Short Circuits in Power Systems

Short Circuits in Power Systems

A Practical Guide to IEC 60909-0

Ismail Kasikci

Second Edition

Author

Ismail Kasikci
Biberach University of Applied Sciences
Karlstraße 11
88400 Biberach
Germany

Cover credit Siemens

All books published by **Wiley-VCH** are carefully produced. Nevertheless, authors, editors, and publisher do not warrant the information contained in these books, including this book, to be free of errors. Readers are advised to keep in mind that statements, data, illustrations, procedural details or other items may inadvertently be inaccurate.

Library of Congress Card No.: applied for

British Library Cataloguing-in-Publication Data
A catalogue record for this book is available from the British Library.

Bibliographic information published by the Deutsche Nationalbibliothek
The Deutsche Nationalbibliothek lists this publication in the Deutsche Nationalbibliografie; detailed bibliographic data are available on the Internet at http://dnb.d-nb.de.

© 2018 Wiley-VCH Verlag GmbH & Co. KGaA, Boschstr. 12, 69469 Weinheim, Germany

All rights reserved (including those of translation into other languages). No part of this book may be reproduced in any form – by photoprinting, microfilm, or any other means – nor transmitted or translated into a machine language without written permission from the publishers. Registered names, trademarks, etc. used in this book, even when not specifically marked as such, are not to be considered unprotected by law.

Print ISBN: 978-3-527-34136-8
ePDF ISBN: 978-3-527-80336-1
ePub ISBN: 978-3-527-80338-5
Mobi ISBN: 978-3-527-80339-2
oBook ISBN: 978-3-527-80337-8

Cover Design Adam-Design, Weinheim, Germany
Typesetting SPi Global, Chennai, India
Printing and Binding C.O.S. Printers Pte Ltd Singapore

Printed on acid-free paper

Contents

Preface *xi*
Acknowledgments *xiii*

1   **Definitions: Methods of Calculations** *1*
1.1   Time Behavior of the Short-Circuit Current *2*
1.2   Short-Circuit Path in the Positive-Sequence System *3*
1.3   Classification of Short-Circuit Types *5*
1.4   Methods of Short-Circuit Calculation *7*
1.4.1   Superposition Method *7*
1.4.2   Equivalent Voltage Source *10*
1.4.3   Transient Calculation *11*
1.4.4   Calculating with Reference Variables *12*
1.4.4.1   The Per-Unit Analysis *12*
1.4.4.2   The %/MVA Method *14*
1.4.5   Examples *14*
1.4.5.1   Characteristics of the Short-Circuit Current *14*
1.4.5.2   Calculation of Switching Processes *14*
1.4.5.3   Calculation with pu System *14*
1.4.5.4   Calculation with pu Magnitudes *16*
1.4.5.5   Calculation with pu System for an Industrial System *17*
1.4.5.6   Calculation with MVA System *19*

2   **Fault Current Analysis** *23*

3   **The Significance of IEC 60909-0** *29*

4   **Supply Networks** *33*
4.1   Calculation Variables for Supply Networks *34*
4.2   Lines Supplied from a Single Source *35*
4.3   Radial Networks *35*
4.4   Ring Networks *35*
4.5   Meshed Networks *37*

5 Network Types for the Calculation of Short-Circuit Currents 39
5.1 Low-Voltage Network Types 39
5.2 Medium-Voltage Network Types 39
5.3 High-Voltage Network Types 44

6 Systems up to 1 kV 47
6.1 TN Systems 48
6.1.1 Description of the System is Carried Out by Two Letters 48
6.2 Calculation of Fault Currents 49
6.2.1 System Power Supplied from Generators: 50
6.3 TT systems 52
6.3.1 Description of the System 52
6.4 IT Systems 53
6.4.1 Description of the System 53
6.5 Transformation of the Network Types Described to Equivalent Circuit Diagrams 54
6.6 Examples 56
6.6.1 Example 1: Automatic Disconnection for a TN System 56
6.6.1.1 Calculation for a Receptacle 56
6.6.1.2 For the Heater 56
6.6.2 Example 2: Automatic Disconnection for a TT System 57

7 Neutral Point Treatment in Three-Phase Networks 59
7.1 Networks with Isolated Free Neutral Point 63
7.2 Networks with Grounding Compensation 64
7.3 Networks with Low-Impedance Neutral Point Treatment 66
7.4 Examples 69
7.4.1 Neutral Grounding 69

8 Impedances of Three-Phase Operational Equipment 71
8.1 Network Feed-Ins, Primary Service Feeder 71
8.2 Synchronous Machines 73
8.2.1 a.c. Component 78
8.2.2 d.c. Component 78
8.2.3 Peak Value 78
8.3 Transformers 80
8.3.1 Short-Circuit Current on the Secondary Side 81
8.3.2 Voltage-Regulating Transformers 83
8.4 Cables and Overhead Lines 85
8.5 Short-Circuit Current-Limiting Choke Coils 96
8.6 Asynchronous Machines 97
8.7 Consideration of Capacitors and Nonrotating Loads 98
8.8 Static Converters 98
8.9 Wind Turbines 99
8.9.1 Wind Power Plant with AG 100
8.9.2 Wind Power Plant with a Doubly Fed Asynchronous Generator 101

8.9.3	Wind Power with Full Converter *101*	
8.10	Short-Circuit Calculation on Ship and Offshore Installations *102*	
8.11	Examples *104*	
8.11.1	Example 1: Calculate the Impedance *104*	
8.11.2	Example 2: Calculation of a Transformer *104*	
8.11.3	Example 3: Calculation of a Cable *105*	
8.11.4	Example 4: Calculation of a Generator *105*	
8.11.5	Example 5: Calculation of a Motor *106*	
8.11.6	Example 6: Calculation of an LV motor *106*	
8.11.7	Example 7: Design and Calculation of a Wind Farm *106*	
8.11.7.1	Description of the Wind Farm *106*	
8.11.7.2	Calculations of Impedances *111*	
8.11.7.3	Backup Protection and Protection Equipment *116*	
8.11.7.4	Thermal Stress of Cables *118*	
8.11.7.5	Neutral Point Connection *119*	
8.11.7.6	Neutral Point Transformer (NPT) *119*	
8.11.7.7	Network with Current-Limiting Resistor *120*	
8.11.7.8	Compensated Network *124*	
8.11.7.9	Insulated Network *125*	
8.11.7.10	Grounding System *125*	

9   Impedance Corrections   *127*
9.1   Correction Factor K_G for Generators   *128*
9.2   Correction Factor K_{KW} for Power Plant Block   *129*
9.3   Correction Factor K_T for Transformers with Two and Three Windings   *130*

10   Power System Analysis   *133*
10.1   The Method of Symmetrical Components   *136*
10.2   Fundamentals of Symmetrical Components   *137*
10.2.1   Derivation of the Transformation Equations   *139*
10.3   General Description of the Calculation Method   *140*
10.4   Impedances of Symmetrical Components   *142*

11   Calculation of Short-Circuit Currents   *147*
11.1   Three-Phase Short Circuits   *147*
11.2   Two-Phase Short Circuits with Contact to Ground   *148*
11.3   Two-Phase Short Circuit Without Contact to Ground   *149*
11.4   Single-Phase Short Circuits to Ground   *150*
11.5   Peak Short-Circuit Current, i_p *153*
11.6   Symmetrical Breaking Current, I_a *155*
11.7   Steady-State Short-Circuit Current, I_k *157*

12   Motors in Electrical Networks   *161*
12.1   Short Circuits at the Terminals of Asynchronous Motors   *161*
12.2   Motor Groups Supplied from Transformers with Two Windings   *163*
12.3   Motor Groups Supplied from Transformers with Different Nominal Voltages   *163*

13	**Mechanical and Thermal Short-Circuit Strength**	*167*
13.1	Mechanical Short-Circuit Current Strength	*167*
13.2	Thermal Short-Circuit Current Strength	*173*
13.3	Limitation of Short-Circuit Currents	*176*
13.4	Examples for Thermal Stress	*176*
13.4.1	Feeder of a Transformer	*176*
13.4.2	Mechanical Short-Circuit Strength	*178*
14	**Calculations for Short-Circuit Strength**	*185*
14.1	Short-Circuit Strength for Medium-Voltage Switchgear	*185*
14.2	Short-Circuit Strength for Low-Voltage Switchgear	*186*
15	**Equipment for Overcurrent Protection**	*189*
16	**Short-Circuit Currents in DC Systems**	*199*
16.1	Resistances of Line Sections	*201*
16.2	Current Converters	*202*
16.3	Batteries	*203*
16.4	Capacitors	*204*
16.5	Direct Current Motors	*205*
17	**Power Flow Analysis**	*207*
17.1	Systems of Linear Equations	*208*
17.2	Determinants	*209*
17.3	Network Matrices	*212*
17.3.1	Admittance Matrix	*212*
17.3.2	Impedance Matrix	*213*
17.3.3	Hybrid Matrix	*213*
17.3.4	Calculation of Node Voltages and Line Currents at Predetermined Load Currents	*214*
17.3.5	Calculation of Node Voltages at Predetermined Node Power	*215*
17.3.6	Calculation of Power Flow	*215*
17.3.6.1	Type of Nodes	*216*
17.3.6.2	Type of Loads and Complex Power	*216*
17.3.7	Linear Load Flow Equations	*218*
17.3.8	Load Flow Calculation by Newton–Raphson	*219*
17.3.9	Current Iteration	*223*
17.3.9.1	Jacobian Method	*223*
17.3.10	Gauss–Seidel Method	*224*
17.3.11	Newton–Raphson Method	*224*
17.3.12	Power Flow Analysis in Low-Voltage Power Systems	*226*
17.3.13	Equivalent Circuits for Power Flow Calculations	*227*
17.3.14	Examples	*228*
17.3.14.1	Calculation of Reactive Power	*228*
17.3.14.2	Application of Newton Method	*228*
17.3.14.3	Linear Equations	*229*
17.3.14.4	Application of Cramer's Rule	*229*
17.3.14.5	Power Flow Calculation with NEPLAN	*230*

18	**Examples: Calculation of Short-Circuit Currents** *233*
18.1	Example 1: Radial Network *233*
18.2	Example 2: Proof of Protective Measures *235*
18.3	Example 3: Connection Box to Service Panel *237*
18.4	Example 4: Transformers in Parallel *238*
18.5	Example 5: Connection of a Motor *240*
18.6	Example 6: Calculation for a Load Circuit *241*
18.7	Example 7: Calculation for an Industrial System *243*
18.8	Example 8: Calculation of Three-Pole Short-Circuit Current and Peak Short-Circuit Current *244*
18.9	Example 9: Meshed Network *246*
18.10	Example 10: Supply to a Factory *249*
18.11	Example 11: Calculation with Impedance Corrections *250*
18.12	Example 12: Connection of a Transformer Through an External Network and a Generator *253*
18.13	Example 13: Motors in Parallel and their Contributions to the Short-Circuit Current *255*
18.14	Example 14: Proof of the Stability of Low-Voltage Systems *257*
18.15	Example 15: Proof of the Stability of Medium-Voltage and High-Voltage Systems *259*
18.16	Example 16: Calculation for Short-Circuit Currents with Impedance Corrections *269*

Bibliography *273*
Standards *277*
Explanations of Symbols *281*
Symbols and Indices *283*
Indices *286*
Secondary Symbols, Upper Right, Left *287*
American Cable Assembly (AWG) *287*

Index *289*

Preface

This book is the result of many years of professional activity in the area of power supply, teaching at the VDE, as well as at the Technical Academy in Esslingen. Every planner of electrical systems is obligated today to calculate the single-pole or three-pole short-circuit current before and after the project management phase. IEC 60909-0 is internationally recognized and used. This standard will be discussed in this book on the basis of fundamental principles and technical references, thus permitting a summary of the standard in the *simplest* and *most understandable* way possible. The rapid development in all areas of technology is also reflected in the improvement and elaboration of the regulations, in particular in regard to IEC 60909-0. Every system installed must not only be suitable for normal operation, but must also be designed in consideration of fault conditions and must remain undamaged following operation under normal conditions and also following a fault condition. Electrical systems must therefore be designed so that neither persons nor equipment are endangered. The dimensioning, cost effectiveness, and safety of these systems depend to a great extent on being able to control short-circuit currents. With increasing power of the installation, the importance of calculating short-circuit currents has also increased accordingly. Short-circuit current calculation is a prerequisite for the correct dimensioning of operational electrical equipment, controlling protective measures and stability against short circuits in the selection of equipment. Solutions to the problems of selectivity, back-up protection, protective equipment, and voltage drops in electrical systems will not be dealt with in this book. The reduction factors, such as frequency, temperatures other than the normal operating temperature, type of wiring, and the resulting current carrying capacity of conductors and cables will also not be dealt with here.

This book comprises the following sections:

Chapter 1 describes the most important terms and definitions, together with relevant processes and types of short circuits.

Chapter 2 is an overview of the fault current analysis.

Chapter 3 explains the significance, purpose, and creation of IEC 60909-0.

Chapter 4 deals with the network design of supply networks.

Chapter 5 gives an overview of the network types for low, medium and high-voltage network.

Chapter 6 describes the systems (network types) in the low-voltage network (IEC 60364) with the cut-off conditions.

Chapter 7 illustrates the types of neutral point treatment in three-phase networks.

Chapter 8 discusses the impedances of the three-phase operational equipment along with relevant data, tables, diagrams, and characteristic curves.

Chapter 9 presents the impedance corrections for generators, power substation transformers, and distribution transformers.

Chapter 10 is concerned with the power system analysis and the method of symmetrical components. With the exception of the three-pole short-circuit current, all other fault currents are unsymmetrical. The calculation of these currents is not possible in the positive-sequence system. The method of symmetrical components is therefore described here.

Chapter 11 is devoted to the calculation of short-circuit types.

Chapter 12 discusses the contribution of high-voltage and low-voltage motors to the short-circuit current.

Chapter 13 deals with the subject of mechanical and thermal stresses in operational equipment as a result of short-circuit currents.

Chapter 14 gives an overview of the design values for short-circuit current strength.

Chapter 15 is devoted to the most important overcurrent protection devices, with time–current characteristics.

Chapter 16 gives a brief overview of the procedure for calculating short-circuit currents in DC systems.

Chapter 17 gives an introduction into power flow analysis.

Chapter 18 represents a large number of examples taken from practice which enhance the understanding of the theoretical foundations. A large number of diagrams and tables that are required for the calculation simplify the application of the IEC 60909 standard as well as the calculation of short-circuit currents and therefore shorten the time necessary to carry out the planning of electrical systems.

I am especially indebted to Dr.-Ing. Waltraud Wüst, Dr. Martin Preuss from Wiley-VCH and Kishore Sivakolundu from SPI for critically reviewing the manuscript and for valuable suggestions.

At this point, I would also like to express my gratitude to all those colleagues who supported me with their ideas, criticism, suggestions, and corrections. My heartiest appreciation is due to Wiley Press for the excellent cooperation and their support in the publication of this book.

Furthermore, I welcome every suggestion, criticism, and idea regarding the use of this book from those who read the book.

Finally, without the support of my family this book could never have been written. In recognition of all the weekends and evenings I sat at the computer, I dedicate this book to my family.

Weinheim
21.07.2017

Ismail Kasikci

Acknowledgments

I would like to thank the companies Siemens and ABB for their help with figures, pictures, and technical documentation. In particular, as a member, I am also indebted to the VDE (Association for Electrical, Electronic and Information Technologies) for their support and release of different kinds of tables and data.

Additionally, I would like to thank Wiley for publishing this book and especially Dr.-Ing. Waltraud Wüst, Dr. Martin Preuss from Wiley-VCH and Kishore Sivakolundu from SPI for their assistance in supporting me and checking the book for clarity.

Finally, I appreciate the designers and planners for their feedbacks, the students for their useful recommendations, and the critics.

1

Definitions: Methods of Calculations

The following terms and definitions correspond largely to those defined in IEC 60909-0. Refer to this standard for all the terms not used in this book.

The terms *short circuit* and *ground fault* describe faults in the isolation of operational equipment, which occur when live parts are shunted out as a result.

1) *Causes*:
   - Overtemperatures due to excessively high overcurrents;
   - Disruptive discharges due to overvoltages; and
   - Arcing due to moisture together with impure air, especially on insulators.
2) *Effects*:
   - Interruption of power supply;
   - Destruction of system components; and
   - Development of unacceptable mechanical and thermal stresses in electrical operational equipment.
3) *Short circuit*: According to IEC 60909-0, a short circuit is the accidental or intentional conductive connection through a relatively low resistance or impedance between two or more points of a circuit that are normally at different potentials.
4) *Short-circuit current*: According to IEC 60909-0, a short-circuit current results from a short circuit in an electrical network.

   It is necessary to differentiate between the short-circuit current at the position of the short circuit and the transferred short-circuit currents in the network branches.
5) *Initial symmetrical short-circuit current*: The effective value of the symmetrical short-circuit current at the moment at which the short circuit arises, when the short-circuit impedance has its value from the time zero.
6) *Initial symmetrical short-circuit apparent power*: The short-circuit power represents a fictitious parameter. During the planning of networks, the short-circuit power is a suitable characteristic number.

Short Circuits in Power Systems: A Practical Guide to IEC 60909-0, Second Edition. Ismail Kasikci.
© 2018 Wiley-VCH Verlag GmbH & Co. KGaA. Published 2018 by Wiley-VCH Verlag GmbH & Co. KGaA.

7) *Peak short-circuit current*: The largest possible momentary value of the short circuit occurring.
8) *Steady-state short-circuit current*: Effective value of the initial symmetrical short-circuit current remaining after the decay of all transient phenomena.
9) *Direct current (d.c.) aperiodic component*: Average value of the upper and lower envelope curve of the short-circuit current, which slowly decays to zero.
10) *Symmetrical breaking current*: The effective value of the short-circuit current that flows through the contact switch at the time of the first contact separation.
11) *Equivalent voltage source*: The voltage at the position of the short circuit, which is transferred to the positive-sequence system as the only effective voltage and is used for the calculation of the short-circuit currents.
12) *Superposition method*: Considers the previous load of the network before the occurrence of the short circuit. It is necessary to know the load flow and the setting of the transformer step switch.
13) *Voltage factor*: Ratio between the equivalent voltage source and the network voltage, U_n, divided by $\sqrt{3}$.
14) *Equivalent electrical circuit*: Model for the description of the network by an equivalent circuit.
15) *Far-from-generator short circuit*: The value of the symmetrical alternating current (a.c.) periodic component remains essentially constant.
16) *Near-to-generator short circuit*: The value of the symmetrical a.c. periodic component does not remain constant. The synchronous machine first delivers an initial symmetrical short-circuit current, which is more than twice the rated current of the synchronous machine.
17) *Positive-sequence short-circuit impedance*: The impedance of the positive-sequence system as seen from the position of the short circuit.
18) *Negative-sequence short-circuit impedance*: The impedance of the negative-sequence system as seen from the position of the short circuit.
19) *Zero-sequence short-circuit impedance*: The impedance of the zero-sequence system as seen from the position of the short circuit. Three times the value of the neutral point to ground impedance occurs.
20) *Short-circuit impedance*: Impedance required for the calculation of the short-circuit currents at the position of the short circuit.

1.1 Time Behavior of the Short-Circuit Current

Figure 1.1 shows the time behavior of the short-circuit current for the occurrence of far-from-generator and near-to-generator short circuits.

The d.c. aperiodic component depends on the point in time at which the short circuit occurs. For a near-to-generator short circuit, the subtransient and the transient behaviors of the synchronous machines are important. Following the decay of all transient phenomena, the steady state sets in.

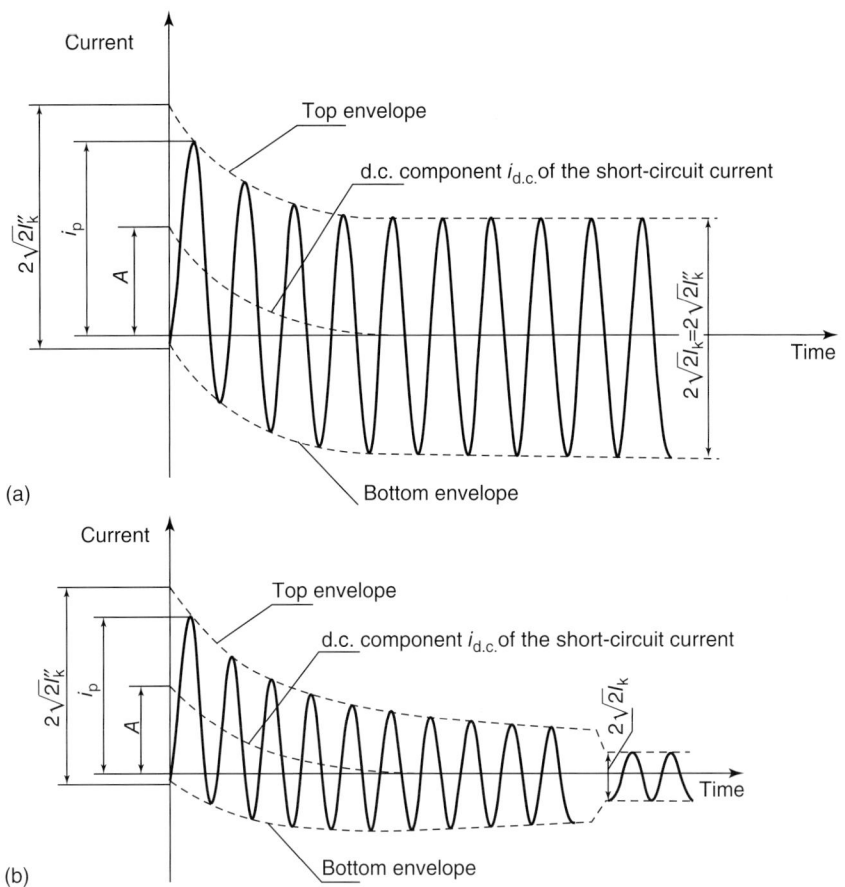

Figure 1.1 Time behavior of the short-circuit current (see Ref. [1]). (a) Far-from-generator short circuit and (b) near-to-generator short circuit. I_k'': initial symmetrical short-circuit current; i_p: peak short-circuit current; $i_{d.c.}$: decaying d.c. aperiodic component; and A: initial value of d.c. aperiodic component.

1.2 Short-Circuit Path in the Positive-Sequence System

For the same external conductor voltages, a three-phase short circuit allows three currents of the same magnitude to develop among the three conductors. Therefore, it is only necessary to consider one conductor in further calculations. Depending on the distance from the position of the short circuit from the generator, it is necessary to consider near-to-generator and far-from-generator short circuits separately. For far-from-generator and near-to-generator short circuits, the short-circuit path can be represented by a mesh diagram with an a.c. voltage source, reactances X, and resistances R (Figure 1.2). Here, X and R replace all components such as cables, conductors, transformers, generators, and motors.

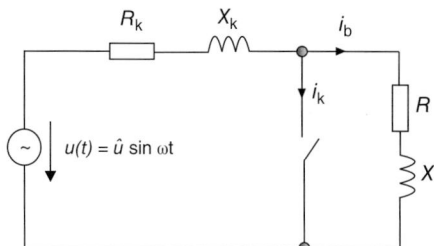

Figure 1.2 Equivalent circuit of the short-circuit current path in the positive-sequence system.

The following differential equation can be used to describe the short-circuit process:

$$i_k \cdot R_k + L_k \frac{di_k}{dt} = \hat{u} \cdot \sin(\omega t + \psi) \tag{1.1}$$

where ψ is the phase angle at the point in time of the short circuit. The inhomogeneous first-order differential equation can be solved by determining the homogeneous solution i_k and a particular solution I_k''.

$$i_k = i_{k\sim}'' + i_{k-} \tag{1.2}$$

The homogeneous solution, with the time constant $\tau_g = L/R$, yields the following:

$$i_k = \frac{-\hat{u}}{\sqrt{(R^2 + X^2)}} e^{t/\tau_g} \sin(\psi - \varphi_k) \tag{1.3}$$

For the particular solution, we obtain the following:

$$i_k'' = \frac{-\hat{u}}{\sqrt{(R^2 + X^2)}} \sin(\omega t + \psi - \varphi_k) \tag{1.4}$$

The total short-circuit current is composed of both the components:

$$i_k = \frac{-\hat{u}}{\sqrt{(R^2 + X^2)}} [\sin(\omega t + \psi - \varphi_k) - e^{t/\tau_g} \sin(\psi - \varphi_k)] \tag{1.5}$$

The phase angle of the short-circuit current (short-circuit angle) is then, in accordance with the above equation,

$$\varphi_k = \psi - \nu = \arctan \frac{X}{R} \tag{1.6}$$

Figure 1.3 shows the switching processes of the short circuit.

For the far-from-generator short circuit, the short-circuit current is, therefore, made up of a constant a.c. periodic component and the decaying d.c. aperiodic component. From the simplified calculations, we can now reach the following conclusions:

1) The short-circuit current always has a decaying d.c. aperiodic component in addition to the stationary a.c. periodic component.
2) The magnitude of the short-circuit current depends on the operating angle of the current. It reaches a maximum at $\gamma = 90°$ (purely inductive load). This case serves as the basis for further calculations.
3) The short-circuit current is always inductive.

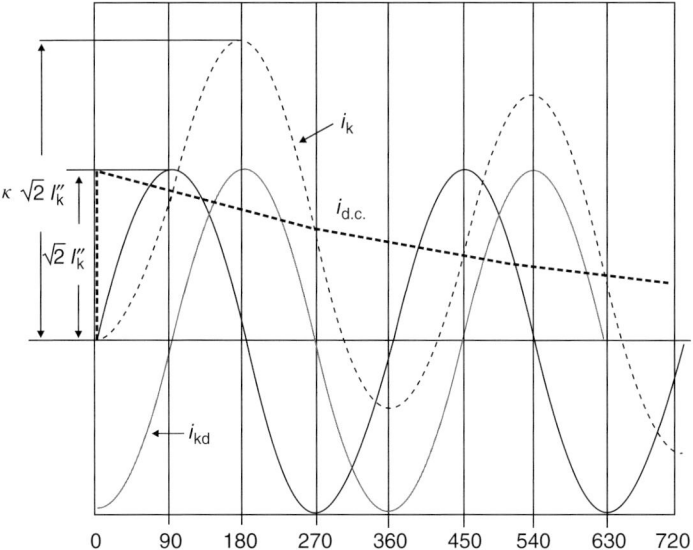

Figure 1.3 Switching processes of the short circuit.

1.3 Classification of Short-Circuit Types

For a three-phase short circuit, three voltages at the position of the short circuit are zero. The conductors are loaded symmetrically. Therefore, it is sufficient to calculate only in the positive-sequence system. The two-phase short-circuit current is less than that of the three-phase short circuit, but largely close to synchronous machines. The single-phase short-circuit current occurs most frequently in low-voltage (LV) networks with solid grounding. The double ground connection occurs in networks with a free neutral point or with a ground fault neutralizer grounded system.

For the calculation of short-circuit currents, it is necessary to differentiate between the far-from-generator and the near-to-generator cases.

1) Far-from-generator short circuit
   When double the rated current is not exceeded in any machine, we speak of a far-from-generator short circuit.
   $$I_k'' < 2 \cdot I_{rG} \tag{1.7}$$
   or when
   $$I_k'' = I_a = I_k \tag{1.8}$$

2) Near-to-generator short circuit
   When the value of the initial symmetrical short-circuit current I_k'' exceeds double the rated current in at least one synchronous or asynchronous machine at the time the short circuit occurs, we speak of a near-to-generator short circuit.
   $$I_k''2 > I_{rG} \tag{1.9}$$

or when

$$I_k'' > I_a > I_k \tag{1.10}$$

Figure 1.4 schematically illustrates the most important types of short circuits in three-phase networks.

1) Three-phase short circuits:
   - connection of all conductors with or without simultaneous contact to ground;
   - symmetrical loading of the three external conductors;
   - calculation only according to single phase.
2) Two-phase short circuits:
   - unsymmetrical loading;
   - all voltages are nonzero;
   - coupling between external conductors;
   - for a near-to-generator short circuit $I_{k2}'' > I_{k3}''$
3) Single-phase short circuits between phase and PE:
   - very frequent occurrence in LV networks.
4) Single-phase short circuits between phase and N:
   - very frequent occurrence in LV networks.
5) Two-phase short circuits with ground:
   - in networks with an insulated neutral point or with a suppression coil grounded system $I_{kEE}'' < I_{k2E}''$.

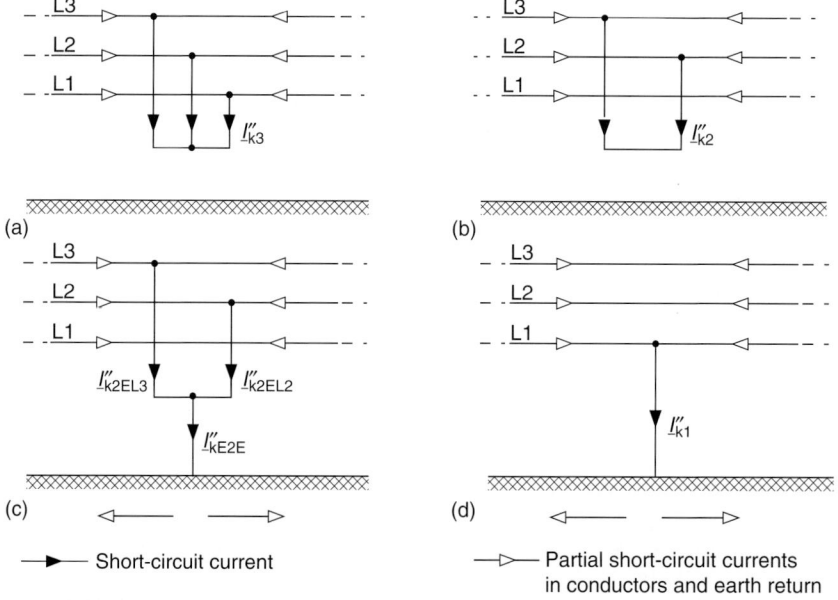

Figure 1.4 Types of short-circuit currents in three-phase networks [1].

With a suppression coil grounded system, a residual ground fault current I_Rest occurs. I_C and I_Rest are special cases of I_k''.

1.4 Methods of Short-Circuit Calculation

The measurement or calculation of short-circuit current in LV networks on final circuits is very simple. In meshed and extensive power plants, the calculation is more difficult because of the short-circuit current of several partial short-circuit currents in conductors and earth return.

The short-circuit currents in three-phase systems can be determined by three different calculation procedures:

1) superposition method for a defined load flow case;
2) calculating with the equivalent voltage source $\frac{c \cdot U_n}{\sqrt{3}}$ at the fault location; and
3) transient calculation.

1.4.1 Superposition Method

The superposition method is an exact method for the calculation of the short-circuit currents. The method consists of three steps. The voltage ratios and the loading condition of the network must be known before the occurrence of the short circuit. In the first step, the currents, voltages, and internal voltages for steady-state operation before onset of the short circuit are calculated (Figure 1.5b). The calculation considers the impedances, power supply feeders, and node loads of the active elements. In the second step, the voltage applied to the fault location before the occurrence of the short circuit and the current distribution at the fault location are determined with a negative sign (Figure 1.5b). This is the only voltage source in the network. The internal voltages are short-circuited. In the third step, both the conditions are superimposed. We then obtain a zero voltage at the fault location. The superposition of the currents also leads to the value zero. The disadvantage of this method is that the steady-state condition must be specified. The data for the network (effective and reactive power, node voltages, and the step settings of the transformers) are often difficult to determine. The question also arises: Which operating state leads to the greatest short-circuit current?

The superposition method assumes that the power flow is known of the network before the fault inception and the setting of the tap changer of the transformer and the voltage set points of the generators.

By the superposition method, the power state is superimposed with an amendments state before the short circuit occurs. For this condition, the consideration of positive sequence is sufficient.

The network consists of $i = 1,\ldots,n$ load nodes and $j = 1,\ldots,m$ generators and power supply applications. With a suitable program, the load flow can be calculated for a network condition. After the changes in the network through the short circuit, there are other values at each node. For a three-phase short circuit, the voltage at the fault point equals zero. This condition is also fulfilled when the

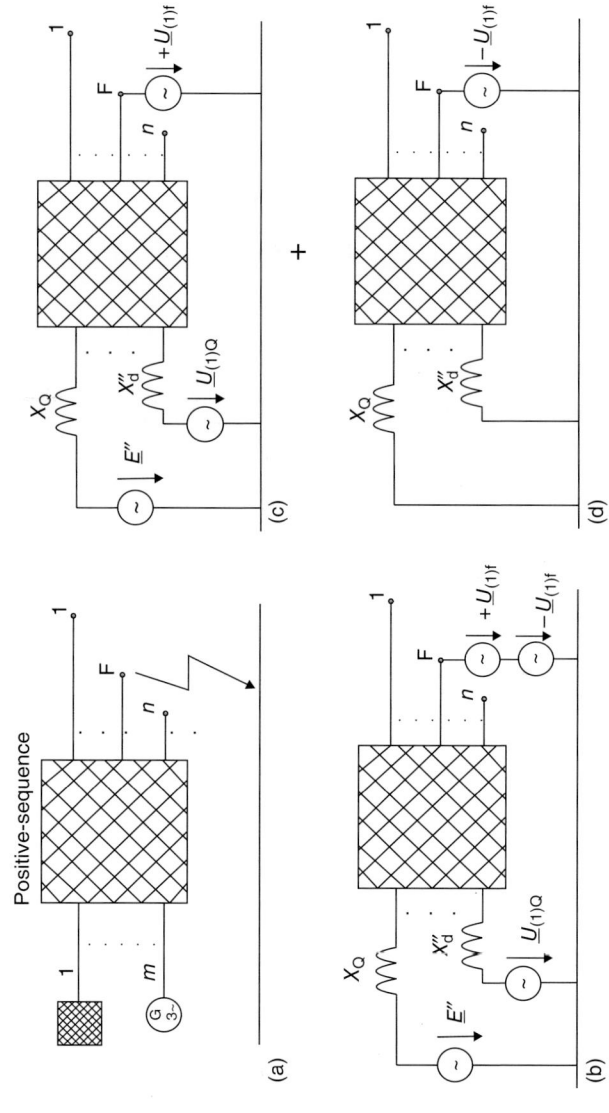

Figure 1.5 Methods for the short-circuit calculation. (a) Single line diagram; (b) voltage source at the fault location; (c) superposition; and (d) equivalent voltage source.

1.4 Methods of Short-Circuit Calculation

same voltage is given at the fault location but with an opposite voltage sign. All network feeders, synchronous, and asynchronous machines are replaced by their internal impedances (Figure 1.5d).

The calculation of a short-circuit current is a linear problem that can be solved easily with linear equations. There is a linear relationship between the node voltages and node currents.

With the help of nodal admittance matrix systems, linear equations can be solved. All impedances are converted to the LV side of the transformers. In contrast to the load flow calculation, an iteration is not required. The equations are obtained at the short-circuit location i in matrix notation.

$$\underline{i} = \underline{Y} \cdot \underline{u} \qquad (1.11)$$

$$\begin{bmatrix} 0 \\ 0 \\ \vdots \\ \underline{I}''_{ki} \\ \vdots \\ 0 \end{bmatrix} = \begin{bmatrix} \underline{Y}_{11} & \underline{Y}_{1n} \\ \underline{Y}_{21} & \underline{Y}_{2n} \\ \vdots & \vdots \\ \underline{Y}_{i1} & \underline{Y}_{in} \\ \vdots & \vdots \\ \underline{Y}_{n1} & \underline{Y}_{nn} \end{bmatrix} \cdot \begin{bmatrix} \underline{U}_1 \\ \underline{U}_2 \\ \vdots \\ -c\dfrac{U_n}{\sqrt{3}} \\ \vdots \\ \underline{U}_n \end{bmatrix} \qquad (1.12)$$

After inversion, we obtain the following:

$$\underline{u} = \underline{Y}^{-1} \cdot \underline{i} \qquad (1.13)$$

$$\begin{bmatrix} \underline{U}_1 \\ \underline{U}_2 \\ \vdots \\ -c\dfrac{U_n}{\sqrt{3}} \\ \vdots \\ \underline{U}_n \end{bmatrix} = \begin{bmatrix} \underline{Z}_{11} & \underline{Z}_{1n} \\ \underline{Z}_{21} & \underline{Z}_{2n} \\ \vdots & \vdots \\ \underline{Z}_{i1} & \underline{Z}_{in} \\ \vdots & \vdots \\ \underline{Z}_{n1} & \underline{Z}_{nn} \end{bmatrix} \cdot \begin{bmatrix} 0 \\ 0 \\ \vdots \\ \underline{I}''_{ki} \\ \vdots \\ 0_n \end{bmatrix} \qquad (1.14)$$

From the ith row of the equation results

$$-c\dfrac{U_n}{\sqrt{3}} = \underline{Z}_{ii} \cdot \underline{I}''_{ki} \qquad (1.15)$$

The initial short-circuit a.c. can be calculated by redirecting the above equation:

$$\underline{I}''_{ki} = -\dfrac{c \cdot U_n}{\sqrt{3} \cdot \underline{Z}_{ii}} \qquad (1.16)$$

For the node voltages, follow:

$$\underline{U}_k = \underline{Z}_{ki} \cdot \underline{I}''_{ki} \qquad (1.16a)$$

1 Definitions: Methods of Calculations

Since the operating voltage $U_{(1)f} = \frac{U_n}{\sqrt{3}}$ is not known at the fault location, for the equivalent voltage source at the fault point can be introduced.

$$-\underline{U}_{(1)f} = \frac{c \cdot U_n}{\sqrt{3}} \tag{1.17}$$

At the short-circuit point, the only active voltage is the Thevenin equivalent voltage source of the system.

1.4.2 Equivalent Voltage Source

Figure 1.6 shows an example of the equivalent voltage source at the short-circuit location F as the only active voltage of the system fed by a transformer with or without an on-load tap changer. All other active voltages in the system are short-circuited. Thus, the network feeder is represented by its internal impedance, $\underline{Z}_{Qt}$, transferred to the LV side of the transformer and the transformer by its impedance referred to the LV side. The shunt admittances of the line, the transformer, and the nonrotating loads are not considered. The impedances of the network feeder and the transformer are converted to the LV side.

The transformer is corrected with K_T, which will be explained later.

The voltage factor c (Table 1.1) will be described briefly as follows:

If there are no national standards, it seems adequate to choose a voltage factor c, according to Table 1.1, considering that the highest voltage in a normal

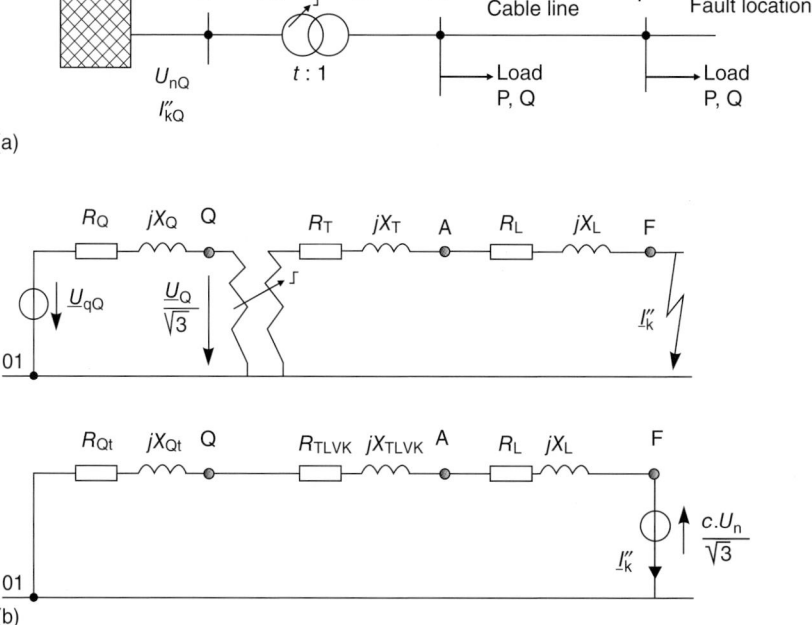

Figure 1.6 Network circuit with equivalent voltage source [2]. (a) System diagram and (b) equivalent circuit diagram of the positive-sequence system.

Table 1.1 Voltage factor c, according to IEC 60909-0: 2016-10 [1].

Nominal voltage, U_n	Voltage factor c for calculation of	
	Maximum short-circuit currents (c_{max})[a]	Minimum short-circuit currents (c_{min})
Low voltage		
100–1000 V	1.05[b]	0.95[b]
(IEC 38, Table I)	1.10[c]	0.9[c]
High voltage[d]		
>1–35 kV	1.10	1.00
(IEC 38, Tables III and IV)		

a) $c_{max} U_n$ should not exceed the highest voltage U_m for equipment of power systems.
b) For LV systems with a tolerance of ±6%, for example, systems renamed from 380 to 400 V.
c) For LV systems with a tolerance of ±10%.
d) If no nominal voltage is defined, $c_{max} U_n = U_m$ or $c_{min} U_n = 0.90 U_m$ should be applied.

(undisturbed) system does not differ, on average, by more than approximately +5% (some LV systems) or +10% (some high-voltage, HV, systems) from the nominal system voltage U_n [3].

1) The different voltage values depending on time and position
2) The step changes of the transformer switch
3) The loads and capacitances in the calculation of the equivalent voltage source can be neglected
4) The subtransient behavior of generators and motors must be considered.

This method assumes the following conditions:

1) The passive loads and conductor capacitances can be neglected
2) The step setting of the transformers need not be considered
3) The excitation of the generators need not be considered
4) The time and position dependence of the previous load (loading state) of the network need not be considered.

1.4.3 Transient Calculation

With the transient method, the individual operating equipment and, as a result, the entire network are represented by a system of differential equations. The calculation is very tedious. The method with the equivalent voltage source is a simplification relative to the other methods. Since 1988, it has been standardized internationally in IEC 60909-0. The calculation is independent of a current operational state. Therefore, in this book, the method with the equivalent voltage source will be dealt with and discussed.

1.4.4 Calculating with Reference Variables

There are several methods for performing short-circuit calculations with absolute and reference impedance values. A few methods are summarized here, and examples are calculated for comparison. To define the relative values, there are two possible reference variables.

For the characterization of electrotechnical relationships, we require the four parameters:

1) voltage U in V;
2) current I in A;
3) impedance Z in Ω; and
4) apparent power S in VA.

Three methods can be used to calculate the short-circuit current:

1) *The Ohm system*: units – kV, kA, V, and MVA.
2) *The per-unit (pu) system*: this method is used predominantly for electrical machines; all four parameters u, i, z, and s are given as per unit (unit = 1). The reference value is 100 MVA. The two reference variables for this system are U_B and S_B. Example: The reactances of a synchronous machine X_d, X'_d, and X''_d are given in pu or in %pu, multiplied by 100%.
3) *The %/MVA system*: this system is especially well suited for the quick determination of short-circuit impedances. As a formal unit, only the % symbol is added.

1.4.4.1 The Per-Unit Analysis

Today, the power system consists of complex and complicated mesh, ring, and radial networks with many transformers, generators, and cables. The calculation of such a circuit can be very tedious and incorrect. The use of sophisticated computer programs is a big help for engineers. On the other hand, for a quick calculation a simple method, per unit system also can be used. However, this method is not accepted worldwide and is not standardized by IEC, EN, or IEEE committees.

The pu method uses the electrical variables $\underline{U}, \underline{I}, \underline{Z}$, and $\underline{S}$. They are based on a dimensionless same references, namely, $U_{base}, I_{base}, Z_{base}$, or S_{base}. The resulting dimensionless quantities are described with the lowercase $\underline{u}, \underline{i}, \underline{z}$, or $\underline{s}$.

A pu system is defined as follows:

$$\text{Per unit value (pu)} = \frac{\text{the actual value (in any unit)}}{\text{the base or reference value (in the same unit)}}$$

$$\underline{u}_{pu} = \frac{\underline{U}}{U_{base}}$$

A reference voltage and a reference apparent power are selected and then reference current and impedance are calculated as follows:

$$Z_{base} = \frac{U_{base}^2}{S_{base}}$$

$$I_{base} = \frac{S_{base}}{U_{base}}$$

Only a single global base value is selected in the short-circuit current calculation. This reference value is then used for all other networks. The choice of reference values can be carried out arbitrarily in principle. However, it is appropriate to select the rated voltage at the short-circuit location as a reference voltage. For example, as reference apparent power is the rated apparent power of the largest transformer in the network or a power of the same selected magnitude (e.g., 100 MVA). The best choice of base can be achieved when the impedances and currents in easily handled orders of magnitude.

It should be noted that related parameters' individual resources, such as the relative short-circuit voltage of a transformer u_{kr} or related subtransient reactance x''_d of the generator, are always relative to a base, which consists of the design parameters of the particular equipment. In a short-circuit current calculation as per pu method, these parameters must first be converted to the selected global basis. If we give an example for voltage and current, the expression is as follows:

$$U_{pu} = \frac{U_{actual}}{U_{base}}$$

$$I_{pu} = \frac{I_{actual}}{I_{base}}$$

Note that the voltage according to the international system of units (SI) is not V, but U. The letter V is a unit in this case. V is used especially in Anglo-Saxon countries.

For other values, we can write for 1 pu impedance (Ω):

$$Z_{base} = \frac{U_{base}}{I_{base}} = \frac{U_{base}}{I_{base}} \quad \text{or in pu} \quad Z_{pu} = \frac{U_{pu}}{I_{pu}}$$

$$I_{base} = \frac{S_{base}}{U_{base}}$$

We convert the values to pu:

$$R_{pu} = \frac{R}{Z_{base}}$$

$$X_{pu} = \frac{X}{Z_{base}}$$

Remember that a symmetrical three-phase system has two voltages, line–line voltage U_L (U_n) and U_{LN} (U_0). By definition:

$$U_{LN} = \frac{U_L}{\sqrt{3}}$$

Now consider:

$$U_{LNpu} = \frac{U_{LN}}{U_{LNbase}}$$

It follows that:

$$U_{LNpu} = \frac{U_{LN}}{U_{LNbase}} = \frac{U_L/\sqrt{3}}{U_{Lbase}/\sqrt{3}} = \frac{U_L}{U_{Lbase}} = U_{Lpu}$$

Consider that the factor $\sqrt{3}$ disappears in the pu equation.

1.4.4.2 The %/MVA Method

The %/MVA method can be considered as a modification of the pu method and designed specifically for the HV network calculation. The impedances of the electrical equipment can be determined easily in %/MVA from the synchronous machine and transformer characteristics. It utilizes the fact that for the pu calculation, apparent power S_{base} is completely arbitrary. Consequently, instead of S_{base}, the dimensionless value 1 is inserted. This has the result that the related sizes of the pu are no longer dimensionless.

The related impedances can be represented in %/MVA. The %/MVA method has the advantage that a conversion with t^2 or $1/t^2$ eliminates the transformation of impedances in the voltage level on the transformers. Furthermore, all impedances of resources and the dimensioning data can be obtained from the nameplate.

$$z = \frac{Z}{U_{\text{base}}^2} \times 100\%$$

$$u = \frac{Z}{U_{\text{base}}} \times 100\%$$

$$i = I \cdot U_{\text{base}}$$

1.4.5 Examples

1.4.5.1 Characteristics of the Short-Circuit Current

The short-circuit current is composed of two parts. The first term describes the a.c. (continuous current) and the second term the compensation process.

$$i_{k(t)} = \hat{i}_k \cdot \sin(\omega t - \Psi + \varphi_u) - \hat{i}_k \cdot \sin(\Psi - \varphi) \cdot e^{-(t/\tau)}$$

The size of the short-circuit current is therefore dependent on the phase angle $\varphi = \arctan \frac{X}{R}$, the time constant $\tau = \frac{L}{R}$, the time t, and the switching angle Ψ.

Thus we can obtain many shifts by changing the sizes (Figure 1.7).

Example: $R_Q/X_Q = 0.176$, $\varphi = -0°$, $\varphi_u = -90°$, $\tau = \infty\,\text{ms}$, $3 \cdot \tau = \infty\,\text{ms}$, $\kappa = 1.02$.

For further considerations, one can write

$$\sin(\omega t - \varphi) = 1, \quad \frac{R}{X} = \frac{1}{\tan \varphi}, \quad \Psi = 0$$

1.4.5.2 Calculation of Switching Processes

Given are the following sizes of short-circuit current (Figure 1.8):

$$\frac{R_Q}{X_Q} = 0.176, \quad \varphi = -80°, \quad \varphi_u = -170°, \quad \tau = 0.018\,\text{ms},$$

$$3 \cdot \tau = 0.054\,\text{ms}, \quad \kappa = 1.597$$

Draw the switching process of the short circuit.

1.4.5.3 Calculation with pu System

Given is a system with 20/6 kV network (Figure 1.9).

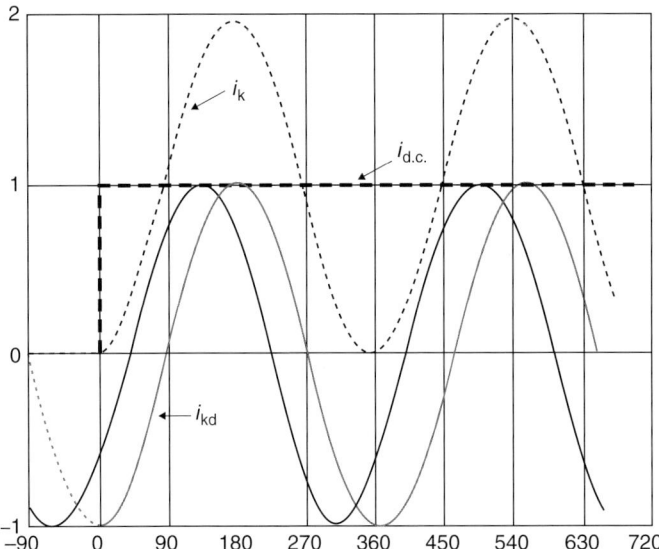

Figure 1.7 Short-circuit current components – switch on.

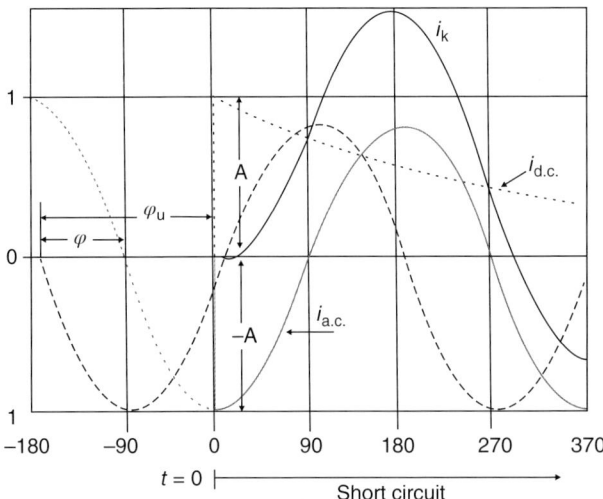

Figure 1.8 Switching process.

Transformer:

$$S_{rT} = 25\,\text{MVA}, \quad u_{krT} = 13\%, \quad 20/6.3\,\text{kV}$$

Motors 1 and 2:

$$2 \times P_{rM} = 2.3\,\text{MW}, \quad U_{rM} = 6\,\text{kV}, \quad \cos\varphi_{rM} = 0.86$$
$$p = 2, \quad I_a/I_{rM} = 5, \quad \eta = 0.97$$

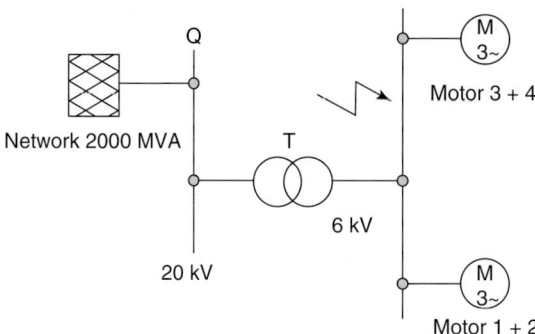

Figure 1.9 Impact of engines on the current.

Motors 3 and 4:

$$2 \times P_{rM} = 0.36\,\text{MW}, \quad U_{rM} = 6\,\text{kV}, \quad \cos\varphi_{rM} = 0.87$$
$$p = 1, \quad I_a/I_{rM} = 5.5, \quad \eta = 0.98$$

1.4.5.4 Calculation with pu Magnitudes

Given: $U_B = U_n = 6\,\text{kV}$ bzw. $20\,\text{kV}$, $S_B = 100\,\text{MVA}$. Calculate example 1.4.5.3 using pu magnitude.

$$U_* = \frac{U}{U_B} \quad I_* = \frac{I \cdot U_B}{SB} \quad Z_* = \frac{Z \cdot S_B}{U_B^2} \quad S_* = \frac{S}{S_B}$$

Transformer translation in pu system:

$$t_r^* = \frac{U_{rTOS}}{U_{rTUS}} \cdot \frac{U_{B,6\,\text{kV}}}{U_{B,20\,\text{kV}}} = \frac{20\,\text{kV}}{6.3\,\text{kV}} \cdot \frac{6\,\text{kV}}{20\,\text{kV}} = 0.9524$$

Power supply:

$$Z_{Qt}^* = \frac{c \cdot U_{nQ}^{2*}}{S_{kQ}''^*} \cdot \frac{1}{t_r^{2*}} = \frac{1.1 \cdot (1 \cdot \text{pu})^2}{10\,\text{pu}} \cdot \frac{1}{0.9524^2} = 0.1212\,\text{pu}$$

Transformer:

$$Z_T^* = \frac{u_{krT}}{100\%} \cdot \frac{U_{rTUS}^2}{S_{rT}} \cdot \frac{S_B}{U_{B,6\,\text{kV}}^2} = \frac{13\%}{100\%} \cdot \frac{(6.3\,\text{kV})^2}{25\,\text{MVA}} \cdot \frac{100\,\text{MVA}}{(6\,\text{kV})^2} = 0.5733\,\text{pu}$$

Impedance:

$$Z_k^* = Z_{Qt}^* + Z_T^* = 0.6945\,\text{pu}$$

$I_k''^*$ without motors.

$$I_k''^* = \frac{c \cdot U_n^*}{\sqrt{3} \cdot Z_k^*} = \frac{1.1 \cdot 1\,\text{pu}}{\sqrt{3} \cdot 0.6945\,\text{pu}} = 0.9144\,\text{pu}$$

Current in kA:

$$I_k'' = I_k^* \cdot \frac{S_B}{U_{B,6\,\text{kV}}} = 0.9144\,\text{pu} \cdot \frac{100\,\text{MVA}}{6\,\text{kV}} = 15.24\,\text{kA}$$

Impedances of motors in pu systems:

$$Z^*_{m1} = \frac{1}{2} \frac{\eta \cdot \cos\varphi}{I_{an}/I_{rM}} \cdot \frac{U^2_{rM}}{P_{rM}} \cdot \frac{S_B}{U^2_{B,6kV}} = \frac{1}{2} \frac{\eta \cdot \cos\varphi}{I_{an}/I_{rM}} \cdot \frac{S_B}{P_{rM}}$$

$$= \frac{1}{2} \cdot \frac{0.86 \cdot 0.97}{5} \cdot \frac{100\,\text{MVA}}{2.3\,\text{MVA}} = 3.63\,\text{pu}$$

$$Z_{m2} = \frac{1}{2} \cdot \frac{0.87 \cdot 0.98}{5.5} \cdot \frac{100\,\text{MVA}}{0.36\,\text{MVA}} = 21.5\,\text{pu}$$

Partly current:

$$I''^*_{km1} = \frac{c \cdot U^*_n}{\sqrt{3} \cdot Z^*_{m1}} = \frac{1.1 \cdot 1\,\text{pu}}{\sqrt{3} \cdot 3.63\,\text{pu}} = 0.175\,\text{pu}$$

$$I''_{km1} = 2.92\,\text{kA}$$

$$I''^*_{km2} = 0.0295\,\text{pu}$$

$$I''_{km2} = 0.492\,\text{kA}$$

1.4.5.5 Calculation with pu System for an Industrial System

Given is Figure 1.10. Calculate the short circuit power and short-circuit currents of an industrial power plant with the pu method.

First, the equivalent circuit diagram is drawn (Figure 1.11).

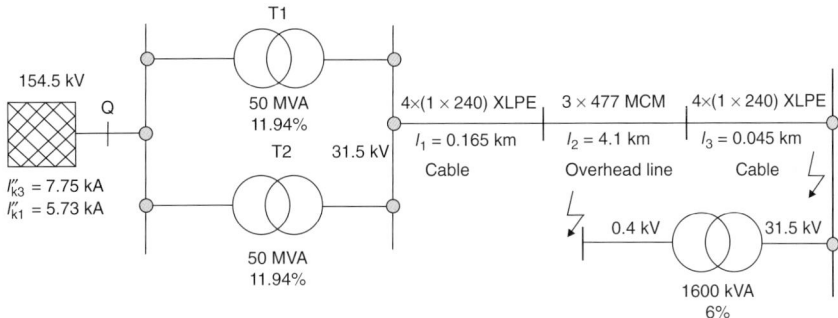

Figure 1.10 Supply for an industrial company.

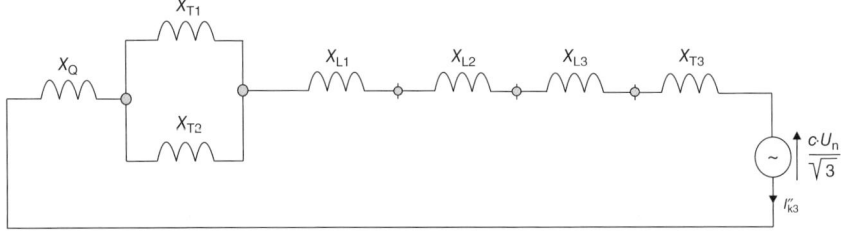

Figure 1.11 Equivalent circuit diagram in positive sequence.

Impedance of the power supply:

$$Z_Q = \frac{U_n}{\sqrt{3} \cdot I''_{k3}} = \frac{154\,\text{kV}}{\sqrt{3} \cdot 7.75\,\text{kA}} = 11.486\,\Omega$$

Short-circuit power of the supply:

$$S''_{kQ} = \sqrt{3} \cdot U_n \cdot I''_{k3} = \sqrt{3} \cdot 154\,\text{kV} \cdot 7.75\,\text{kA} = 2064.75\,\text{MVA}$$

Reactance of the feed:

$$X_Q = \frac{1 \cdot 100}{S''_{kQ}} = \frac{1 \cdot 100}{2064.75\,\text{MVA}} = 0.0484\,\text{pu}$$

Reactance of transformers:

$$X_T = \frac{X_{T1} \cdot X_{T2}}{X_{T1} + X_{T2}} = 5.97\%$$

$$X_T = \frac{100}{50\,\text{MVA} + 50\,\text{MVA}} \cdot 5.97\% = 0.0597\,\text{pu}$$

Total reactance:

$$X_G = X_Q + X_T = 0.0484\,\text{pu} + 0.0597\,\text{pu} = 0.10813\,\text{pu}$$

$$I_{pu} = \frac{U}{X_G} = \frac{1}{0.10813\,\text{pu}} = 9.248\,\text{pu}$$

Short-circuit power:

$$X_{kQ} = 1\,\text{pu} \cdot 100 = 9.248 \cdot 100 = 924.8\,\text{MVA}$$

$$X_{Hpu} = \frac{X_H}{X_B} = \frac{X_F L}{X_B}$$

$$X_B = \frac{U^2}{100} = \frac{31.5\,\text{kV}^2}{100} = 9.922\,\Omega$$

$$X_H = X_{H1} + X_{H1}$$

$$X_T = 0.0161 + 1.5088 = 1.528\,\text{pu}$$

$$X_{H1} = l_1 \cdot X_{H1} = 0.21\,\text{km} \cdot 0.0754\,\Omega/\text{km} = 0.0161\,\Omega$$

$$X_{H2} = l_2 \cdot X_{H2} = 4.1\,\text{km} \cdot 0.368\,\Omega/\text{km} = 1.5088\,\Omega$$

$$X_{Hpu} = \frac{1.526}{9.922} = 0.154\,\text{pu}$$

$$X_{G2} = X_{G1} + X_H = 0.10813 + 0.154 = 0.26213\,\text{pu}$$

$$I_B = \frac{S}{\sqrt{3} \cdot U} = \frac{100 \cdot 10^3}{\sqrt{3} \cdot 31.5\,\text{kV}} = 1835\,\text{A}$$

$$I_{pu} = \frac{1}{X_{G2}} = \frac{1}{0.26213\,\text{pu}} = 3.8\,\text{pu}$$

On distribution transformer:

$$I''_k = I_{pu} \cdot I_B = 3.8\,\text{pu} \cdot 1835\,\text{A} = 6.973\,\text{kA}$$

1.4 Methods of Short-Circuit Calculation

Short-circuit power on the primary side of the transformer:

$$S_k'' = \sqrt{3} \cdot U \cdot I_k'' = \sqrt{3} \cdot 31.5\,\text{kV} \cdot 6.973\,\text{kA} = 380\,\text{MVA}$$

Short-circuit power on the secondary side of the transformer:

$$X_{TR} = \frac{100}{1.6} \cdot 0.06 = 3.75\,\text{pu}$$

$$X_{G3} = X_{G2} + X_{G1} = 0.26213 + 3.75 = 4.10213\,\text{pu}$$

$$I_{pu} = \frac{1}{X_{G3}} = \frac{1}{4.10213\,\text{pu}} = 0.24924\,\text{pu}$$

$$I_{k(0.4\,\text{kV})}'' = I_{pu} \cdot I_B = 0.24924\,\text{pu} \cdot 1.835\,\text{kA} = 0.457\,\text{kA}$$

1.4.5.6 Calculation with MVA System

Figure 1.12 shows a power plant with auxiliary power system and power supply. Calculate using the %/MVA system at the busbar SS circuit power, the peak short-circuit current and the breaking current.

The total reactance at the fault is calculated with the aid of the equivalent circuit shown in Figure 1.13 by gradual power conversion.

1) Calculation of the reactance of the individual resources
   Network reactance:

$$X_Q = \frac{1.1 \cdot 100}{S_{kQ}''} = \frac{1.1 \cdot 100}{8000\,\text{MVA}} = 0.0138\%/\text{MVA}$$

   Transformer 1:

$$X_{T1} = \frac{u_{kr}}{S_{rT1}} = \frac{13}{100\,\text{MVA}} = 0.1300\%/\text{MVA}$$

   Generator:

$$X_G = \frac{x_d''}{S_{rG}} = \frac{11.5}{93.7\,\text{MVA}} = 0.1227\%/\text{MVA}$$

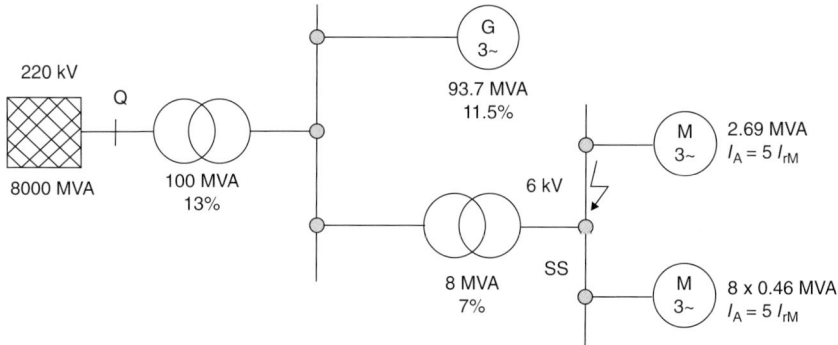

Figure 1.12 Power plant with auxiliary power system.

1 Definitions: Methods of Calculations

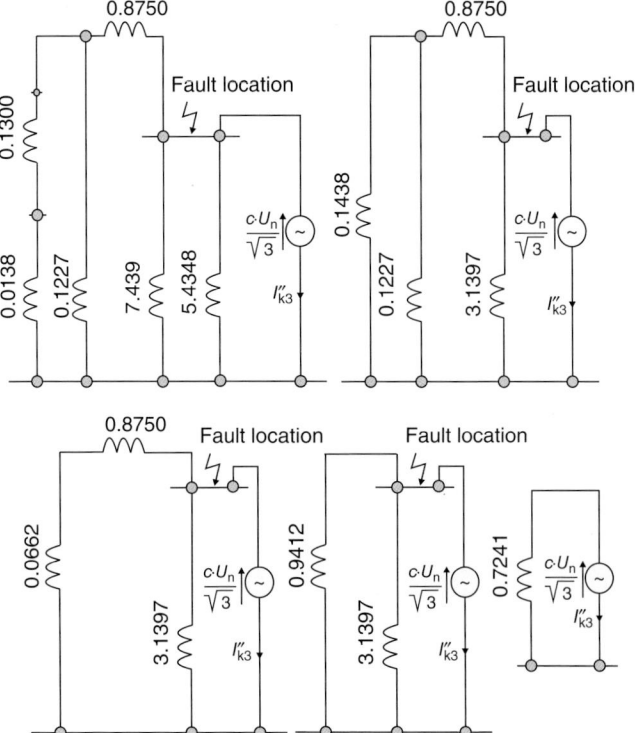

Figure 1.13 Equivalent circuit diagram in positive sequence.

Transformer 2:

$$X_{T2} = \frac{u_{kr}}{S_{rT2}} = \frac{7}{8\,\text{MVA}} = 0.8750\%/\text{MVA}$$

Asynchronous motor:

$$X_{M1} = \frac{I_{rM}/I_A}{S_{rM}} \cdot 100 = \frac{1}{5 \cdot 2.69\,\text{MVA}} \cdot 100 = 7.439\%/\text{MVA}$$

Asynchronous motor group:

$$X_{M2} = \frac{I_{rM}/I_A}{S_{rM}} \cdot 100 = \frac{1}{5 \cdot 8 \cdot 0.46\,\text{MVA}} \cdot 100 = 5.4348\%/\text{MVA}$$

Total reactance at the fault point:

$$S''_k = \frac{1.1 \cdot 100\%}{X_k} = \frac{1.1 \cdot 100\%}{0.7225\%/\text{MVA}} = 152\,\text{MVA}$$

2) Parts of each feed on the short-circuit power.
   With the total reactance, we can determine the circuit power.

$$S''_k = \frac{1.1 \cdot 100\%}{X_G} = \frac{1.1 \cdot 100\%}{0.7241} = 152\,\text{MVA}$$

Thus, the proportions of the individual feeds to the short-circuit power:

Part of each motor gives:

$$S''_{kM1} = \frac{0.1345}{1.381} \cdot 152\,\text{MVA} = 14.8\,\text{MVA}$$

Parts of the motor group is then:

$$S''_{kM1} = \frac{0.184}{1.381} \cdot 152\,\text{MVA} = 20.3\,\text{MVA}$$

Part of transformer 2:

$$S''_{kM1} = \frac{1.0625}{1.381} \cdot 152\,\text{MVA} = 116.9\,\text{MVA}$$

3) For the 220-kV grid:
   Part of the generator:

$$S''_{kG} = \frac{8.150}{15.104} \cdot 116.9\,\text{MVA} = 63.1\,\text{MVA}$$

   Part of the 220-kV network

$$S''_{kQ} = \frac{6.954}{15.104} \cdot 116.9\,\text{MVA} = 53.8\,\text{MVA}$$

4) Determining the μ and q factors
   From Figure 13.3, we get the μ factors for $t_v = 0.1\,\text{s}$.
   For each motor:

$$\frac{S''_{kM1}}{S_{rM1}} = \frac{14.8}{2.69} = 5.50 \Rightarrow \mu = 0.77$$

   For motor groups:

$$\frac{S''_{kM2}}{S_{rM2}} = \frac{20.3}{8 \cdot 0.46} = 5.52 \Rightarrow \mu = 0.76$$

   For the generator:

$$\frac{S''_{kG}}{S_{rG}} = \frac{63.1}{93.7} = 0.67 \Rightarrow \mu = 1$$

   The q factors are determined from the engine power/phase pair, according to Figure 13.3 for $t_v = 0.1\,\text{s}$.
   For each motor:

$$\frac{\text{Rated power}}{\text{Pol-pair}} = \frac{2.3}{2} = 1.15 \Rightarrow q = 0.55$$

   For motor groups:

$$\frac{\text{Rated power}}{\text{Pol-pair}} = \frac{0.36}{3} = 0.12 \Rightarrow q = 0.3$$

5) Determination of the individual feeds to the switching capacity
   For each motor:

$$S_{aM1} = \mu \cdot q \cdot S''_{kM1} = 0.77 \cdot 0.55 \cdot 14.8\,\text{MVA} = 6.3\,\text{MVA}$$

   For motor groups:

$$S_{aM2} = \mu \cdot q \cdot S''_{kM2} = 0.76 \cdot 0.3 \cdot 20.3\,\text{MVA} = 64.6\,\text{MVA}$$

For the generator:

$$S_{aG} = \mu \cdot S''_{kG} = 1 \cdot 63.1\,\text{MVA} = 63.1\,\text{MVA}$$

220-kV network:

$$S_{aQ} = \mu \cdot S''_{kQ} = 1 \cdot 53.8\,\text{MVA} = 53.8\,\text{MVA}$$

6) Calculation of short-circuit currents
   Initial short-circuit a.c.:

$$I''_k = \frac{S''_k}{\sqrt{3} \cdot U_n} = \frac{152\,\text{MVA}}{\sqrt{3} \cdot 6\,\text{kV}} = 14.63\,\text{kA}$$

Peak short-circuit current:

$$i_p = \kappa \cdot \sqrt{3} \cdot I''_k = 1.8 \cdot \sqrt{3} \cdot 14.63\,\text{kA} = 37.2\,\text{kA}$$

Breaking current:

$$I_a = \frac{S_a}{\sqrt{3} \cdot U_n} = \frac{127.8\,\text{MVA}}{\sqrt{3} \cdot 6\,\text{kV}} = 12.3\,\text{kA}$$

The sample was calculated for comparison with NEPLAN. As we can see, the results are the same (Figure 1.14).

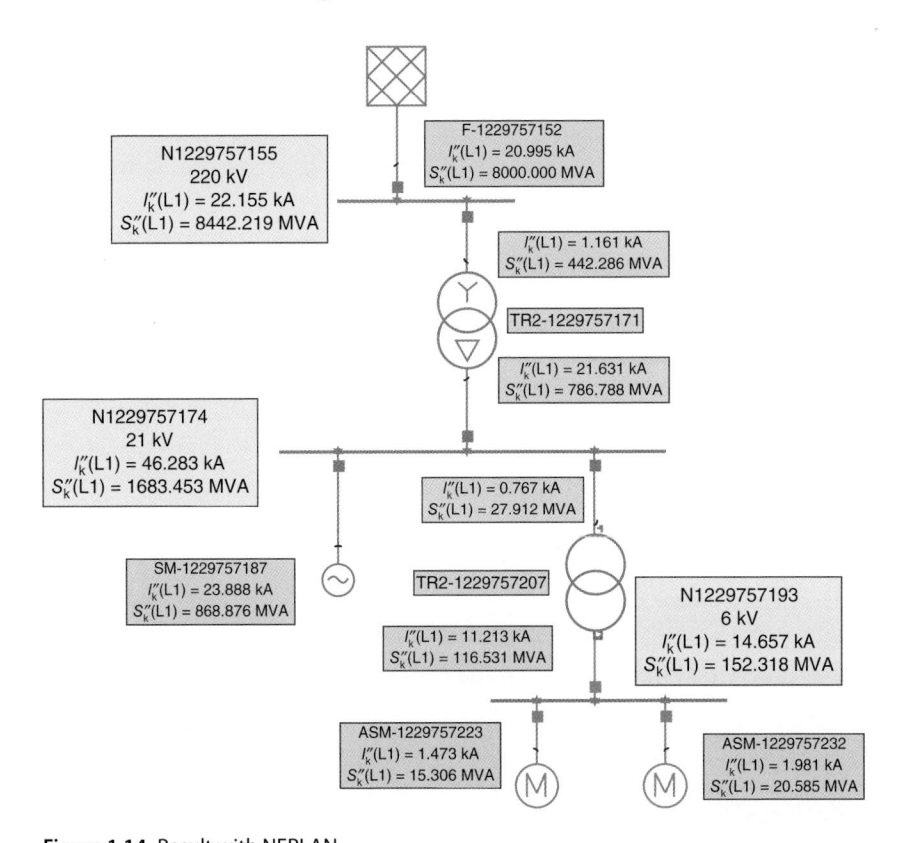

Figure 1.14 Result with NEPLAN.

2

Fault Current Analysis

Every electrical engineer and designer is committed to perform before and after the project is constructed, especially the calculation of the single- or three-phase short-circuit current in electric power systems to check the protection and short-circuit strength of the electrical systems for the selection of equipment and to adjust the protective devices.

This book deals with the calculation of short circuits in electrical installations and the load flow in low-voltage and high-voltage networks to the highest industry standards and regulations (IEC 60909-0). The purpose of the review of fundamentals of power systems and short-circuit calculation by using the standards is to obtain a better understanding of the basic complexities involved in a.c. systems.

The presentation of the regulation is maintained and summarized in a simple and understandable way, so that the reader can do his or her job without much trouble.

Each electrical system must be designed not only for the normal operation, but also for abnormal conditions such as short circuits. Therefore, electrical systems are to be dimensioned, so that neither people nor property values are jeopardized. The design for economy and safety is strongly dependent on the calculation of short circuits.

The knowledge of the magnitude of the short-circuit currents occurring is a decisive factor in the design and selection of system components in electrical power grids. The three-phase short-circuit currents cause a few exceptions to the strongest mechanical and thermal stresses of the equipment.

The initial short-circuit current of the three-phase short circuit, I''_{k3}, is the central value of any other relevant information for the design characteristic short-circuit current. All other short circuits can be calculated based on the standardized factors.

For the calculation of the initial short-circuit a.c. in three-phase power systems, there is an exact calculation using the superposition method and the other standardized method according to IEC 60909-0, the equivalent voltage source to the short circuit at the fault location in the system.

It is assumed that there is always a perfect short circuit in both the calculation methods (e.g., no arcing or contact resistances provided).

Short Circuits in Power Systems: A Practical Guide to IEC 60909-0, Second Edition. Ismail Kasikci.
© 2018 Wiley-VCH Verlag GmbH & Co. KGaA. Published 2018 by Wiley-VCH Verlag GmbH & Co. KGaA.

According to IEC 60909-0, the standardized short-circuit calculation is a very simple method, which has the advantage of minimum data and equipment parameters to get sufficiently precise results.

The method calculates the maximum and minimum short-circuit currents, irrespective of the load flow condition based on standardized correction factors for generators and transformers. On the other hand, the superposition method requires the complete data of the system.

The load flow, the voltages at all the feeds, the transformer tap positions, the consumer loads, and so on, must be taken into consideration. However, all of these do not necessarily lead to the maximum short-circuit current.

For the calculation of the short-circuit current, the voltage at the short-circuit node before the occurrence of a short circuit is required. By replacing this value with an equivalent voltage source, the short-circuit current can be determined approximately without taking into account the power flow calculation only with the condition of reverse feed.

A reasonable assumption for the voltage source in the "reverse feed" gives the equivalent voltage at the short-circuit location, $\frac{c \cdot U_n}{\sqrt{3}}$, where c is the voltage factor and U_n is the network nominal system voltage. The equivalent voltage source is the only active voltage of the system. All network feeders and synchronous and asynchronous motors are replaced by their internal impedances. Whereas the network's generators are simplified or replaced by their subtransient X_d'' that can be found or obtained from the data sheet of the generators.

Wind turbines with a full converter (a full-scale power converter) are limited to the rated current and do not contribute, or only marginally contribute, to the short-circuit current. According to IEC 60909-0, asynchronous generators are modeled by the equivalent impedance Z_M.

The increasing amount of installed capacity in the high-voltage level increases the short-circuit power in the networks, whereby higher fault currents arise. Even the choice of neutral point treatment, as well as the network, forms a massive influence on the protection concept.

Networks with isolated neutral or resonant-grounded phase faults in overhead networks are usually self-extinguishing. The protection relays should not intervene in this case, but only when standing shorts, as they occur in cable networks or overhead line conductors. For these special arrangements, earth fault relays or watt-metric relays are required.

Low-impedance grounded systems and single-phase and multiphase short circuits are detected by the same electric protective device. Each short circuit causes a high-fault current, which is detectable and switched off selectively by the protective device. Typically, a multiphase reclosure tries to give the chance of self-healing and, on the other side, the network users are affected by the disorder. If it fails, short interruption is switched off three-phase.

For the protection of medium-voltage networks, which are not usually designed as a mesh system, an independent maximum time overcurrent protection is used. The setting of the overcurrent relay of trip factor k must be known, which links the breaking current and the current setting. For all of these, the short-circuit current calculation is essential.

A three-phase system can be temporarily or permanently disturbed by errors, especially short-circuit measures or consumers. To calculate, the operating variable computational models and algorithms are needed for solving systems of power generation, transmission, and distribution of a comprehensive tool for planning, design, analysis, optimization, and management of any network of energy supply.

Due to the liberalization of energy markets and, in particular, the rapid expansion of renewable energy, the demands on the network planning and management processes are becoming increasingly complex. Complex network technologies and network topologies are impossible today without calculation programs.

The scope of power calculations and system planning may be as follows:

- balanced and unbalanced load flow calculations in coupled and meshed a.c./d.c. systems, including any power plant and network control functions;
- short-circuit current calculation, according to IEC 60909-0, IEEE 41, ANSI C37, G74, and IEC 61363, as well as the complete overlay method taking into account the voltage support of inverters and multiple-error-calculating any fault;
- fast power failure calculation with the support of parallel computing structures;
- network state estimation (as for supervisory control and data acquisition, SCADA, applications);
- protection coordination and distance protection devices, including protection simulation;
- arc fault calculation, according to IEEE 1584-2002 and NFPA 70E-2012;
- calculation of power quality, including harmonic load flow (IEC 61000-3-6); flicker calculation, according to IEC 61400-21 and IEC 61000-4-15; and filter sizing;
- separation points optimization in medium-voltage systems and optimization of compensation devices;
- optimization of transformer tap changers in directional power flow;
- reliability calculations, including optimal restocking strategies;
- optimal load flow calculation for active and reactive power optimizations (optimal power flow, OPF);
- stability calculations (root mean squares (RMS)), which include power plants, consumers, and protection devices;
- calculation of electromagnetic transient (EMT) phenomena (e.g., overvoltages and ferroresonances by transformer saturation);
- computing eigenvalues, eigenvectors, and participation factors; and
- modeling virtual power plants.

Appropriate programs can provide the assessment and selection of electrical equipment, the calculation of the mechanical and thermal short-circuit strengths, the calculation of short-circuit currents, selectivity, and backup protection for the selection of overcurrent protective devices, as well as the calculation of the temperature that increases in the control cabinets.

Transformers with or without medium voltage, generators, and neutral mains can supply the networks. A neutral power supply can be displayed by setting the impedances of the loop impedance or short-circuit currents. Circuits can be optionally charged to the hedging of parallel cables through a protective device, a single hedge parallel cable with several protective devices, and dimensioned. The chosen feeds can be connected to each other through directed or undirected couplings. Thus, it can be explained by the resulting possibility of defining the various required operating modes (e.g., normal operation and emergency operation), represent the main supply practical and be included in the calculation.

A parallel network operation can be displayed by combining two nonsimilar feeds via a nondirectional clutch. Feeds on the subdistribution level can also be included in the calculation, and a presentation of separate networks is possible.

In the case of selecting a transformer with medium voltage, a medium-voltage switchgear is required and is an important factor to size the transformer's feeder elements.

Transformer, generator, or neutral network in feed according to the selection of feeder types, switchgear to the transformer or generator or neutral network supply, which consists of a switching device, a circuit breaker, disconnector with fuse, fuse switch disconnectors or fuse with base, cable or busbar connection with switching device before the feed point, which stands as a type of the circuit breaker, a circuit breaker, air circuit breakers, switch disconnectors with fuse, fuse switch disconnectors or fuse with base to choose from.

The distributors are subdistribution, group switch, busway, busway center feed, or distribution with equivalent impedances for selection. Concerning selecting these elements, there are specific requirements to meet the design again (e.g., whether the connection line is to be performed as a busbar or as a cable and which and how many switching devices are to be used). A cable selection section is also provided for laying so that the influenced values of the current-carrying capacity are taken into account in the dimensioning. The distributors are always inserted on a busbar in the graphic. This can be the busbar, which symbolizes the entry point, the busbar of an already connected distributor, or the representation of a current rail track so that, in this way, the network can be branched as a radial network.

Final circuits are available as elements consumers with fixed connection, socket circuits, motors, loading units, capacitors, and dummy loads as items to choose from. These, in turn, are connected to the busbar of existing subdistributors or the representation of a current rail track, or directly to the entry point symbolizing busbar. There are also various options in the placement of these elements in the network diagram.

In practice, a selectivity detection is often required (e.g., the equipment of emergency power supply). The switching device selection and possibly backup protection are taken into account (e.g., the switching capacity of a downstream switch can be increased as the upstream circuit breaker trips simultaneously and, thereby, limits the current).

IEC 60909-0 includes a standard procedure for the calculation of short-circuit currents in low-voltage and high-voltage networks up to 550 kV at 50 or 60 Hz [1]. The purpose of this procedure is to define a brief, general, and easy-to-handle calculation procedure, which is intended to lead with sufficient accuracy to results on the safe side. For this purpose, we calculate with an equivalent voltage source at the position of the short circuit. It is also possible to use the superposition method here.

A complete calculation of the time behavior for far-from-generator and near-to-generator short circuits is not required here. In most cases, it is sufficient to calculate the three-phase and the single-phase short-circuit currents, assuming that for the duration of the short circuit no change takes place in the type of short circuit, the step switch of the variable-ratio transformers is set to the principal tapping and arc resistances can be neglected.

The short-circuit currents and short-circuit impedances can always be determined by the following methods:

- calculation by hand,
- calculation using a PC,
- using field tests, and
- measurements on network models.

The short-circuit currents and short-circuit impedances can be measured in low-voltage networks by measuring instruments directly at the assumed position of the short circuit.

For the dimensioning and the choice of operational equipment and overcurrent protective equipment, the calculation of short-circuit currents in three-phase networks is of great importance, since the electrical systems must be designed not only for the normal operational state, but also to withstand fault situations.

IEC 60909-0 describes the basis for calculation, which consists of the following parts:

1) Main Part I describes the application areas and the definitions.
2) Main Part II explains the characteristics of short circuits and their currents and the calculation method of equivalent voltage source.
3) Main Section III deals with the short-circuit impedances of electrical equipment, the impedance correction factors of generators, power transformers, and power stations.
4) Main Section IV provides the calculation of the individual short-circuit currents.

Summary of IEC 60909-0:

- Restructuring of calculations.
- The supplementary pages with examples and conversion factors supplement the theoretical part.
- The high-voltage and low-voltage networks are treated in the same way.
- The rules for calculating the smallest and largest short circuits are equally valid for high-voltage and low-voltage networks.
- The corrections to the impedances of generators and power plant blocks do not depend on the time behavior of the short-circuit current.
- In low-voltage networks, a temperature rise of 20–80 °C is assumed for the single-phase short circuit. This increases the resistance of the cable or conductor by a factor of 1.24.
- The indices for symmetrical components (0, 1, 2) are internationally standardized.

- The short-circuit currents are determined with the equivalent voltage source method, in accordance with IEC 60909-0. For this, the internal voltages in the network are short-circuited. The only effective voltage at the position of the short circuit is then $\frac{c \cdot U_n}{\sqrt{3}}$, where c is the voltage factor.
- The superposition method is the more accurate method. However, this requires the knowledge of the network conditions before the occurrence of the short circuit.

3

The Significance of IEC 60909-0

The short circuit is an undesired network operating state. This state can cause overloading of the operational equipment (transformers, transmission lines, cables, and generators) as well as damage to the insulation. The transition from normal operation to operation under fault conditions takes place through electromagnetic and electromechanical transient phenomena, which influence the magnitude and temporal behavior of the short-circuit currents. These processes depend on the current sources, the position of the short circuit, and the time from the onset of the short circuit until it decays.

The most common type of short circuit is the "dead" short circuit, that is, the impedance at the faulty location is negligibly small. Short-circuit currents are as a rule much larger than the load currents. The thermal and dynamic stresses resulting from these short circuits can destroy the operational equipment and endanger persons. During the planning and project management of electrical systems, the smallest short-circuit current I''_{k1min} must therefore be determined and taken into account for configuring the overcurrent protection equipment and the largest short-circuit current I''_{k3max} for dimensioning the operational equipment. Only in this way can electrical systems be correctly dimensioned and protected, allowing their safe and economic operation. Otherwise, unpleasant consequences can be expected. A few of these are listed below:

- impairment of safety and reliability of the power supply
- interruption of the power supply
- destruction of system components
- emergence of mechanical and thermal stresses in the operational electrical equipment
- emergence of overvoltages.

Until 1962, VDE 0670 switchgear regulations were the standard for short-circuit calculations. VDE 0102 was released in 1971 and revised in 1975, so that in Germany the calculations for low-voltage and high-voltage networks were made uniform. In the meantime, further developments in electrical power systems have taken place and various software has appeared on the market. In order to meet the requirements and developments, in 1985 both parts,

Short Circuits in Power Systems: A Practical Guide to IEC 60909-0, Second Edition. Ismail Kasikci.
© 2018 Wiley-VCH Verlag GmbH & Co. KGaA. Published 2018 by Wiley-VCH Verlag GmbH & Co. KGaA.

"Calculation of Three-Phase Networks" in accordance with DIN VDE 0102, were extended to include the newly summarized information about operational equipment.

In 1988, based on this draft version, the IEC publication "Short Circuit Current Calculation in Three-Phase AC Systems" appeared. In 1990, the present standard IEC 60909-0 "Calculation of Short Circuit Currents in Three-Phase Networks" was released.

The method of symmetrical components is used for both symmetrical and asymmetrical short circuits. The capacitances of conductors and the shunt admittances of passive loads are neglected here. In this method, motors are treated as generators in high-voltage networks and are neglected in low-voltage networks. For a double ground connection, only the voltage of the short-circuit current source is used as the effective voltage.

For the assessment of electrical systems, such as breaking conditions, protective measures, thermal and mechanical short-circuit strengths, selectivity, and voltage drop, comprehensive calculations are performed.

For the calculation of short circuits, the following are important:

- power draw and documentation of result
- short-circuit currents
- transferred short-circuit currents
- impedance protection, maximum current-dependent time relays, and current-independent time relays
- examination of breaking conditions
- proof of stability of switchgear, switching devices, cables, and conductors against short circuits.

Figure 3.1 makes clear the importance and the range of applicability of short-circuit current calculations and additional calculations relating to other regulations.

In medium-voltage networks, the type of the smallest fault current that must be considered depends on the type of neutral point design. This is decisive for the type of network protection required.

The maximum calculated short-circuit current (I''_{k3}) is responsible for the selection and rating of equipment regarding the mechanical and thermal stresses and the determination of the time–current coordination of protective relays. The system remains symmetrical. The balanced three-phase short-circuit current can be performed using a single-phase equivalent voltage source method that has only line-to-neutral voltage and impedance in positive-sequence components.

The minimum short-circuit current (I''_{k1}) has to be calculated for the selection of the system protection. According to the IEC 60364-4-41, a protective device interrupts the supply to the line conductor in the event of a fault between the line conductor and an exposed-conductive-part or a protective conductor in the circuit within the disconnection time required.

Table 3.1 provides an overview of the design criteria and short-circuit current according to IEC 60909-0, which are considered for the planning and design of electrical power systems.

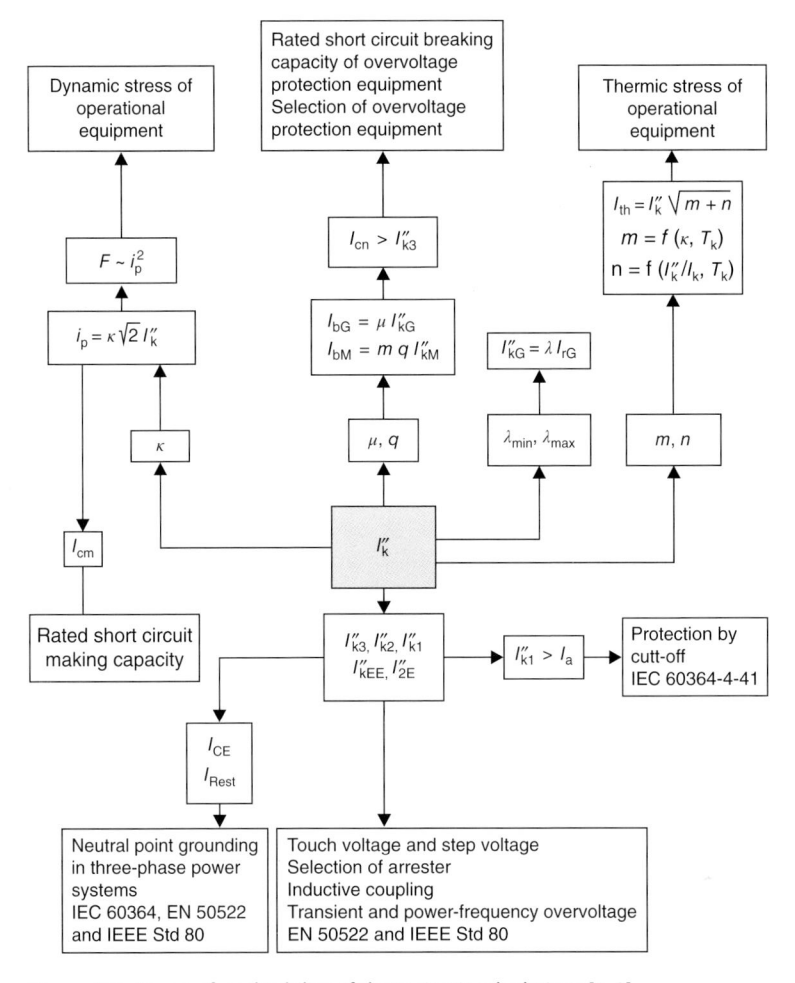

Figure 3.1 Range of applicability of short-circuit calculations [1, 2].

Table 3.1 Selection of short-circuit currents.

Design criteria	Physical effects	Short circuit	Limitation
Dynamic stress for components	Forces, F	Peak short-circuit current, i_p	Instantaneous value
Thermal stress for components and lines	Temperature increase	Steady-state short-circuit current	RMS value
Short-circuit breaking capacity for switching devices	Thermal equivalent short-circuit current, I^{th}	Breaking current, I_b	RMS value
Switching devices Protection setting Tripping of relays	Protective measures	Initial symmetrical short-circuit current Steady-state short-circuit current, I_k	Single-phase short-circuit current, I''_{k1}

4

Supply Networks

All electrical appliances that are used in household and commercial and industrial applications work with low voltage. High voltage is used not only for the transmission of electrical energy over very large distances, but also for regional distribution to the load centers. According to international rules, there are only two voltage levels:

- Low voltage: up to and including 1 kV a.c. (or 1500 V d.c.)
- High voltage: >1 kV a.c. (or 1500 V d.c.).

However, as different high-voltage levels are used for transmission and regional distribution and because the tasks and requirements of the switchgear and substations are also very different, the term *medium voltage* has come to be used for the voltages required for regional power distribution that are part of the high-voltage range from 1 kV a.c. up to and including 52 kV a.c. Most operating voltages in medium-voltage systems are in the 3–40.5 kV a.c. range. The electrical transmission and distribution systems connect power plants and electricity consumer and consist of (Figure 4.1) the following:

- Power plants, for generators and station supply systems.
- Transformer substations of the primary distribution level (public supply systems or systems of large industrial companies), in which power supplied from the high-voltage system is transformed to medium voltage.
- Local supply, transformer, or customer transfer substations for large consumers (secondary distribution level), in which the power is transformed from medium to low voltage and distributed to the consumer.

The following stages of calculating and dimensioning circuits and equipment are of a great interest. They can be worked out efficiently using modern dimensioning tools, so that there is more freedom left for the creative planning stage of finding conceptual solutions.

Concept finding:

- analysis of the supply task
- selection of the network configuration

Short Circuits in Power Systems: A Practical Guide to IEC 60909-0, Second Edition. Ismail Kasikci.
© 2018 Wiley-VCH Verlag GmbH & Co. KGaA. Published 2018 by Wiley-VCH Verlag GmbH & Co. KGaA.

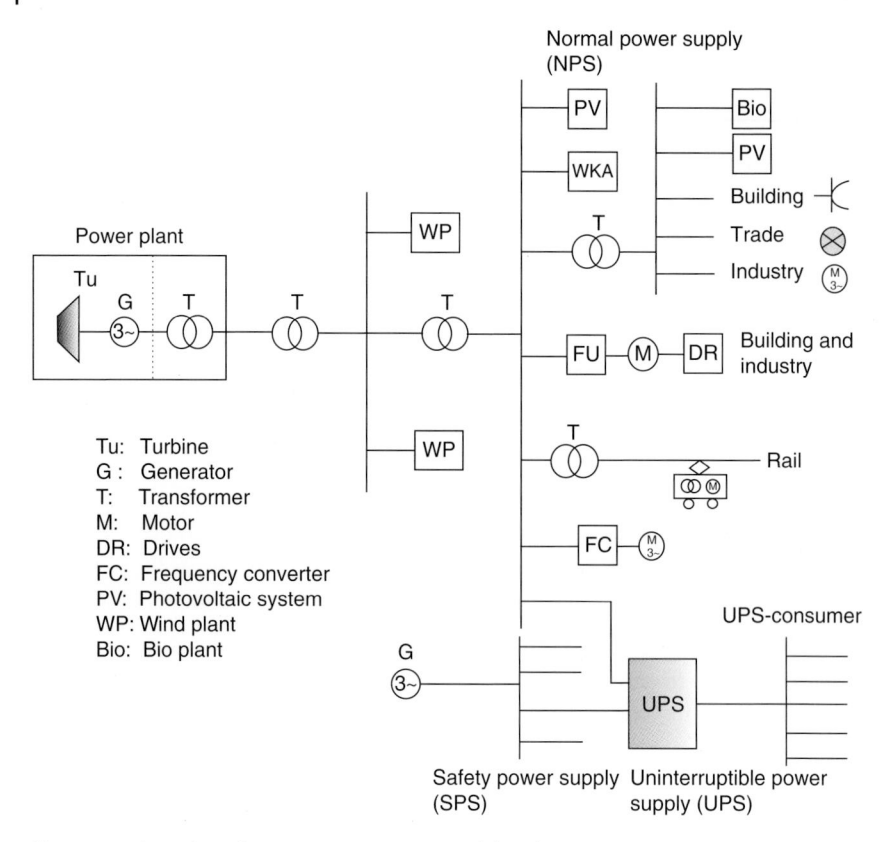

Figure 4.1 Overview of a power transmission and distribution system.

- selection of the type of power supply system
- definition of the technical features.

Calculation:

- energy balance
- load flow (normal/fault)
- short-circuit currents (uncontrolled/controlled).

Dimensioning:

- selection of equipment, transformers, cables, protective and switching devices, and so on
- requirements according to selectivity and back-up protection.

Electrical supply system configurations that can be found in practice will be briefly explained here.

4.1 Calculation Variables for Supply Networks

- Short-circuit currents in accordance with IEC 60909-0
- Ground loop impedance

Figure 4.2 Line supplied from a single source.

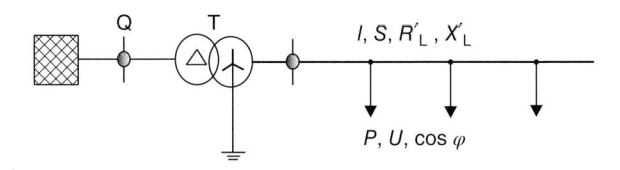

I, S, R'_L, X'_L

$P, U, \cos \varphi$

- Peak short-circuit current
- Initial symmetrical short-circuit power
- Load flow
- Load distribution for the network.

4.2 Lines Supplied from a Single Source

All consumers are centrally supplied from one power source. Each connecting line has an unambiguous direction of energy flow. A feed-in supplies any number of distributed loads along the line (e.g., bus distributor, Figure 4.2).
Characteristics of this input:

- No security of supply
- High network losses.

4.3 Radial Networks

All consumers are centrally supplied from two to n power sources. They are rated as such that each of it is capable of supplying all consumers directly connected to the main power distribution system (stand-alone operation with open couplings). If one power source fails, the remaining sources of supply can also supply some consumers connected to the other power source. In this case, any other consumer must be disconnected (load shedding). A feed-in supplies a large number of branched lines (e.g., industrial network, Figure 4.3). As with the simple line-fed, all consumers are switched off after the fault.
Characteristics of this input:

1) Advantages
   - Very clearly arranged
   - Simple network protection
   - Easily calculated.
2) Disadvantages
   - Single source supply
   - Low security of supply
   - Poor voltage stabilization
   - High voltage drop.

4.4 Ring Networks

The ring network is usually fed from two sources (e.g., industrial network, Figure 4.4). After shutdown of the error, the network can be put back into

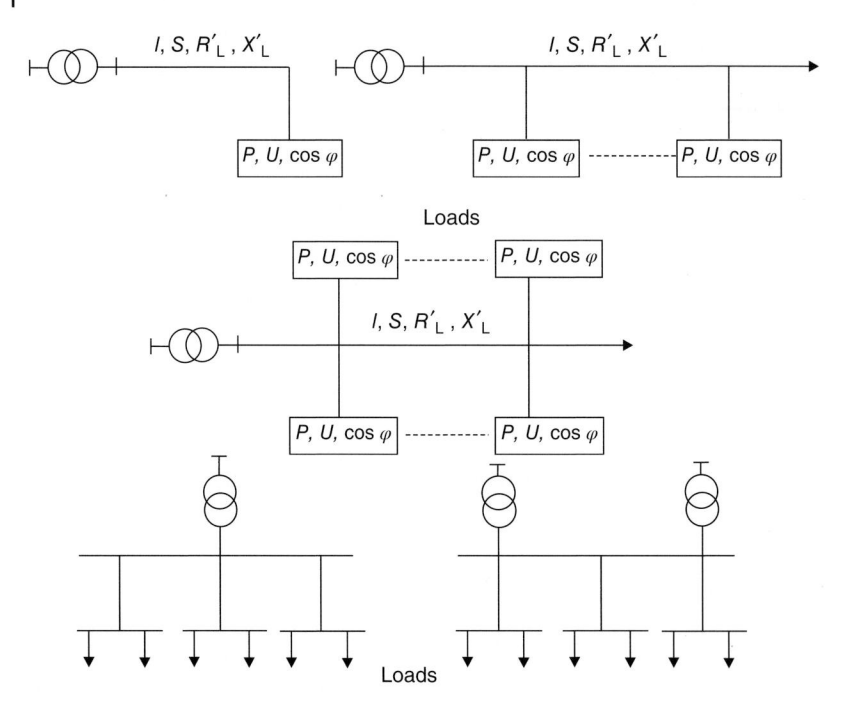

Figure 4.3 Radial network.

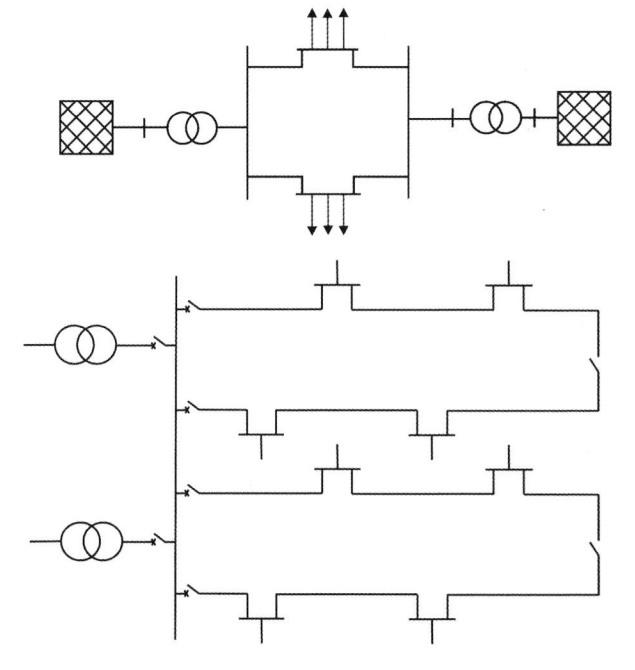

Figure 4.4 Ring network.

operation. The ring networks are usually open-operated networks. When a short circuit occurs, the power may be fed from both sides while the affected line on both sides will be unlocked.

Ring system in an interconnected network individual radial systems, in which the connected consumers are centrally supplied by one power source, is additionally coupled electrically with other radial systems by means of coupling connections. All couplings are normally closed.

Characteristics of this input:

1) Advantages
    - Increased voltage security
    - Better load balancing
    - Better voltage stability
    - Very high voltage quality.
2) Disadvantages
    - Network protection difficult.

4.5 Meshed Networks

Radial system with power distribution via busbars. In this special case of radial systems that can be operated in an interconnected network, busbar trunking systems are used instead of cables. The supply of each load is ensured by the linking of several supply lines and in part by several feed-ins. The failure of one line or one feed-in can normally be compensated by the remaining part of the network (e.g., computer centers and chemical industry, Figure 4.5).

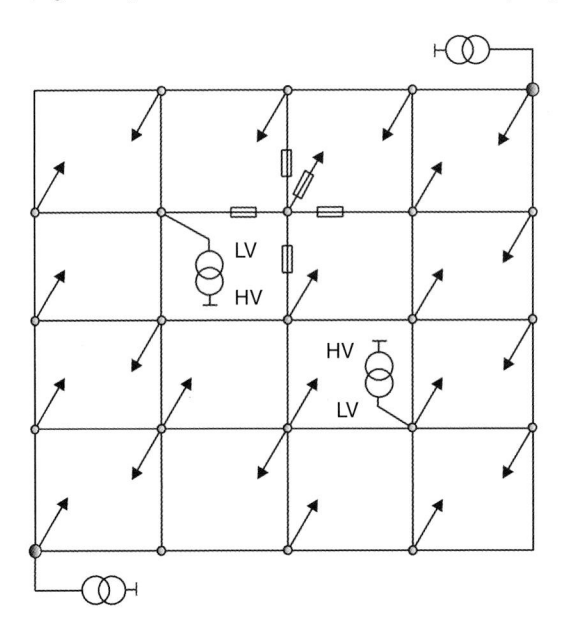

Figure 4.5 Meshed network.

Characteristics of this input:

1) Advantages
    - High security of supply
    - Good voltage stability
    - Good load balancing
    - Low network losses.
2) Disadvantages
    - Selectivity is very poor
    - Impedance protection is complicated
    - High short-circuit currents in the system
    - Extensive short-circuit and load flow calculations needed.

5

Network Types for the Calculation of Short-Circuit Currents

5.1 Low-Voltage Network Types

In this section other types of networks are shown, which are sometimes encountered in practice and in which the short-circuit currents are fed from different sources. The most frequently found network types in the public and the industrial sectors are radial networks. For these networks, the calculation of short-circuit currents is very simple. The medium-voltage and low-voltage sides can be configured arbitrarily, according to the requirements for supplying power (Figure 5.1).

In industrial networks, power supply to the systems must not fail. In the event of a malfunction, switchover can take place from another transformer (Figure 5.2). Radial networks with redundant inputs have a higher security of supply and a high voltage quality (Figure 5.3). The transformers can be loaded uniformly. The meshed network with different inputs is the most widely used network type for electrical distribution in the industry (Figure 5.4). The disadvantages of such networks are high costs of investment and an arrangement that is difficult to oversee.

5.2 Medium-Voltage Network Types

For optimum design of medium-voltage systems, the following points that are not explained in further detail are of great importance:

- Network losses
- Complexity of maintenance
- Investment costs
- Power requirement coverage
- Security of supply
- Ease of operation
- Environmental compatibility.

Figure 5.5 illustrates an industrial load center network that supplies individual large-scale loads and Figure 5.6 a ring network, in which the supply of power is ensured.

Short Circuits in Power Systems: A Practical Guide to IEC 60909-0, Second Edition. Ismail Kasikci.
© 2018 Wiley-VCH Verlag GmbH & Co. KGaA. Published 2018 by Wiley-VCH Verlag GmbH & Co. KGaA.

40 | 5 Network Types for the Calculation of Short-Circuit Currents

Figure 5.1 Simple radial networks with different centers of load distribution.

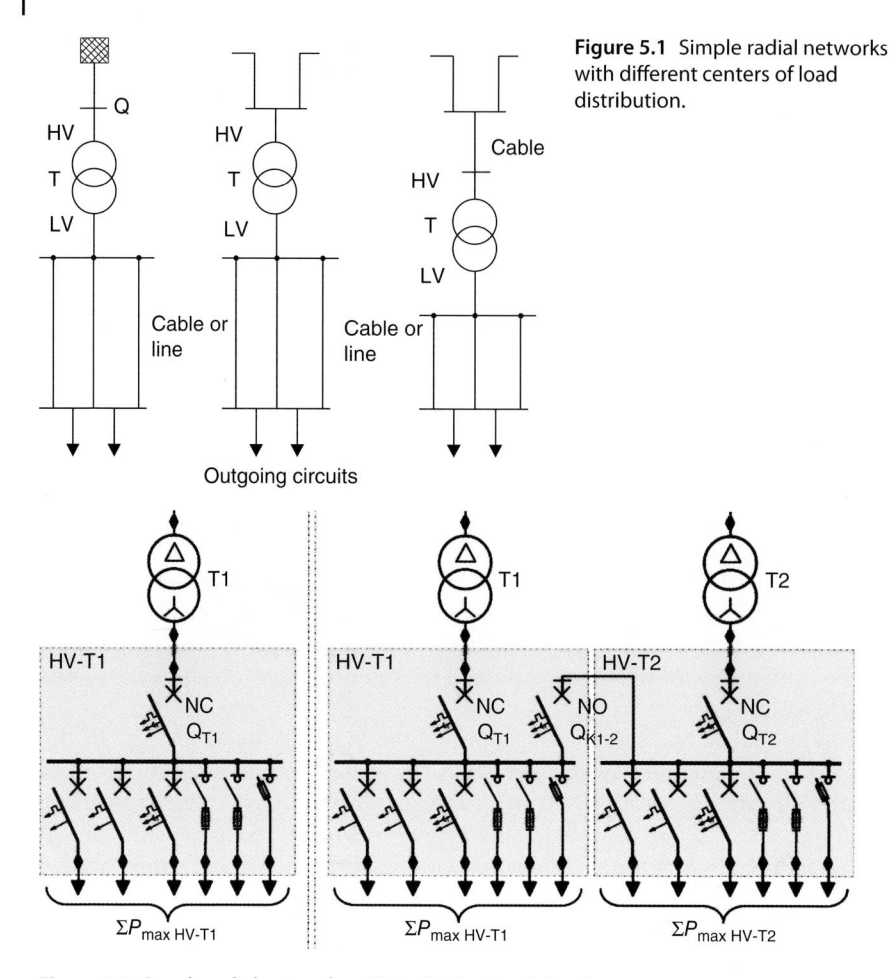

Figure 5.2 Simple radial networks with individual load circuits.

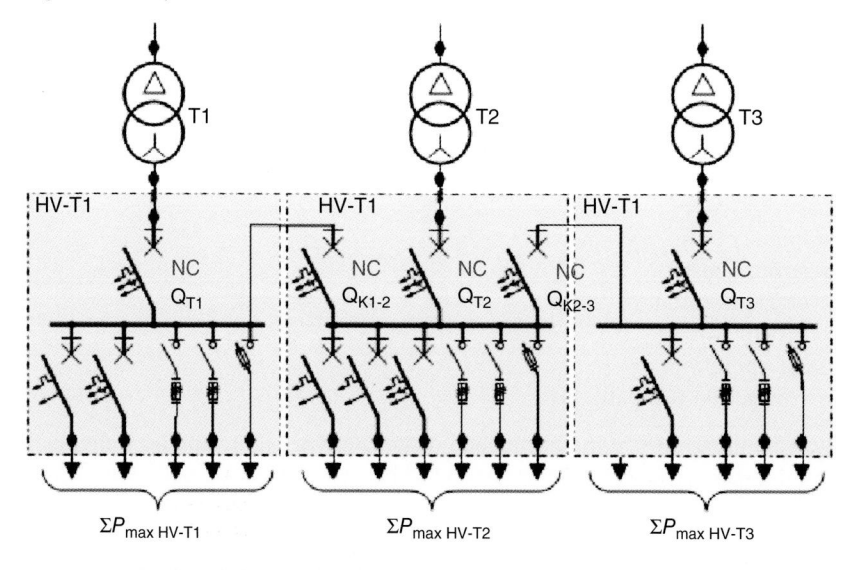

Figure 5.3 Simple radial networks with redundant inputs.

5.2 Medium-Voltage Network Types

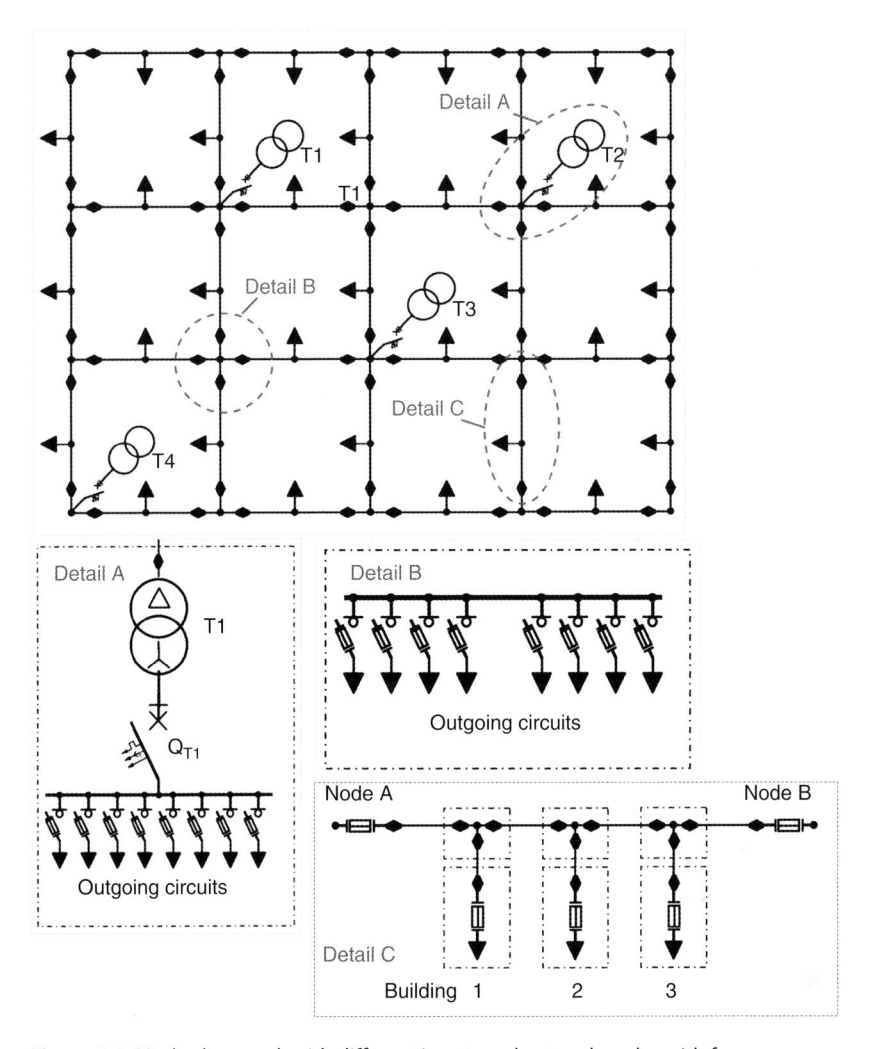

Figure 5.4 Meshed network with different inputs and network nodes with fuses.

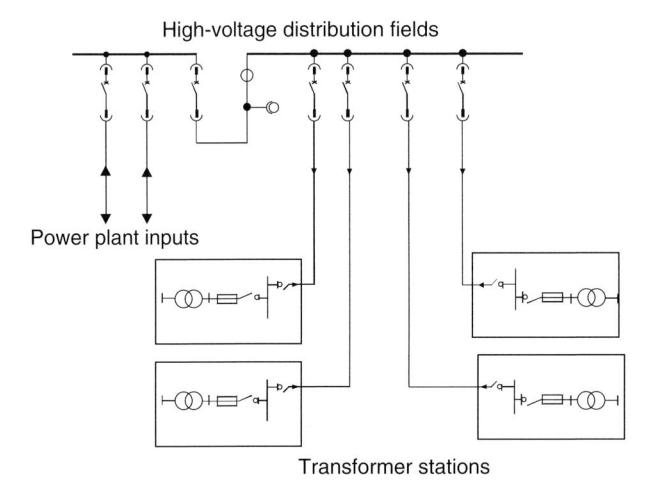

Figure 5.5 Industrial load center network.

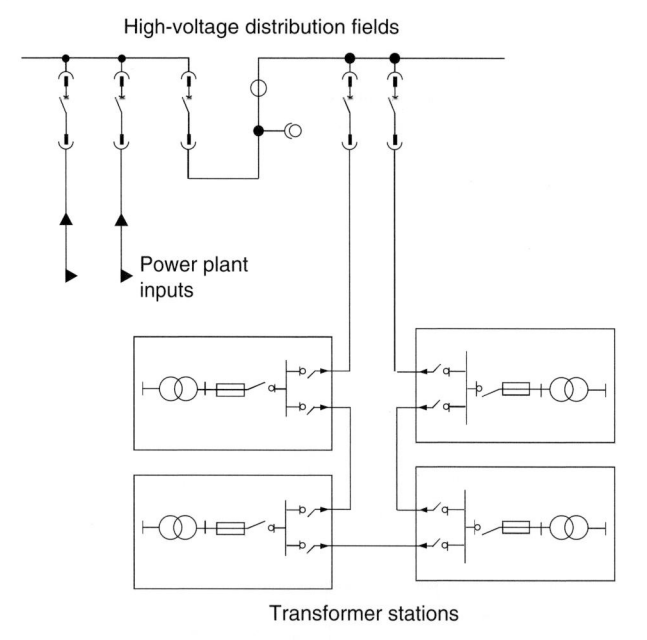

Figure 5.6 Industrial ring network.

Figures 5.5–5.9 depict medium-voltage network types with different possible structures. The network with open rings (Figure 5.5) is connected through circuit breakers to the busbar. The ring can be opened and closed by the load switch disconnector. The network with remote station (Figures 5.6 and 5.7) with network supporting structure is connected through several input cables to the busbar of the transformer substation. An industrial area can also be supplied from several transformer stations (Figure 5.8). The short-circuit current can also be fed from different sources, as shown in Figure 5.9.

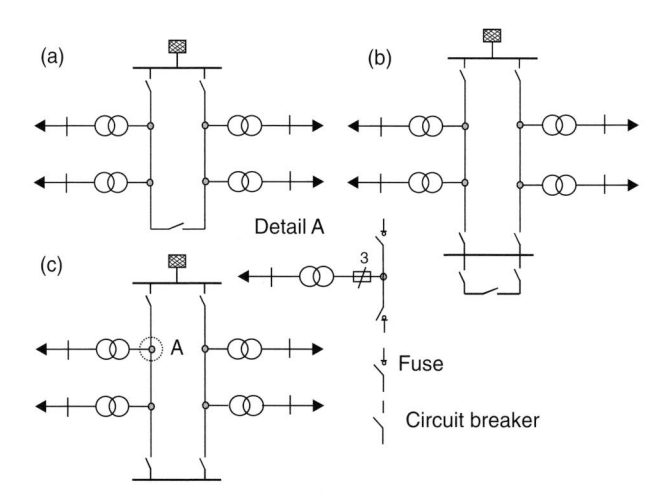

Figure 5.7 (a) Ring network, (b) network with remote station, and (c) network supporting structure.

5.2 Medium-Voltage Network Types | 43

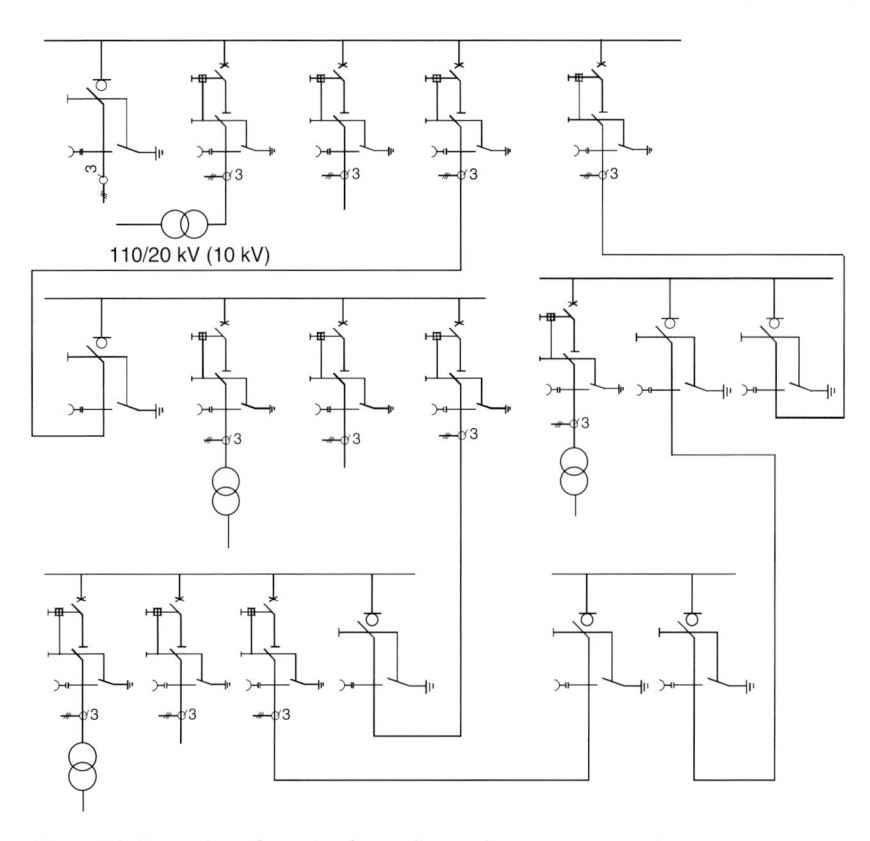

Figure 5.8 Network configuration for medium-voltage systems.

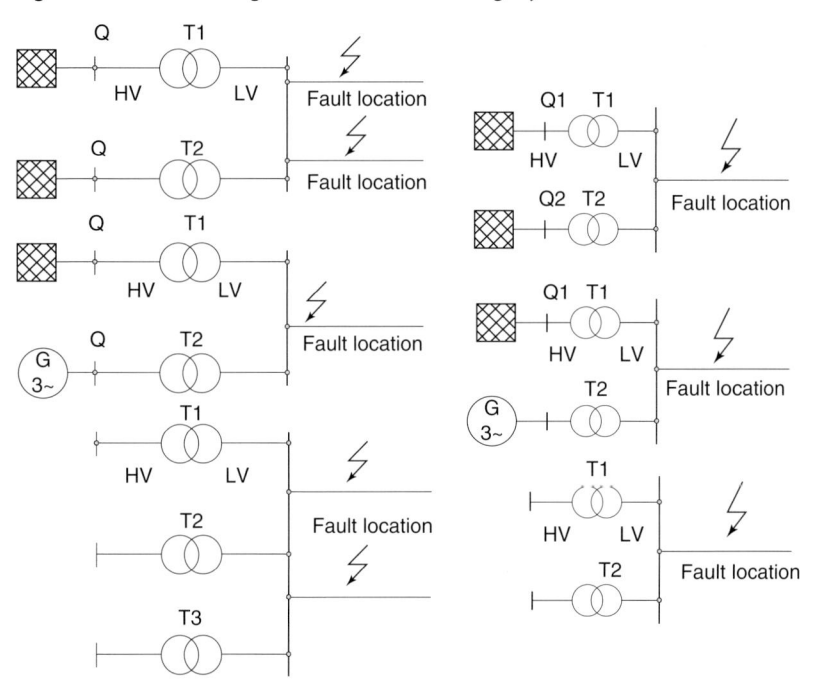

Figure 5.9 Short circuit with simple inputs.

5.3 High-Voltage Network Types

High-voltage substations are points in the power system where power can be pooled from generating sources, distributed and transformed, and delivered to the load points. Substations are interconnected with each other, so that the power system becomes a meshed network. This increases the reliability of the power supply system by providing alternate paths for flow of power to take care of any contingency, so that power delivery to the loads is maintained and the generators do not face any outage.

The high-voltage substation is a critical component in the power system, and the reliability of the power system depends upon the substation. Therefore, the circuit configuration of the high-voltage substation has to be selected carefully and designed. Busbars are that part of the substation where all the power is concentrated from the incoming feeders and distributed to the outgoing feeders.

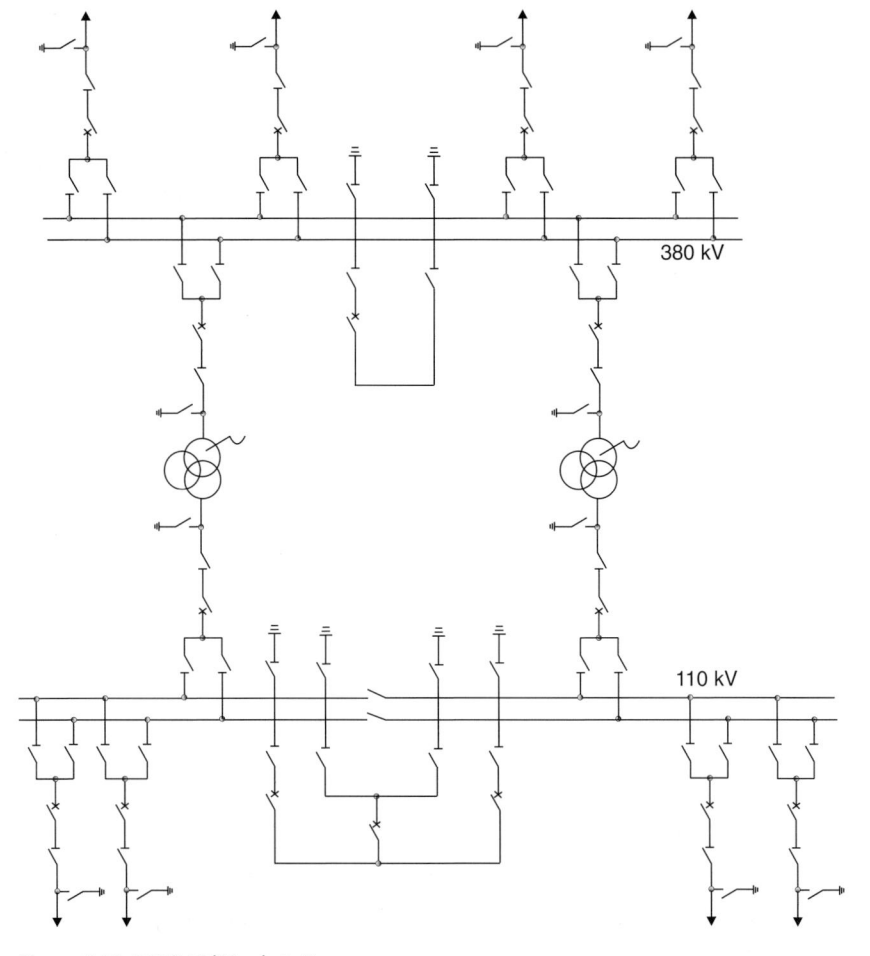

Figure 5.10 380/110 kV-substation.

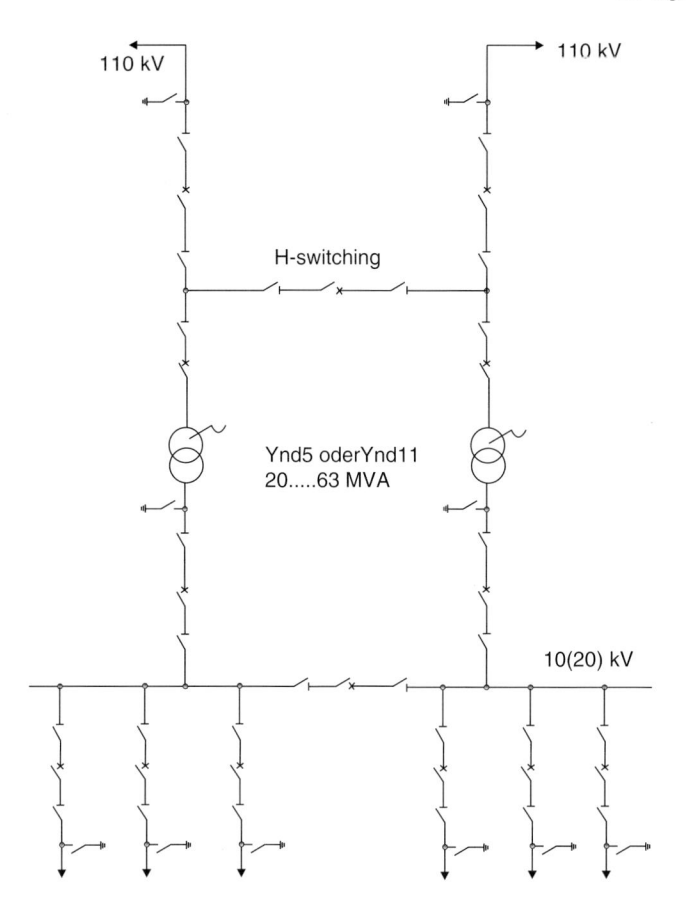

Figure 5.11 High-voltage transmission line.

That means that the reliability of any high-voltage substation depends on the reliability of the busbars present in the power system. An outage of any busbar can have dramatic effects on the power system. An outage of a busbar leads to the outage of the transmission lines connected to it.

As a result, the power flow shifts to the surviving healthy lines that are now carrying more power than they are capable of. This leads to tripping of these lines, and the cascading effect goes on until there is a blackout or similar situation. The importance of busbar reliability should be kept in mind when taking a look at the different busbar systems that are prevalent.

The three-phase high-voltage systems are found in the combined operation, in cities and in industrial centers. The voltage level is determined by the transmission and short-circuit power. The switchgear is designed as interior room or outdoor switchyard. For the configuration and calculation of switchgear of the scope of the system and the number of the busbars and their equipment is very important.

Figures 5.10–5.12 show different types of high-voltage power systems such as generation, transmission, and distribution.

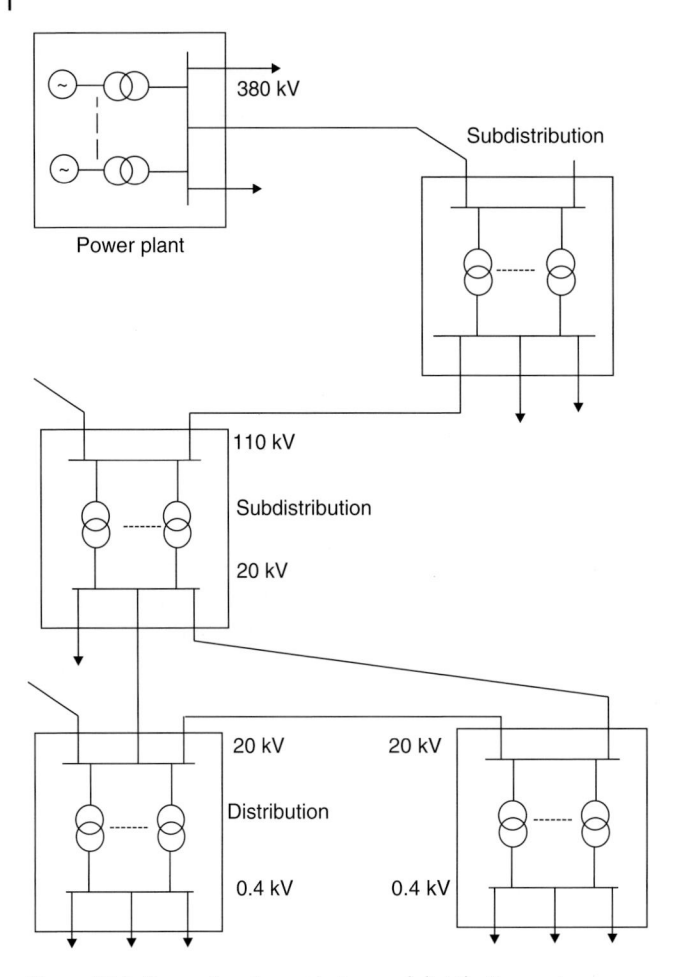

Figure 5.12 Generation, transmission, and distribution system.

6

Systems up to 1 kV

In low-voltage power system, the power supply systems according to the type of connection to earth are described in IEC 60364-1 and NEC 250. The type of connection to earth must be selected carefully for the low-voltage network, as it has a major impact on the expense required for protective measures. Earthing (grounding) influences the system's electromagnetic compatibility (EMC).

In a TN system, in the event of a short circuit to an exposed conductive part, a considerable part of the single-phase short-circuit current is not fed back to the power source via a connection to earth but via the protective conductor (PE). The comparatively high single-phase short-circuit current allows for the use of simple protective devices such as fuses or miniature circuit breakers, which trip in the event of a fault within the permissible tripping time.

TN systems are preferably used today in low-voltage networks. When using a TN-S system, residual currents in the building and thus an electromagnetic interference by galvanic coupling can be prevented in normal operation because the operating currents flow back exclusively via the separately laid isolated N conductor. In the case of a central arrangement of the power sources, we always recommend the TN system as a rule. In that, the system earthing is implemented at one central earthing point (CEP), for example, in the low-voltage main distribution system, for all sources. If a PEN conductor is used, it is to be insulated over its entire course – this includes the distribution system.

The magnitude of the single-phase short-circuit current directly depends on the position of the CEP. Four-phase switches must be used if two TN-S subsystems are connected to each other. In TN-S systems, only one earthing bridge may be active. Therefore, it is not permitted that two earthing bridges be interconnected via two conductors.

Today, networks with Terra–Terra (TT) systems are only used in rural supply areas and in a few countries. In this context, the stipulated independence of the earthing systems must be observed. Networks with an IT system are preferably used for rooms with medical applications in accordance with IEC 60364-7-710 in hospitals and in production, where no supply interruption is to take place upon the first fault, for example, in the cable and optical waveguide production.

Short Circuits in Power Systems: A Practical Guide to IEC 60909-0, Second Edition. Ismail Kasikci.
© 2018 Wiley-VCH Verlag GmbH & Co. KGaA. Published 2018 by Wiley-VCH Verlag GmbH & Co. KGaA.

6 Systems up to 1 kV

The TT system as well as the IT system requires the use of residual current protective devices (RCDs) for almost all circuits [4].

Earthing systems are classified according to the type of ground connection of the power source (input type) and the type of exposed conductive parts of the electrical system (IEC 60364 part 41). In this chapter, the three different types of systems and protective measures are briefly described.

6.1 TN Systems

According to IEC 60364, the TN system is preferred not only in the area of public low-voltage networks but also in the industrial sector. In TN systems, the grounding of the operational equipment is implemented by connection to the PEN conductor or for small cross-sections to the protective ground conductor (PE) (Figure 6.1).

6.1.1 Description of the System is Carried Out by Two Letters

First letter: describes the grounding conditions of the power source

T: direct grounding of a point
I: isolated to the earth.

Second letter: describes the grounding conditions of the exposed conductive parts of the electrical system

N: direct connection of exposed conductive parts through PEN or PE to system grounding, continuing in TN system
S: neutral conductor (N) and protective ground conductor (PE), as separate and separated conductors
C: neutral conductor and protective ground conductor combined in a single conductor (PEN).

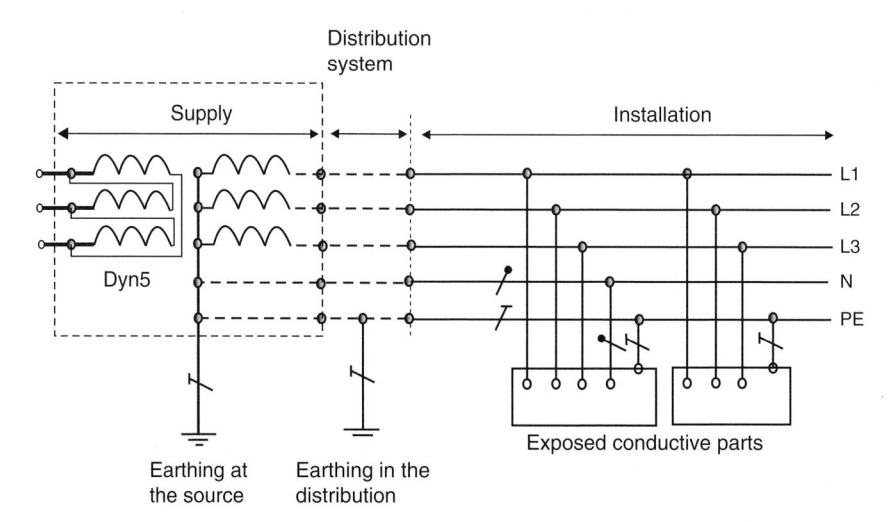

Figure 6.1 Circuitry of the TN–C–S system.

Requirements on fault protection in case of short circuit:

- All equipment must be connected to a PE.
- For distribution circuits, a turn-off time not longer than 5 s is allowed.
- For final circuits, an RCD/30 mA must be provided.

Permissible overcurrent protective equipment for TN systems are as follows:

- Fuses
- Line protection circuit breakers (B or C type)
- Circuit breakers
- RCDs
- Arc fault circuit interrupter.

The following conditions must be satisfied for the ground loop impedance:

$$Z_S \leq \frac{U_0}{I_a} \tag{6.1}$$

where Z_S is the sum of impedances of network feed-ins and power source (loop resistance), U_0 is the conductor-to-ground voltage, and I_a is the breaking current of the overcurrent protective equipment.

The loop impedance of the TN system is required in order to calculate the minimum required fault current at the position of the short circuit (Figure 6.2).

6.2 Calculation of Fault Currents

This section presents and explains the simple fundamental considerations for the calculation of the fault current.

In TN systems, the fault current is calculated in order to ensure protection in case of indirect contact and to guarantee that the protective equipment switches off within the specified time. Figure 6.3 gives an overview of the calculations carried out here.

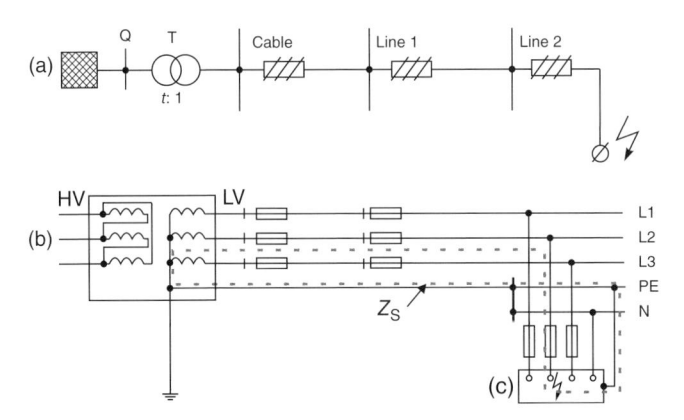

Figure 6.2 Circuitry of the TN system. (a) Distribution systems, (b) loop impedance, and (c) fault location.

6 Systems up to 1 kV

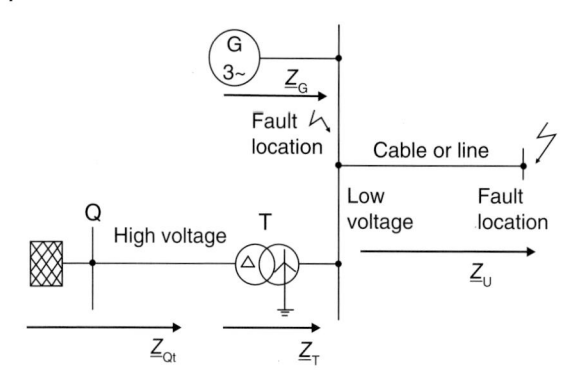

Figure 6.3 Overview of the power supply.

The impedance upstream from the power source is given by:

$$Z_{Qt} = \sqrt{R_{Qt}^2 + X_{Qt}^2} \qquad (6.2)$$

and with the short-circuit power of the high-voltage network:

$$Z_Q = \frac{c \cdot U_n^2}{S_{kQ}''} \qquad (6.3)$$

If exact data are not available for the reactance and resistance, we can then use the following values [5]:

$$R_Q = 0.100 \cdot X_Q \qquad (6.4)$$
$$X_Q = 0.995 \cdot Z_Q \qquad (6.5)$$

Impedance of the transformer:

$$Z_T = \frac{(U_n)^2}{S_{rT}} \cdot \frac{u_{kr}\%}{100} \qquad (6.6)$$

If exact data are not available for the reactance and resistance, we can then use the following values [5]:

$$R_T = 0.31 \cdot Z_T \qquad (6.7)$$
$$X_T = 0.95 \cdot Z_T \qquad (6.8)$$
$$Z_T = \sqrt{R_T^2 + X_T^2} \qquad (6.9)$$

Impedance of the power source:

$$Z_S = Z_Q + Z_T \qquad (6.10)$$

6.2.1 System Power Supplied from Generators:

Transient reactance of generator:

$$X_d' = \frac{U_n^2}{S_{rG}} \cdot \frac{x_d'}{100} \qquad (6.11)$$

Zero-sequence reactance:

$$X_0 = \frac{U_n^2}{S_{rG}} \cdot \frac{x_0}{100} \qquad (6.12)$$

6.2 Calculation of Fault Currents

If exact data are not available for the reactance and resistance, we can then use the following values [5]:

$$X'_d = 30\% \cdot x'_d \tag{6.13}$$
$$X'_d = 6\% \cdot x_0 \tag{6.14}$$

Calculation of the fault current I''_{k1min} (external conductor – protective ground conductor):

In accordance with IEC 60364, protection for indirect contact is ensured if the following equation is satisfied:

$$I''_{k1min} = \frac{\sqrt{3} \cdot c_{min} \cdot U_n}{3 \cdot \sqrt{(2 \cdot l_L + R_v)^2 + (2 \cdot l \cdot X'_L + X_v)^2}} \tag{6.15}$$

For cables and conductors with reduced PEN or protective ground conductor cross-sections, Equation (6.15) can be used if we substitute

$$R'_L = \frac{R'_{L1} + R'_{L2}}{2} \tag{6.16}$$

This protective measure requires coordinating the type of ground connection and the characteristics of the protective ground conductors and protective equipment. An immediate and automatic cut off of the faulty circuit is ensured when the following condition is met:

$$I''_{k1min} \geq I_a \tag{6.17}$$

Calculation of the resistance at a temperature of 80 °C in accordance with IEC 60909 for a minimum single-phase short-circuit current:

$$R_{L80°} = R_{L20°} \cdot \left[1 + 0.004\frac{1}{°C}(\theta_e - 20°C)\right] \tag{6.18}$$

The meanings of the symbols are as follows:

c	voltage factor
S''_{kQ}	short-circuit power of the high-voltage network
X_Q	reactance upstream from power source
X_T	reactance of power source
Z_Q	impedance upstream from power source
R_Q	resistance upstream from power source
R_T	resistance of power source
Z_T	impedance of power source
I''_{k1min}	smallest single-phase short-circuit current
I_a	breaking current of overcurrent protective equipment
l	length of the conductor (half the loop length)
R'_L	resistance per unit length of cable or conductor
X'_L	reactance per unit length of cable or conductor
R_v	ground loop resistance of main network
X_v	ground loop reactance of main network
R'_{L1}	resistance per unit length of external conductor
R'_{L2}	resistance per unit length of PEN or protective ground conductor
Z_S	sum of impedances of network feed-ins and power source.

6.3 TT systems

In TT systems, the neutral conductor does not serve a protective ground conductor function. The connection of the operational equipment takes place through the protective ground conductor to a common grounding system (Figure 6.4). TT systems are of no importance in the industrial sector.

6.3.1 Description of the System

First letter: describes the grounding conditions of the power source

T: direct grounding of a point
I: isolated to the earth.

Second letter: describes the grounding conditions of the exposed conductive parts of the electrical system

T: exposed conductive parts, grounded directly and independently of power source

Permissible overcurrent protective equipments for TT systems are as follows:

- RCDs
- Line protection circuit breakers, for example, with A, B, C, and D characteristic
- Circuit breakers
- Arc fault circuit interrupter
- Fuses.

The following condition must be satisfied for the ground resistance of the exposed conductive parts:

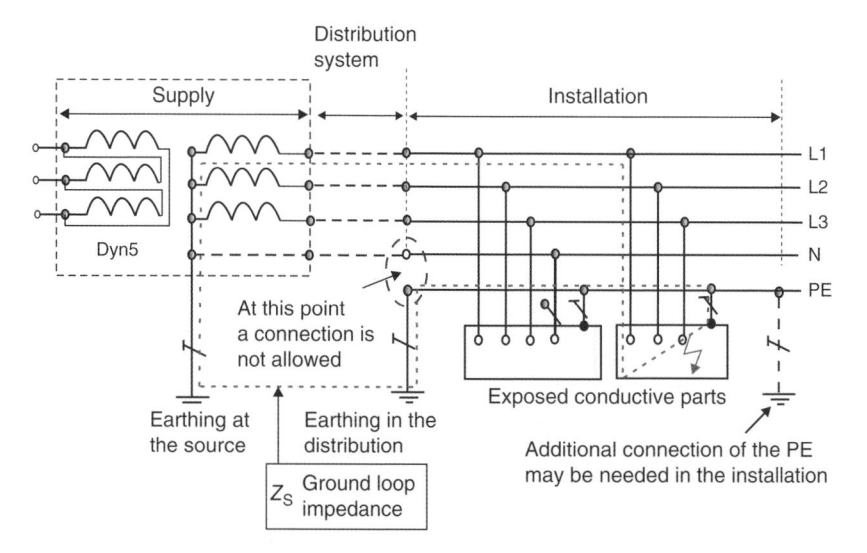

$$R_A \leq \frac{U_T}{I_{\Delta n}} \tag{6.19}$$

$$Z_S \leq \frac{U_0}{I_a} \tag{6.20}$$

Figure 6.4 Circuitry of the TT system.

where R_A is the sum of the resistances of the ground electrode and the protective ground conductor, U_T is the touch voltage, $I_{\Delta n}$ is the rated differential current of the RCD, and Z_S is the ground loop impedance (consisting of source resistance, protection resistance in the installation, and internal resistance).

6.4 IT Systems

The power source for IT systems is isolated. Its application is primarily in the industrial sector and in the operation rooms of hospitals (Figure 6.5). The source may be connected to earth through very high impedance. The neutral conductor may or may not be distributed in the installation.

6.4.1 Description of the System

First letter: describes the grounding conditions of the power source

I: isolation of active parts from ground or connection of active parts to ground through an impedance (indirect grounding).

Second letter: describes the grounding conditions of the exposed conductive parts of the electrical system

T: exposed conductive parts, grounded directly and independently of power source.

The protection for indirect contact is implemented by messages generated in the isolation monitoring, with equipotential bonding or cut off in addition in case of a double fault.

Permissible overcurrent protective equipment for IT systems are as follows:

- Insulation monitoring
- RCDs
- Line protection circuit breakers
- Circuit breakers
- Fuses.

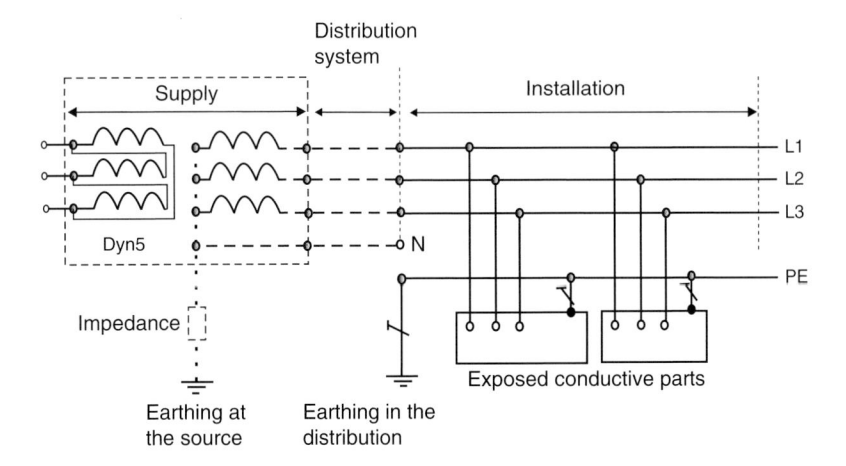

Figure 6.5 Circuitry of the IT system.

The ground resistance of the exposed conductive parts must be sufficiently low to allow the following condition to be satisfied:

$$R_A \leq \frac{U_T}{I_d} \tag{6.21}$$

where R_A is the sum of the resistances of the earth electrode and the PE for each appliances, U_T is the touch voltage, and I_d is the fault current at the first error with a very low impedance between a conductor and an exposed conductive part.

The value of I_d takes into account the leakage currents and the total impedance of the electrical system and ground.

6.5 Transformation of the Network Types Described to Equivalent Circuit Diagrams

There are many possible arrangements of networks. In order to calculate the total impedance at the position of the short circuit, the network topologies in multiple and meshed networks are simplified and transformed in a star–delta or delta–star transformation (Figure 6.6).

With this approach, the network is reduced to a network with simple inputs. The entire short-circuit path is represented by resistances and reactances and the impedance at the position of the short circuit calculated from these. The following relationships then apply:

The impedance is generally:

$$\underline{Z} = R + jX \tag{6.22}$$

The magnitude of the impedance is given by:

$$Z = \sqrt{R^2 + X^2} \tag{6.23}$$

Series circuit (Figure 6.6a):

$$\underline{Z}_{total} = \underline{Z}_1 + \underline{Z}_2 \tag{6.24}$$

Parallel circuit (Figure 6.6b):

$$\underline{Z}_{total} = \frac{\underline{Z}_1 \cdot \underline{Z}_2}{\underline{Z}_1 + \underline{Z}_2} \tag{6.25}$$

Delta–star transformation (Figure 6.6c):

$$\underline{Z}_1 = \frac{\underline{Z}_a \cdot \underline{Z}_c}{\underline{Z}_a + \underline{Z}_b + \underline{Z}_c} \tag{6.26}$$

$$\underline{Z}_2 = \frac{\underline{Z}_a \cdot \underline{Z}_b}{\underline{Z}_a + \underline{Z}_b + \underline{Z}_c}$$

$$\underline{Z}_3 = \frac{\underline{Z}_b \cdot \underline{Z}_c}{\underline{Z}_a + \underline{Z}_b + \underline{Z}_c}$$

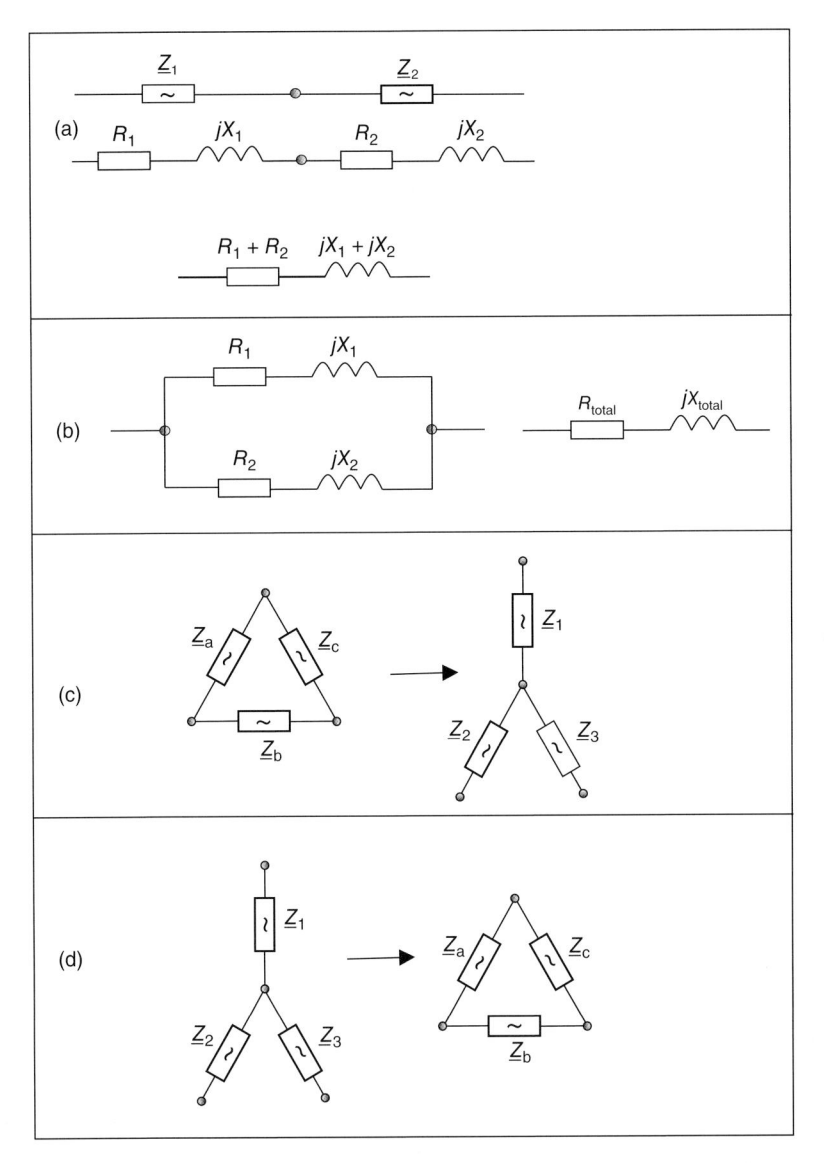

Figure 6.6 Network transformations. (a) Series circuit, (b) parallel circuit, (c) delta–star transformation, and (d) star–delta transformation.

Star–delta transformation (Figure 6.6d):

$$\underline{Z}_a = \frac{\underline{Z}_1 \cdot \underline{Z}_2 + \underline{Z}_1 \cdot \underline{Z}_3 + \underline{Z}_2 \cdot \underline{Z}_3}{\underline{Z}_3} \tag{6.27}$$

$$\underline{Z}_b = \frac{\underline{Z}_1 \cdot \underline{Z}_2 + \underline{Z}_1 \cdot \underline{Z}_3 + \underline{Z}_2 \cdot \underline{Z}_3}{\underline{Z}_1}$$

$$\underline{Z}_c = \frac{\underline{Z}_1 \cdot \underline{Z}_2 + \underline{Z}_1 \cdot \underline{Z}_3 + \underline{Z}_2 \cdot \underline{Z}_3}{\underline{Z}_2}$$

6.6 Examples

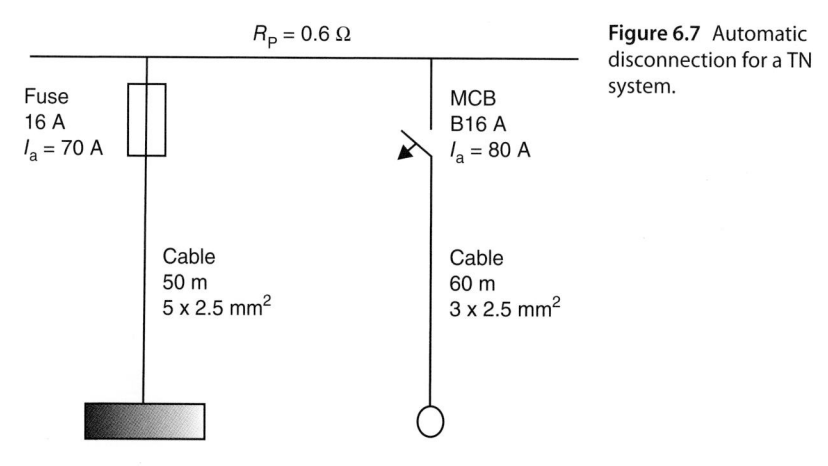

6.6.1 Example 1: Automatic Disconnection for a TN System

In a subdistribution panel (Figure 6.7), a circuit for a heater and the wall receptacle will be calculated in order to prove the automatic disconnection. At the panel, the impedance is given with $R_P = 0.6\,\Omega$.

6.6.1.1 Calculation for a Receptacle
The loop resistance is as follows:

$$R_S = R_P + R_L = 0.6\,\Omega + 1.56\frac{2\cdot 60\,\mathrm{m}}{56\frac{\mathrm{m}}{\Omega\,\mathrm{mm}^2}\cdot 2.5\,\mathrm{mm}^2} = 1.93\,\Omega$$

For a miniature circuit breaker, the disconnection current is $5 \cdot 16\,\mathrm{A} = 80\,\mathrm{A}$. The disconnection time is 0.4 s. The fault current is as follows:

$$I_F = \frac{230\,\mathrm{V}}{1.93\,\Omega} = 119.17\,\mathrm{A}$$

Fault current must be greater than the disconnection current which is

$$I_F > I_a \rightarrow 119.17\,\mathrm{A} > 80\,\mathrm{A}$$

With this current, the shutdown condition is satisfied.

6.6.1.2 For the Heater

$$t_a \leq 0.4\,\mathrm{s}$$

$$R_S = R_P + R_L = 0.6\,\Omega + 1.56\frac{2\cdot 50\,\mathrm{m}}{56\frac{\mathrm{m}}{\Omega\,\mathrm{mm}^2}\cdot 2.5\,\mathrm{mm}^2} = 1.71\,\Omega$$

The disconnection current is for a fuse of gG/16 A 70 A, as shown for the receptacle. It gives

$$I_F = \frac{230\,\mathrm{V}}{1.71\,\Omega} = 134.5\,\mathrm{A}$$

$$I_F > I_a \rightarrow 134.5\,\mathrm{A} > 70\,\mathrm{A}$$

The shutdown condition is satisfied.

Figure 6.7 Automatic disconnection for a TN system.

6.6.2 Example 2: Automatic Disconnection for a TT System

Given is the equipment grounding resistance: $R_A = 5\,\Omega$. In TT system, fault protection can be provided either by overcurrent device or by RCD. The grounding resistance at the consumer must satisfy the disconnection requirements.

We can calculate the disconnection current for this circuit.

$$I_a \leq \frac{U_0}{Z_S} = \frac{230\,\text{V}}{5\,\Omega} = 46\,\text{A}$$

It is difficult to fulfill the safety requirements for this circuit. Therefore, an RCD must be installed with 30 mA:

$$R_A \leq \frac{U_T}{I_{\Delta n}} = \frac{50\,\text{V}}{0.03\,\text{A}} = 1666\,\Omega$$

7

Neutral Point Treatment in Three-Phase Networks

The main faults are the single-phase short circuit and the short circuit to ground. The short circuit to ground is a conductive connection between a point in the network belonging to the operational circuit and ground. Between 80% and 90% of all faults in grounded networks are short circuits to ground. If the short-circuit currents are identical in all three conductors with a three-phase short circuit, then the fault is symmetrical. In all other cases, the fault currents in the three conductors are different and these faults are then asymmetrical. In addition, in three-phase networks various so-called transverse faults are possible. Along with these transverse faults, line interruptions can also occur.

This results in longitudinal faults, which are, however, of no importance for short-circuit current calculations. For ground faults and short circuits to ground, the magnitude of the short-circuit current depends primarily on how the neutral point of the network (transformer) is connected to the ground [2]. Figure 7.1 shows an example of earthing (grounding) of possible transformer at the star point in electrical power systems.

The short-circuit currents are determined by the voltage sources present in the network (generators and motors) and by the network impedances. The requirement of an optimum and inexpensive network can lead to different neutral point treatments. The expense for grounding systems, network protection, network design, operating mode, and size of the network is the determining factor in the choice of neutral point treatment. The neutral point treatment also affects the following parameters (Figure 7.2):

- Touch, step, and grounding electrode voltages
- Single-phase short-circuit currents
- Voltage stress.

For the construction and operation of electrical systems, a knowledge of the grounding measures is indispensable. The most important measure is protection in the event of indirect touching, that is, the safety of human beings and of objects must be ensured. For this, the calculation of the ground potential rise is important. This value characterizes the maximum touch voltage. Thus, in terms of danger to human beings, the touch voltage at which a fault current can flow

Short Circuits in Power Systems: A Practical Guide to IEC 60909-0, Second Edition. Ismail Kasikci.
© 2018 Wiley-VCH Verlag GmbH & Co. KGaA. Published 2018 by Wiley-VCH Verlag GmbH & Co. KGaA.

7 Neutral Point Treatment in Three-Phase Networks

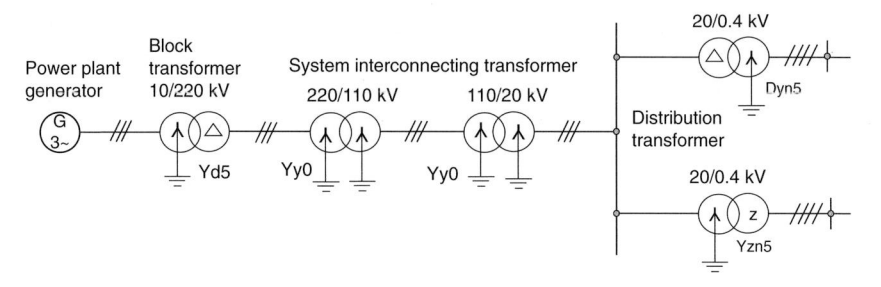

Figure 7.1 Neutral point treatment of transformers.

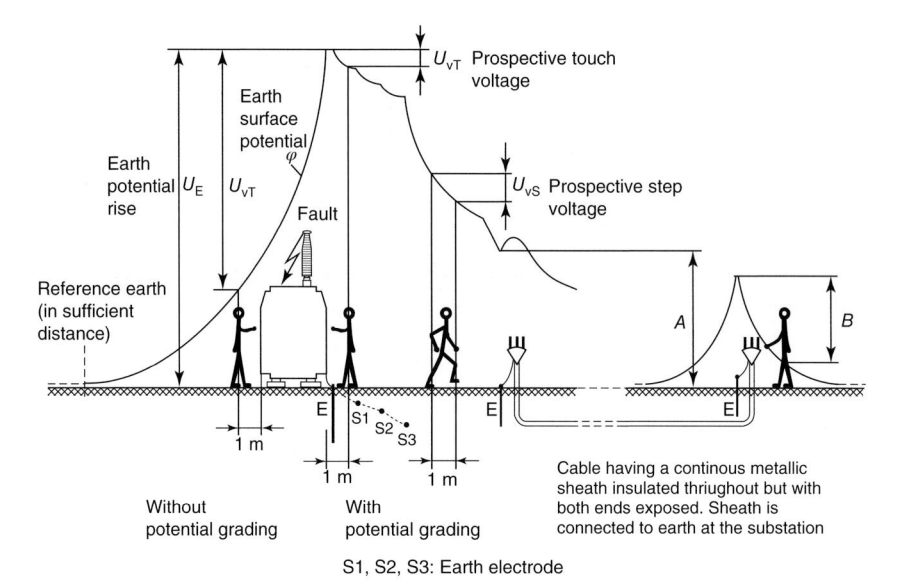

Figure 7.2 Surface potential profile and voltages in case of current-carrying earth electrodes (see also Ref. [2]).

through the body and the heart is especially critical. The touch voltage is part of the ground potential rise in the event of a fault, which can be picked up by human beings. The highest touch voltage that may remain without any limit in time in low-voltage systems is 50 V for a.c. voltages and 120 V for d.c. voltages. As stated in EN 50522, the highest touch voltage for times greater than 5 s in high-voltage systems is 75 V. The magnitude of the permissible touch voltage depends on the duration of the fault and is given in Figure 4 of EN 50522. If the ground potential rise remains less than 150 V, then the condition is satisfied and no further measures are required. Otherwise, additional measures have to be undertaken. For a fault with contact to ground, the type of neutral point grounding determines the magnitude of the line-to-ground voltage and the magnitude of the currents flowing to ground. This section will briefly discuss the principles of grounding systems and then deal with the different methods of neutral point grounding in high-voltage networks.

The project planner has to determine and observe the following parameters for the dimensioning of grounding systems:

1) Magnitude of the fault current (this parameter depends on the neutral point grounding of the high-voltage network. See Table 7.1).
2) Duration of the fault (this parameter depends on the neutral point grounding of the high-voltage network).
3) Characteristics of ground (measurement of the ground resistivity).
4) Ground resistance.
5) Correct choice and dimensioning of materials.

The design of grounding systems must satisfy four requirements:

1) The mechanical strength and resistance to corrosion of the grounding electrode and protective conductor, as well as their connections, must be ensured. These determine the minimum dimensions for the grounding electrodes.
2) The greatest fault current must be calculated and held under control from the thermal point of view.
3) Damage to objects and operational equipment (especially information technology) must be avoided.
4) The safety of persons with respect to voltages on grounding systems (touch voltages and parasitic voltages), which occur at the time of the greatest ground fault current, must be ensured.

The ground potential rises and touch voltages of a grounding system can be calculated from known data. The fault current frequently divides in the system. For the calculation of the grounding system parameters, it is necessary to consider all grounding electrodes and other grounding systems. In accordance with EN 50522, for step voltages, it is not necessary to define permissible values. If a system satisfies the requirements with regard to the touch voltages, then no dangerous step voltages can occur.

Overhead ground wires from overhead lines and metal shielding of grounding cables are involved in leading away fault currents that flow to ground. They assume a part of the ground fault current from the particular circuit. This effect gives rise to the effective relieving of a high-voltage grounding system that is affected by a ground fault. The extent of this relief is described by the reduction factor.

The reduction factor r for an overhead ground wire in a three-phase current conductor is as follows:

$$r = \frac{I_0}{3I_0} = \frac{3I_0 - I_{EW}}{3I_0}$$

The ground potential rise is

$$\underline{U}_E = \underline{Z}_E \cdot \underline{I}_E$$

In the event of a fault, the ground potential rise is

$$\underline{I}_E = r \cdot \sum 3 \underline{I}_0$$

Table 7.1 Decisive currents for the dimensioning of grounding systems [2].

Type of high-voltage network			Decisive for thermal loading		Decisive for ground potential rise and touch voltage
			Ground electrodes	Ground conductor	
Networks with isolated neutral point			–	I''_{kEE}	$I_E = r \cdot I_C$
Networks with ground fault neutralizer grounded system	In systems with ground fault neutralizer grounded system		–	I''_{kEE}	$I_E = r \cdot \sqrt{I_L^2 + I_{Res}^2}$
	In systems without ground fault neutralizer grounded system				$I_E = r \cdot I_{Res}$
Networks with low-resistance neutral point grounding			I''_{k1}	I''_{k1}	I_E
Networks with ground fault neutralizer grounded system and temporary low-resistance neutral point grounding	In systems in which temporary grounding takes place		I''_{k1}	I''_{k1}	I_E
	In all other systems	With ground fault coil	–	I''_{kEE}	$I_E = r \cdot \sqrt{I_L^2 + I_{Res}^2}$
		Without ground fault coil		$I_E = r \cdot I_{Res}$	

I_C, calculated or measured capacitive ground fault current; I_{Res}, residual ground fault current. When the exact value is not known, we can assume 10% of I_C; I_L, sum of the rated currents of parallel ground fault coils for the system under discussion; I''_{k1}, initial symmetrical short-circuit current for a single-phase ground fault, calculated in accordance with IEC 60909; I''_{kEE}, double line to ground fault, calculated in accordance with IEC 60909 or HD533 (for I''_{kEE} 85% of the three-phase initial symmetrical short-circuit current can be used as the greatest value); I_E, grounding current; r, reduction factor.

The meanings of the symbols are as follows:

I_{EW} current in the overhead ground wire in A
U_E ground potential rise in V
I_E grounding current in A
$3I_0$ sum of the zero-sequence currents in A
Z_E grounding impedance in Ω.

We can differentiate between the three types of neutral point treatment.

7.1 Networks with Isolated Free Neutral Point

The short-circuit current to ground flows through the capacitances to ground C_E of the uninterrupted conductors (Figure 7.3). The short-circuit currents to

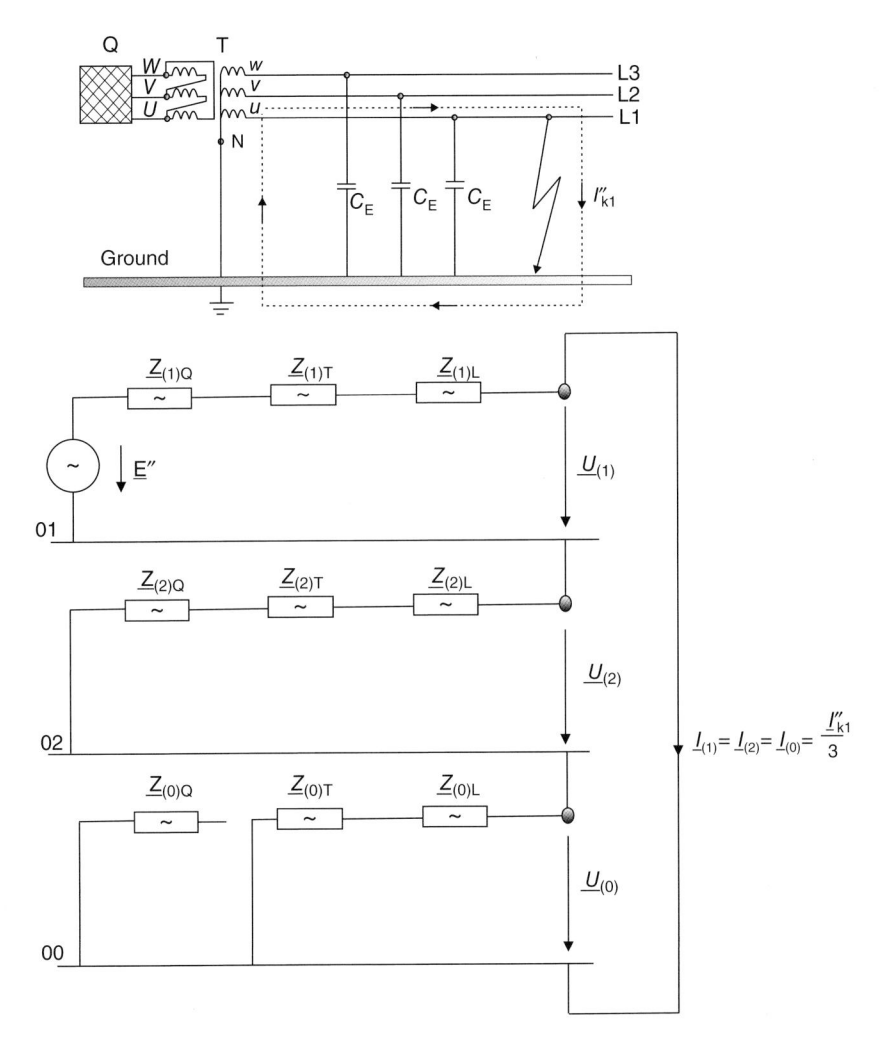

Figure 7.3 Isolated network and equivalent circuit diagram.

Table 7.2 Summary of the isolated network.

Advantages	Disadvantages
No supply interruption	Voltage increase in healthy conductors
Automatic deletion of fault	Stronger current arc
Grid operation of ground fault possible	Limitation to small networks
Low demands on grounding systems	Full isolation in the network
Voltage delta on the primary side	Risk of subsequent errors
Fault location difficulty maintaining	

ground are small in this case and in small networks are usually self-quenching, although large transient overvoltages can occur. The potential of the neutral point relative to ground is determined by the capacitances C_E. The short-circuit current to ground I_{CE} at the position of the short circuit is given by:

$$I_{CE} = 3 \cdot \omega \cdot C_E \frac{c \cdot U_n}{\sqrt{3}} \tag{7.1}$$

The short-circuit current to ground increases with the length of the conductor, so that the operation of this type of network is restricted to smaller networks (up to 30 kV). The limiting value of the short-circuit current to ground is around 35 A, since otherwise the arcing is no longer self-quenching.

Table 7.2 summarizes the advantages and disadvantages of isolated networks.

7.2 Networks with Grounding Compensation

A network with grounding compensation is present when the neutral point is grounded through ground fault coils in such a way that their inductance is matched to the capacitance to ground (Figure 7.4). For the matching condition $I_C = I_L$:

$$I_C = 3 \cdot \omega \cdot C_E \frac{c \cdot U_n}{\sqrt{3}} \quad \text{and} \quad I_L = \frac{c \cdot U_n}{\sqrt{3} \cdot \omega \cdot L} \tag{7.2}$$

Here, we refer to the ground fault current I_F as the unbalanced residual current. The capacitive short circuit to ground I_C is compensated by the inductive coil current I_L of the ground fault quenching coils apart from a residual current. If the short-circuit current to ground exceeds 35 A, then the network must be operated with grounding compensation. The residual current should not exceed 60 A for medium-voltage networks and 130 A for high-voltage networks, in order to ensure the self-quenching of the arcing and to keep the thermal stress under control. If this is not the case, then low-impedance neutral point grounding must be used. Here too, overvoltages occur as for networks with an isolated free neutral point.

7.2 Networks with Grounding Compensation

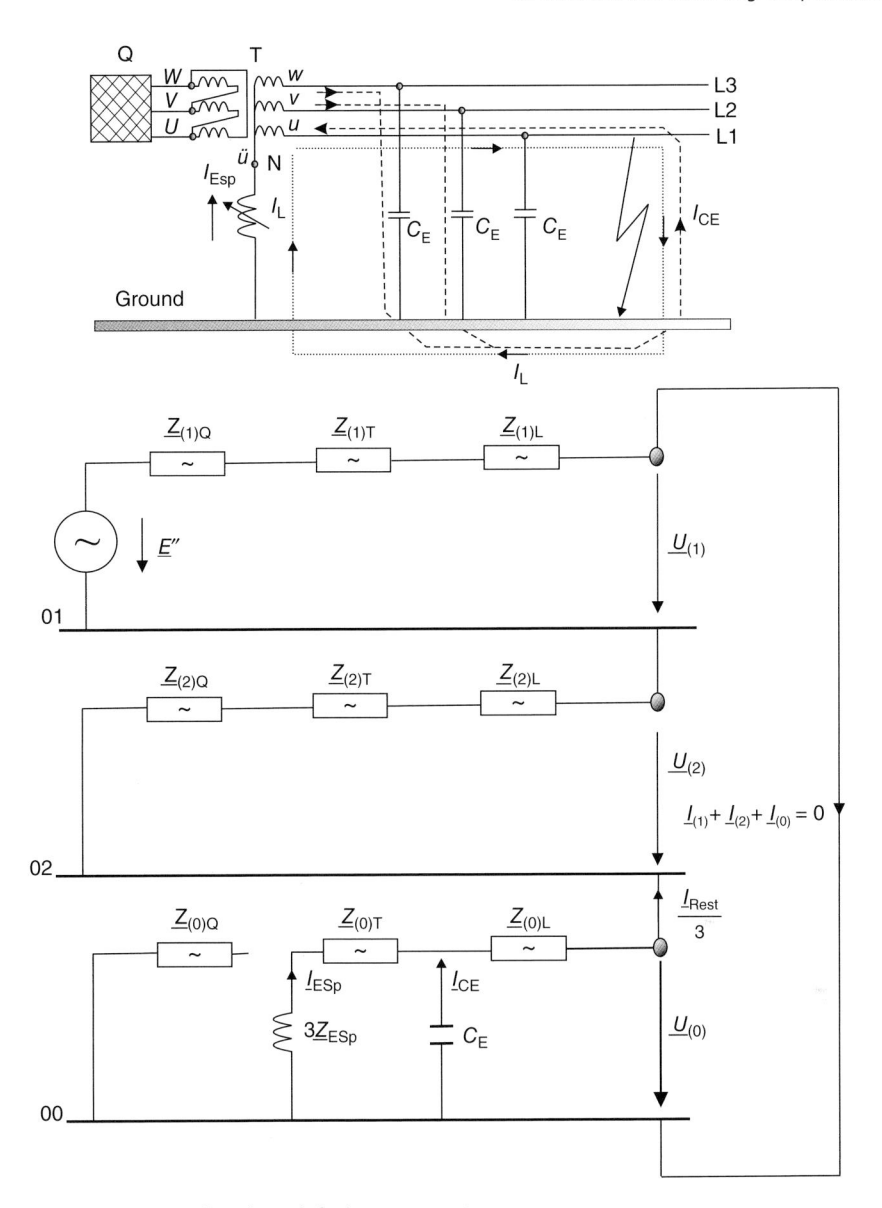

Figure 7.4 Network with earth fault compensation.

After the compensation, a residual fault current flows through the fault earth location. These results are found to be

$$I_r = I_{CE} \cdot \sqrt{d^2 + v^2} \tag{7.3}$$

Table 7.3 summarizes the advantages and disadvantages of networks with earth fault compensation.

Table 7.3 Summary of the compensated network.

Advantages	Disadvantages
No supply interruption	Voltage increase in healthy conductors
Automatic deletion of fault	Stronger current arc
Grid operation of ground fault possible	Limitation to small networks
Low demands on grounding systems	Full isolation in the network
Voltage delta on the primary side	Risk of subsequent errors
Fault location difficulty maintaining	Resonance danger
Residual ground fault	Risk of multiple errors
Extinction if current and voltage in phase	Cost increases in insulation
	Investment for coil

7.3 Networks with Low-Impedance Neutral Point Treatment

In accordance with HD 63751, a network with low-impedance neutral point grounding (Figure 7.5) is present only when the neutral point of one or more transformers is directly grounded and the network protection is designed so that in the event of a short circuit to ground at any arbitrary fault position automatic cut off must take place (protection by cut off).

The fault to ground is described as a short circuit to ground and the fault current as short-circuit current to ground or single-phase short-circuit current. The short-circuit currents to ground are, however, limited by the neutral point impedance ($Z_S = 20,\ldots, 60\,\Omega$) to values less than 5 kA. The ground fault factor $\delta = \frac{U_{LE}}{U/\sqrt{3}}$ will be introduced here to describe the voltage conditions for the neutral point treatment, where U_{LE} is the conductor-to-ground voltage for a fault and U is the operating voltage before the fault occurs. For a single-phase short-circuit current, then

$$I''_{k1} = \frac{\sqrt{3} \cdot c_{min} \cdot U_n}{Z_1 + Z_2 + Z_0} \tag{7.4}$$

Table 7.4 summarizes the advantages and disadvantages of low-resistance grounded network.

Table 7.5 gives an overview of the neutral point arrangement on fault behavior in three-phase high-voltage networks.

Table 7.6 summarizes an overview of the application of neutral point arrangements.

7.3 Networks with Low-Impedance Neutral Point Treatment

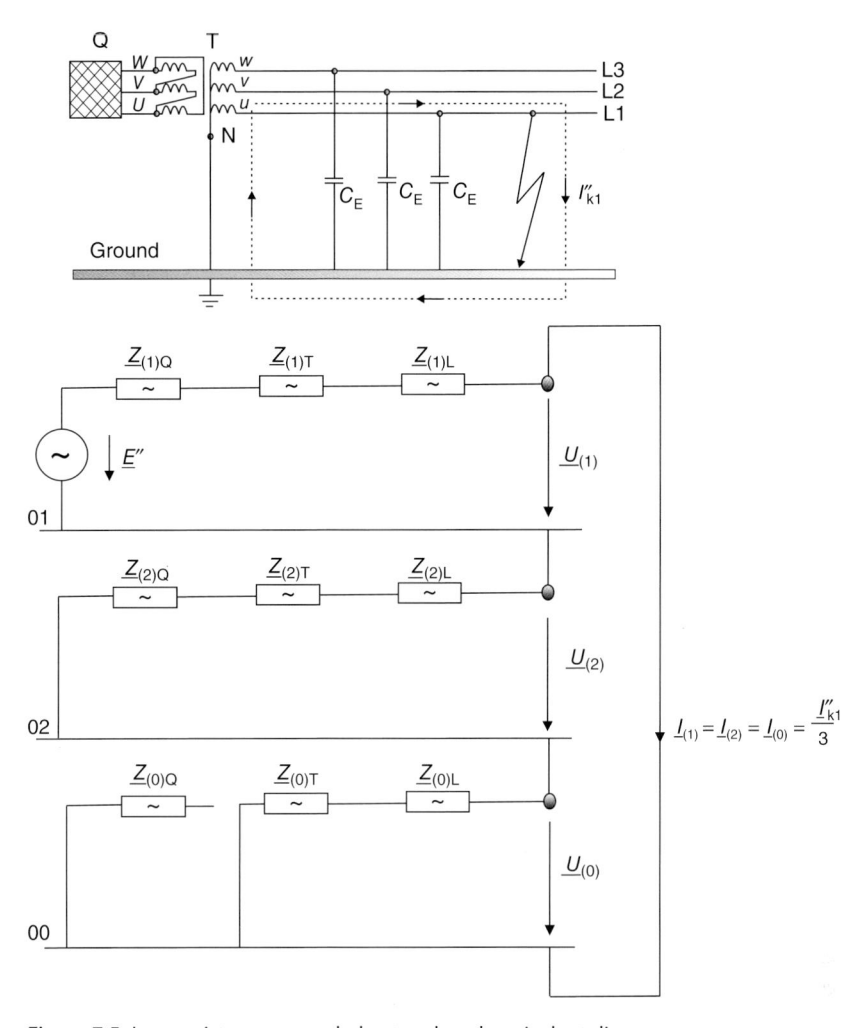

Figure 7.5 Low-resistance grounded network and equivalent diagram.

Table 7.4 Summary of the low-resistance grounded network.

Advantages	Disadvantages
Supply interruption	Every short circuit must be shut down
Selective disconnection	Fault currents over earth
Small overvoltages	Ground fault is short circuit
Low level of isolation	High breaking capacity
No traveling wave hazards	High demands on earthing systems
No limitation on network expansion	A three-phase power protection needed

Table 7.5 Arrangement of neutral point.

Arrangement of neutral point (only one phase is shown)	Isolated	With arc suppression coil	Current limiting	Low-resistance ground
Examples of use	Power plant auxiliaries	Overhead line	Cable network	High voltage
Fault current	$I_E < 40$ A	$I_{Rest} < 60\text{–}120$ A	For example, $I''_{k1} = 2$ kA	High
Fault duration	$t > 2$ h	$t < 2\text{–}3$ h	$t \leq 1\text{–}3$ s	$t \leq 1\text{–}3$ s
Ground fault factor $\delta = U_{LE}/(U_{LL}/\sqrt{3})$	$\delta \sim \sqrt{3}$	$\delta \sim \sqrt{3}$	$\delta \sim 1.4\text{–}1.8$	$\delta \leq 1.4$
Overvoltage	$k \sim 2.5$	$k \sim 3.0$	$k < 2.5$	$k < 2.5$
Voltage rise	Yes	Yes	No	No
Ground fault arc	Self-quenching up to several A	Self-quenching	Usually sustained	Sustained
Detection of fault	Location by disconnection	Location by disconnection	Selective disconnection	Short-circuit protection

δ, ground fault factor; I''_{kEE}: double ground fault; U_{LE}, conductor to ground voltage at fault occurrence; I''_{kEE}, two phase to ground fault; U_{LL}, operating voltage before fault occurrence; I''_{kEE}, phase to ground fault; I_E, ground fault current; I_{Rest}, residual current; C_E, ground capacitance.

Table 7.6 Application of neutral point arrangements.

Neutral point selection	Ground fault factor (δ)	Application	Results
Direct grounding	≤ 1.4	In low-voltage power systems < 1 kV In high-voltage power systems > 110 kV	Short-circuit protection selective disconnection Saving of insulation
Low-resistance ground	0.87–1.4	Medium-voltage power systems for cable networks 10–30 kV	Short-circuit protection Selective disconnection
With arc suppression coil	> 1.4	Overhead and cable networks up to 110 kV	Location by disconnection
Short circuited with suppression coil	First > 1.4 after short grounding < 1.4	Cable network 10–110 kV	First location by disconnection then disconnection

7.4 Examples

7.4.1 Neutral Grounding

Calculate the earth fault in a 110-kV insulated power at 70 km line length and conductor earth capacitance $C'_E = 5\,\text{nF/km}$.

$$C_E = C'_E \cdot l = 5\,\text{nF/km} \cdot 70\,\text{km} = 0.350\,\mu\text{F}$$

This gives the earth fault current:

$$I_E = 3 \cdot \omega \cdot C_E \cdot U_0 = 3 \cdot \omega \cdot 0.350\,\mu\text{F} \cdot 110\,\text{km}/\sqrt{3} = 20.9\,\text{A}$$

The network will be operated with arc suppression coil. Determine in this case the values of the coil.

$$Q_{rD} = \frac{U_n}{\sqrt{3}} \cdot I_{rD} = \frac{110\,\text{kV}}{\sqrt{3}} \times 20.9\,\text{A} = 1.32\,\text{Mvar}$$

8

Impedances of Three-Phase Operational Equipment

For the calculation of short-circuit currents, it is necessary to know conductor-specific equivalent data and impedances of electrical operational equipment, which are usually given by the respective manufacturer. The calculation of short-circuit currents is based on the use of equivalent circuits for the operational equipment. In principle, the equivalent resistances and reactances must be determined for all equipment. The impedances of generators, network transformers, and power plant blocks should take account of the impedance corrections for calculating the short-circuit currents. For generators, transformers, and choke coils, the impedances and reactances are given in the per unit (pu) or in the %/MVA system. Cables and lines are, however, assigned Ω/km values. The impedances of operational equipment are described in detail in the following section [1].

8.1 Network Feed-Ins, Primary Service Feeder

The input is from a network, usually designated "Q" for source, and not from a generator. The calculation of this network is performed with the initial symmetrical short-circuit power, S''_{kQ}, or the initial symmetrical short-circuit current, I''_{kQ}, at the feeder connection point Q should be determined without transformer (Figure 8.1a) and with transformer on the low-voltage (LV) side (Figure 8.1b).

The internal impedance of a high-voltage (HV) network or a medium-voltage (MV) network can then be determined according to the following equation:

$$\underline{Z}_Q = R_Q + jX_Q$$

$$Z_Q = \frac{c \cdot U_{nQ}}{\sqrt{3} \cdot I''_{kQ}} \tag{8.1}$$

If the short circuit is fed through transformers, it is possible to further extend the above relationships:

$$Z_{Qt} = \frac{c \cdot U_{nQ}}{\sqrt{3} \cdot I''_{kQ}} \cdot \frac{1}{t_r^2}$$

Short Circuits in Power Systems: A Practical Guide to IEC 60909-0, Second Edition. Ismail Kasikci.
© 2018 Wiley-VCH Verlag GmbH & Co. KGaA. Published 2018 by Wiley-VCH Verlag GmbH & Co. KGaA.

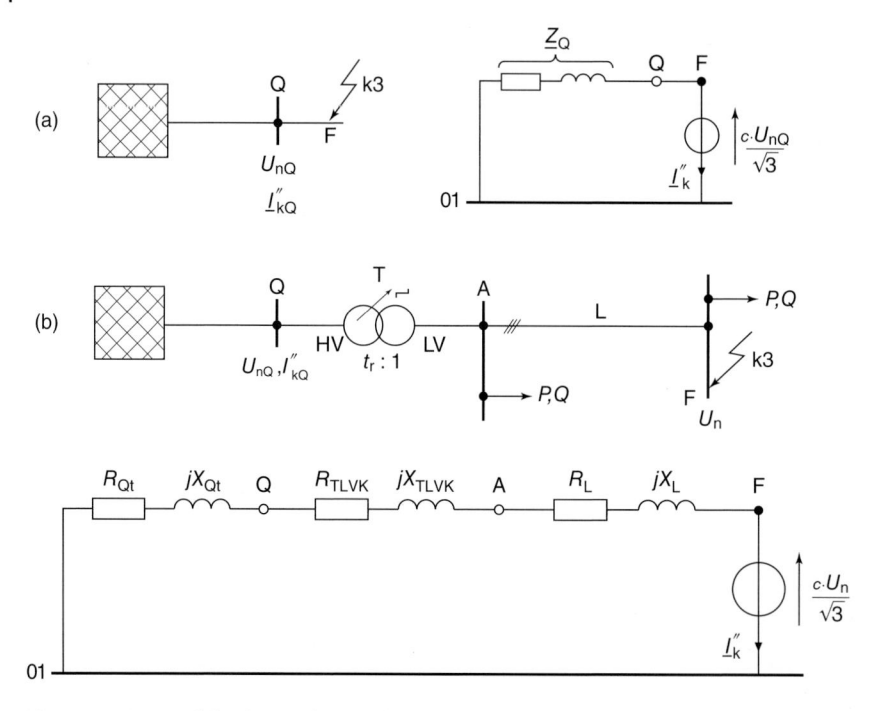

8 Impedances of Three-Phase Operational Equipment

Figure 8.1 Network feed-in and equivalent circuit. (a) System diagram and equivalent circuit diagram without transformer and (b) system diagram and equivalent circuit diagram with transformer.

$$Z_{Qt} = \frac{c \cdot U_{nQ}^2}{S_{kQ}''} \cdot \frac{1}{t_r^2} \tag{8.2}$$

$$I_{kQ}'' = \frac{c \cdot U_{nQt}}{\sqrt{3} \cdot Z_{Qt}} \tag{8.3}$$

It is sufficient to calculate only with reactances in HV networks with a voltage of greater than 35 kV, that is, $\underline{Z} = 0 + jX_Q$. In all other cases, the calculation proceeds as follows:

$$X_Q = 0.995 \cdot Z_Q$$
$$R_Q = 0.1 \cdot X_Q$$
$$S_{kQ}'' = \sqrt{3} \cdot U_{nQ} \cdot I_{kQ}'' \tag{8.4}$$

The meaning of the symbols is as follows:
U_{nQ} nominal voltage of the network at the interface Q
S_{kQ}'' initial symmetrical short-circuit power
I_{kQ}'' initial symmetrical short-circuit current at the feeder connection point
c voltage factor for the nominal voltage
t_r rated transformation ratio at which the on-load tap changer is in the main position

Z_Q positive-sequence impedance of short circuit
Z_{Qt} positive-sequence impedance relative to LV side of transformer
R_Q resistance of power supply feeder
X_Q reactance of power supply feeder.

8.2 Synchronous Machines

A synchronous machine is the main power generator in the electric power system and, therefore, is mainly used as a generator. After a winding arrangement, one differentiates between equipment inner-phase and outer-phase machines. In the outer phase, the armature winding is in the rotor, wherein the inner machines are located in the stator.

Outer-phase machines can only be operated with small power, because with them the net power must be removed via slip rings. Inner-phase machines need much smaller excitation power to be transferred to the rotor. Finally, there are slip-ring inner-phase machines in which the magnetic excitation power is transmitted to the rotor. The exciter field winding is a d.c. coil. The armature winding is performed in one or three phases. The damper windings are short-circuit windings that are intended to calm the nonstationary processes. After the execution of the rotor, one distinguishes further between turbo generators and salient-phase generators.

The rotor of a turbo generator is also referred to as an inducer and is designed cylindrically. The excitation winding will be introduced into the grooves, which are distributed over almost the entire circumference. The rotor of a salient-phase generator consists of individual phases and is, therefore, also called magnet wheel.

Figure 8.2 shows a turbo generator and a salient-phase generator.

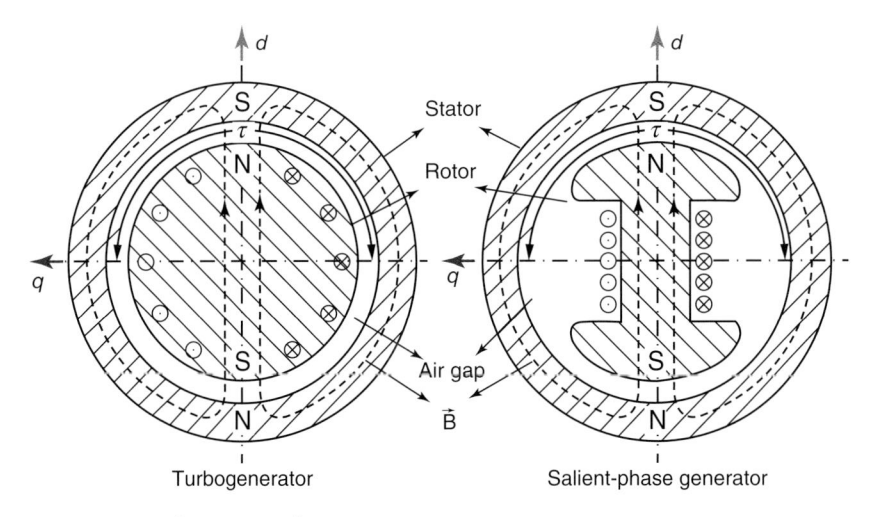

Figure 8.2 Synchronous machine.

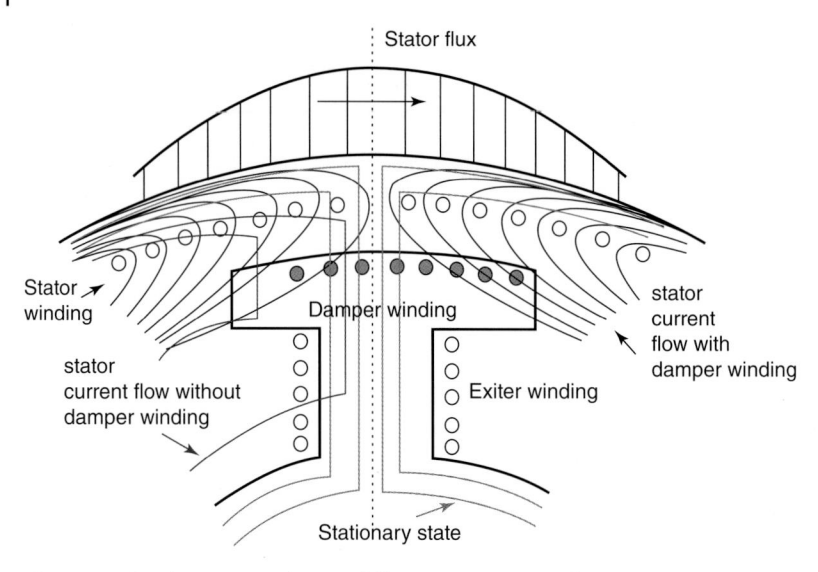

Figure 8.3 Synchronous machine in different operative conditions.

The knowledge of the short-circuit operation is important for the understanding of the generator behavior at all sudden disturbances of the stationary operation (Figure 8.3).

During the nonstationary operation, the stator winding reacts on the exciter and damper winding. In this case, a wide variety of operations occur. Before fault inception, only the excitation field is present. This field forms a constant flux linkage for the excitation and damper winding and for the stator coil, a sinusoidal alternating u_x linkage, which induces the rotor voltage.

No voltage arises in the exciter and damper winding during stationary operation. In dynamic processes, such as short circuits, the behavior of the generator is determined by the transition reactance. Sudden short circuits have an equalization process result in which the magnetic u_x, which is caused by the ow of the armature reaction, largely closes via magnetic stray paths in the air. The magnetic flux is distributed in different time intervals to different dispersions. In A compensating current occurs in the excitation field coil, which also temporarily urges the stator flux. The magnetic resistance is large at the beginning of this process, which is called the transient process, but decreases continuously until a stationary state occurs again. A similar behavior is also exhibited by a transient reactance X'_d. It is at the beginning of the equalization process and is much smaller than the synchronous reactance X_d. The reactance X_d occurs continually greater during the transient event, and if it is a stationary state, X_d becomes effective again. The reactance X'_d substantially includes the leakage reactance of the stator and rotor windings. The synchronous machine is the main power generator in the electric power generation and is, therefore, mainly used as a generator. Figure 8.4 shows an equivalent circuit and phasor diagram of the synchronous machine in the positive sequence in the longitudinal direction (d-direction) with reactances and their effective internal voltages. For three-phase short circuit, only the two inductive reactances X_h and X_σ

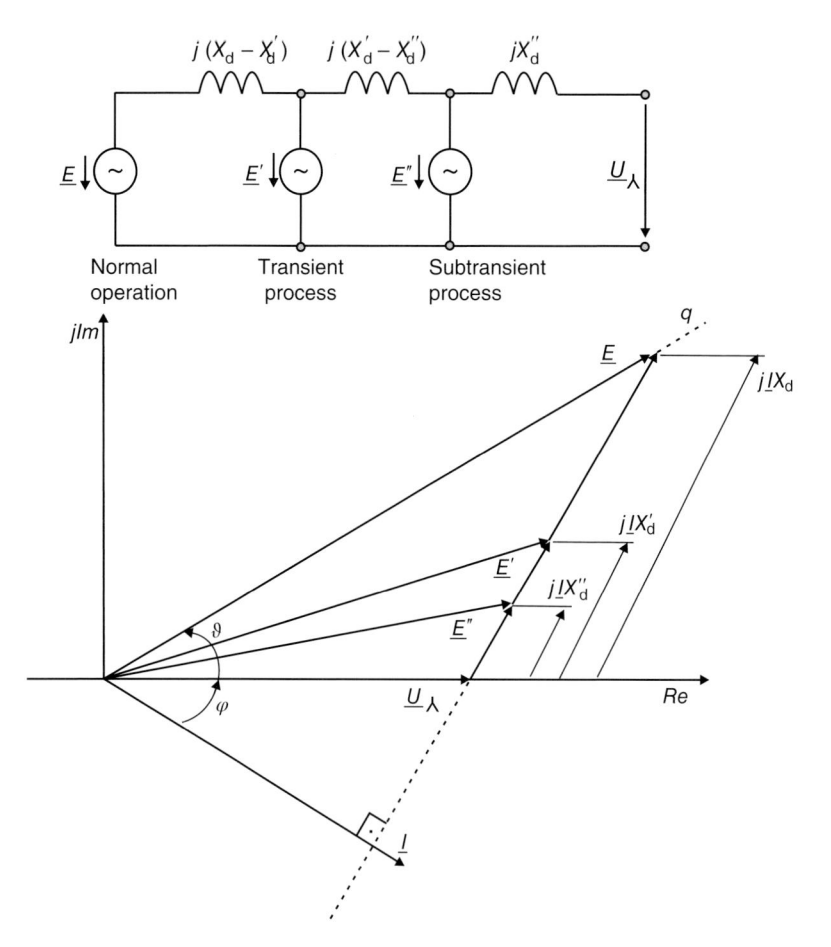

Figure 8.4 Equivalent circuit and phasor diagram of the synchronous machine in the positive sequence.

occur. The amount of short-circuit current, therefore, depends only on this reactance. At first, the largest peak short-circuits current, which then decays to a steady-state short-circuit current. The d.c. component occurs unchanged and decays with constant time. The altered magnetic conditions in various parts of the rotor voltages are induced, which in turn react on the stator. The currents in the damper winding subside quickly, as the effective resistances are very large. These operations are called subtransient (fast transient).

Figure 8.5 shows the maximum d.c. shift of full-time course of the short-circuit current. For the fault inception, the following expression can be stated mathematically:

$$i_k = \sqrt{2} \cdot \left[(I_k'' - I_k') \cdot e^{-\frac{t}{T_d''}} \cdot \sin(\omega t - \alpha) + (I_k' - I_k) \cdot e^{-(t/T_d')} \cdot \sin(\omega t - \alpha) \right.$$
$$\left. + I_k \cdot \sin(\omega t - \alpha) + I_k'' \cdot e^{-(t/T_{dc})} \cdot \sin \alpha \right] \quad (8.5)$$

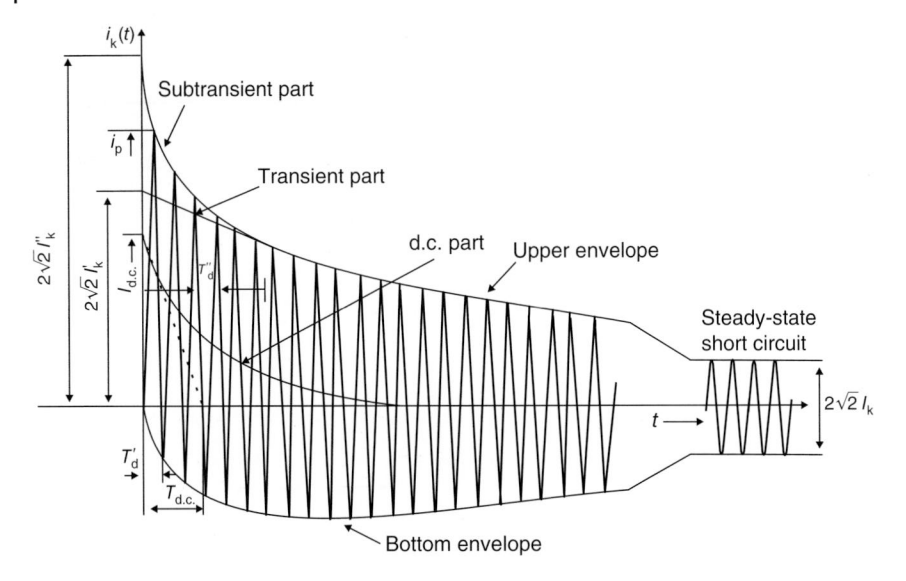

Figure 8.5 Generator near the short-circuit current.

This relationship describes the subtransient part, the transient part, the short-circuit current, and the d.c. link.

The course of time can be calculated for the dimensioning of the generator and all devices located in the short circuit and the peak short-circuit current i_p, which is important for the dynamic stress, the breaking current i_a, which is decisive for the dimensioning of the switch, and the short-circuit current I_k, the basis, for the thermal stress of resources.

The graph of the generator near the initial short-circuit current is determined by the beginning of the reactance:

$$I_k'' = \frac{U_E}{X_d''} \tag{8.6}$$

The currents in the rotor winding decay slower because of the small ohmic resistance. This process is called transient (temporarily) and appointed to the transition reactance (transient reactance):

$$I_k' = \frac{U_E}{X_d'}$$

Lastly, there are steady-state parts still in the time domain, which is conditionally indicated by the following equation:

$$I_k = \frac{U_E}{X_d}$$

The initial reactance (subreactance) X_d'' of the synchronous machine determines the size of the initial short-circuit alternating current. Figure 8.6 shows the equivalent circuit diagram of a synchronous machine.

For LV generators, the following can be given:

$$\underline{Z}_k = R_G + jX_d'', \quad R_G \approx 0.15 \cdot X_d''$$

8.2 Synchronous Machines

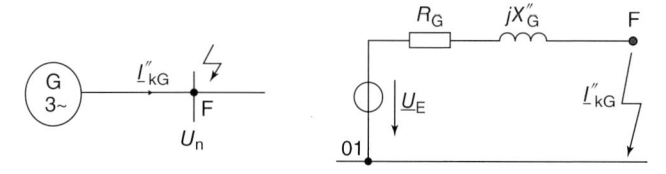

Figure 8.6 Synchronous machine and the equivalent circuit diagram.

For HV generators, $U_{rG} > 1$ kV and $S_{rG} \geq 100$ MVA, the following can be given:

$$R_G \approx 0.05 \cdot X_d'' \quad \text{for } S_{rG} < 100 \text{ MVA} \quad R_G \approx 0.07 \cdot X_d''$$

The subtransient reactance of the generator is

$$X_d'' = \frac{x_d'' \cdot U_{rG}^2}{S_{rG}}$$

The factors 0.05, 0.07, and 0.15 consider the decay of the short-circuit current during the first half period. If the manufacturer indicated otherwise, the following can be assumed.

For the single-phase short-circuit current:

$$I_{k1}'' \approx 5 \cdot I_{rG}$$

For the three-phase short-circuit current:

$$I_{k3}'' \approx 3 \cdot I_{rG}$$

In the following section, a generator near a short circuit will be described. These currents are mainly caused by the reactance of the synchronous generators (SGs).

The impedance of the SG is compared with the relatively large line impedance so that the severity of the current is heavily influenced by the generator. The reactance is described by different phases and is discussed in more detail later. The reactance increases with each phase so that very high short-circuit currents occur, but it will soon disappear.

The impedances are dependent on the machine's operating conditions and are based on evaluating the envelope of the maximum values of the generator's actual time-dependent short circuit. This envelope is a function of power, impedances and voltages.

For normal operations, three functions will be calculated from this envelope. The subtransient reactance, X_d'', determines the level of the short-circuit current I_{kd}'' and the transient reactance, X_d', determines the level of the short-circuit current I_{kd}' and is called a breaking current.

Finally, the compensation process proceeds in the short-circuit current I_{kd}. The more distant a short circuit occurs from the generator, the less pronounced the three stages are.

As described below, the short-circuit current contains both a.c. and d.c. components.

$$i_k(t) = \sqrt{2} \cdot I_{a.c.}(t) + i_{d.c.}(t)$$

8.2.1 a.c. Component

The a.c. component time function is characterized by the subtransient I''_{kd}, transient I'_{kd}, and steady-state currents I_{kd} during the time periods T''_d and T'_d.

$$I_{a.c.}(t) = (I''_{kd} - I'_{kd}) \cdot e^{-(t/T''_d)} + (I'_{kd} - I_{kd}) \cdot e^{-(t/T'_d)} + I_{kd}$$

$$I''_{kd} = \frac{E''_q}{Z''_d} = \frac{E''_q}{\sqrt{R_a^2 + X''^2_d}}$$

$$I'_{kd} = \frac{E'_q}{Z'_d} = \frac{E'_q}{\sqrt{R_a^2 + X'^2_d}}$$

$$I_{kd} = \frac{E_q}{Z_d} = \frac{E_q}{\sqrt{R_a^2 + X_d^2}}$$

The active subtransient voltage can be evaluated using the following equation:

$$E''_q = \sqrt{\left(\frac{U_{rG}}{\sqrt{3}} \cdot \cos\varphi + R_a \cdot I_{rG}\right)^2 + \left(\frac{U_{rG}}{\sqrt{3}} \cdot \sin\varphi + X''_d \cdot I_{rG}\right)^2}$$

And transient voltage:

$$E'_q = \sqrt{\left(\frac{U_{rG}}{\sqrt{3}} \cdot \cos\varphi + R_a \cdot I_{rG}\right)^2 + \left(\frac{U_{rG}}{\sqrt{3}} \cdot \sin\varphi + X'_d \cdot I_{rG}\right)^2}$$

where

$$Z''_d = R_a + jX''^2_d$$
$$Z'_d = R_a + jX'^2_d$$

8.2.2 d.c. Component

The d.c. component can be calculated from the following equation:

$$i_{d.c.}(t) = \sqrt{2} \cdot (I'_{kd} - I_{rG} \cdot \sin\varphi) \cdot e^{-\frac{t}{T_{d.c.}}}$$

8.2.3 Peak Value

The peak value is calculated at the half cycle of the short-circuit condition. The exact time depends on the preload conditions, the generator impedance, and the time constants.

$$i_p(t) = \sqrt{2} \cdot I_{a.c.}(t) + i_{d.c.}(t)$$

The practical short-circuit current calculation in networks, which is ultimately based on the behavior of SGs when the short circuit occurs, does not deal with mathematical functions and their instantaneous values, but only with effective values, the specific be assigned to typical time ranges of the short-circuit current.

The following assumptions can be made to simplify the system calculations that lead to conservative results. These assumptions are as follows:

1) Stator resistance is ignored. The error is negligible.
2) Preload conditions are ignored. This gives a lower value of 10%.
3) Subtransient and transient decays are ignored.
4) Time constants can be calculated from the machine open-circuit time constants and impedance.

All calculating formulae become

$$I_{a.c.}(t) = (I''_{kd} - I'_{kd}) \cdot e^{-(t/T''_d)} + I'_{kd}$$

$$I_{a.c.}(t) = \frac{U_n}{\sqrt{3} \cdot X''_{kd}}$$

$$i_{d.c.} = \sqrt{2} \cdot I_{a.c.}$$

$$i_p = \sqrt{2} \cdot I_{a.c.} + i_{d.c.} = 2 \cdot \sqrt{2} \cdot I_{a.c.}$$

where X''_d is the subtransient reactance, X'_d is transient reactance, X_d is synchron reactance, X_0 is zero reactance, Z_G is impedance of the generator, S_{rG} is rated power of the generator, R_G is resistance of the generator, U_{rG} is the rated voltage of the generator, T is the time constant.

Table 8.1 shows the various characteristics (subtransient, transient, and synchronous longitudinal reactances) for calculating the reactances in synchronous machines. They are effective in short entry X''_d for I''_{k3}, during the decay of the short-circuit current X'_d and the synchronous operation of the generator X_d for the continuous short-circuit current.

Table 8.1 Reactances of a synchronous machine [6].

Generator type	Turbo generators	Salient-phase generators	
		With damper winding[a]	Without damper winding
Subtransient reactance, X''_d (%)	9...22[b]	12...30[c]	20...40[c]
Transient reactance, X'_d (%)	14...35[d]	20...45	20...40
Synchronous reactance,[e] X_d (%)	140...300	80...180	80...180
Negative-sequence reactance,[f] X_2 (%)	9...22	10...25	30...50
Zero-sequence reactance,[g] X_0 (%)	3...10	5...20	5...25
Subtransient time constant, T''_d (s)	0.06...0.10	0.04...0.08	—
Transient time constant, T'_d (s)	0.5...1.8	0.9...2.5	0.7...2.5
d.c. time constant, T_{dc} (s)	0.05...0.3	0.1...0.3	0.15...0.5

a) Valid for laminated-phase shoes and complete damper winding; also for solid-phase shoes with strap connections.
b) Values increase with machine rating; low values for LV generators.
c) Higher values are for low-speed rotors ($n < 375$ min^{-1}).
d) For very large machines (>1000 MVA) as much as 40–45%.
e) Saturated values are 5–20% lower.
f) $x_2 = 0.5 (x''_d + x''_q)$; also valid for transients.
g) Depends on winding pitch.

8.3 Transformers

To transport the electrical energy from the generator (power plant) to the end user (households, industry, and public institutions), a chain of different electrical installations is needed. The last element before the consumer is usually the distribution transformer. A transformer is an alternator – an alternating voltage and current, electromagnetic by induction, that is between two or more windings at the same frequency. It transmits different values of voltage and current. Thus, it is a device for the transmission and transport of electrical energy (Figure 8.7).

At this point, it is useful to explain the transformer and its equivalent circuit for the case of a short circuit (Figure 8.8).

The short-circuit voltage, U_k, is the primary voltage at which a transformer with short-circuited secondary winding already takes up its primary rated current. U_k is usually expressed as a relative short-circuit voltage in percent of the primary voltage. It is a measure for the loading of the voltage change occurring. The following condition applies

$$u_k = \frac{U_k \cdot 100\%}{U_{nHV}} \tag{8.7}$$

When a short circuit occurs during the operation of a transformer on the secondary side, the peak short-circuit current, i_p, flows first, which then gradually decays to the steady-state short-circuit current. The magnitude of i_p depends on the momentary value of the voltage and the magnetic state of the iron core. The

Figure 8.7 Overview of a transformer.

Figure 8.8 Transformer and equivalent circuit.

value of the steady-state short-circuit current, I_k, depends on the short-circuit voltage, U_k, and the internal resistance, Z.

$$I_{kd} = \frac{U_{nHV}}{Z} \tag{8.8}$$

$$Z = \frac{U_k}{I_{nLV}} \tag{8.9}$$

$$U_{nHV} = \frac{U_k \cdot 100\%}{u_k} \tag{8.10}$$

$$I_k = \frac{I_{nLV} \cdot 100\%}{u_k} \tag{8.11}$$

8.3.1 Short-Circuit Current on the Secondary Side

The equivalent circuit of the positive-sequence, negative-sequence, and zero-sequence systems is given by the number and the circuitry of the windings. Due to the phase angle, the negative-sequence impedance is identical to the positive-sequence impedance. The positive-sequence impedance of the transformer is calculated as follows:

$$Z_T = \frac{u_{kr}}{100\%} \frac{U_{rT}^2}{S_{rT}} \tag{8.12}$$

$$R_T = \frac{u_{Rr}}{100\%} \frac{U_{rT}^2}{S_{rT}} = \frac{P_{krT}}{3 \cdot I_{rT}^2} \tag{8.13}$$

$$X_T = \sqrt{Z_T^2 - R_t^2} \tag{8.14}$$

For LV transformers, the equivalent resistances and the inductive reactances in the zero-sequence and positive-sequence systems are as follows (Figure 8.9):
For the connection symbol $Dyn5$:

$$Z_{2T} = Z_{1T} \tag{8.15}$$
$$R_{0T} = R_T \tag{8.16}$$
$$X_{0T} = 0.95 \cdot X_T \tag{8.17}$$

For the connection symbols $Dzn0$ and $Yzn11$:

$$R_{0T} = 0.4 \cdot R_T \tag{8.18}$$
$$X_{0T} = 0.1 \cdot X_T \tag{8.19}$$

For the connection symbol $YYn6$:

$$R_{0T} = R_T \tag{8.20}$$
$$X_{0T} = 7 \cdots 100 X_T \tag{8.21}$$

Transformers with three windings are employed in auxiliary service for the internal requirements of power stations, in the industrial sector or as network transformers. The short-circuit impedances of transformers with three windings in the positive-sequence system can be calculated as follows (in accordance with Figure 8.10):

$$Z_{AB} = \frac{u_{krAB}}{100\%} \frac{U_{rTA}^2}{S_{rTAB}} \tag{8.22}$$

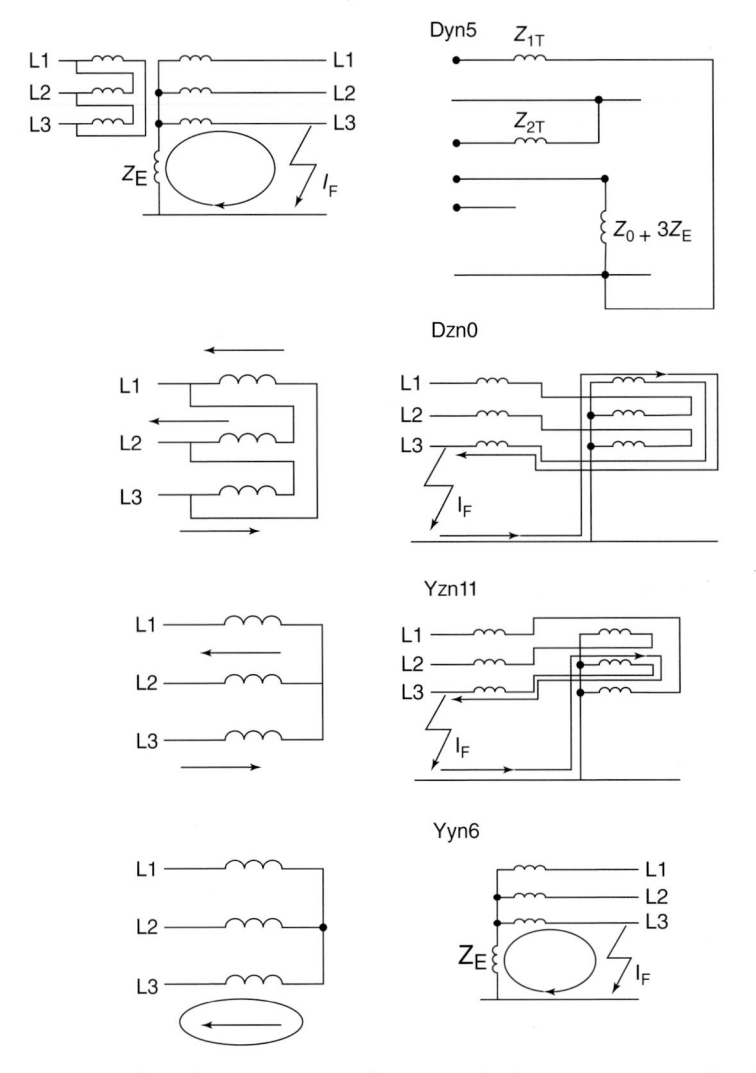

Figure 8.9 Equivalent resistances and reactances in the zero-sequence and positive-sequence systems for LV transformers.

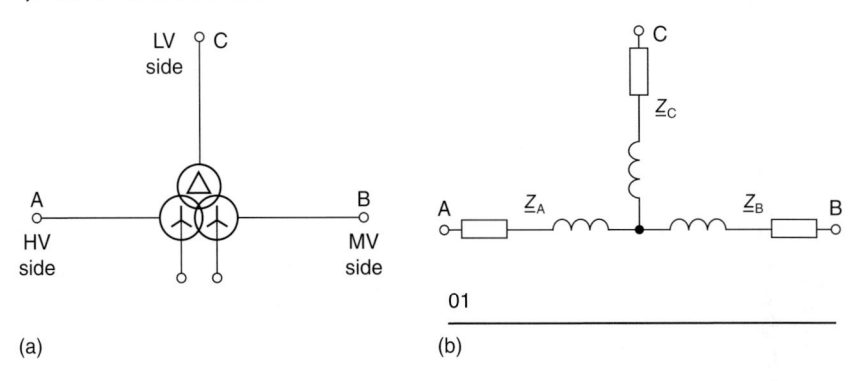

(a)

(b)

Figure 8.10 (a) Circuit diagram for a transformer with three windings and (b) equivalent circuit with side C open.

With side B open:

$$Z_{AC} = \frac{u_{krAC}}{100\%} \frac{U_{rTA}^2}{S_{rTAC}} \tag{8.23}$$

With side A open:

$$Z_{BC} = \frac{u_{krBC}}{100\%} \frac{U_{rTA}^2}{S_{rTBC}} \tag{8.24}$$

With the positive-sequence short-circuit impedances:

$$Z_A = \frac{1}{2} \cdot (Z_{AB} + Z_{AC} - Z_{BC}) \tag{8.25}$$

$$Z_B = \frac{1}{2} \cdot (Z_{BC} + Z_{AB} - Z_{AC}) \tag{8.26}$$

$$Z_C = \frac{1}{2} \cdot (Z_{AC} + Z_{BC} - Z_{AB}) \tag{8.27}$$

The meaning of the symbols is as follows:

U_{rT} rated voltage of the transformer on HV or LV side
I_{rT} rated current of the transformer on HV or LV side
U_{nHV} nominal voltage on HV side
I_{nLV} nominal voltage on LV side
U_k short-circuit voltage
S_{rT} rated apparent power of transformer
P_{krT} total winding losses of transformer at rated current
u_{kr} rated value of short-circuit voltage in %
u_{Rr} rated value of resistive voltage drop in %
R_{0T} zero-phase equivalent resistance of transformer
R_T equivalent resistance of transformer
X_{0T} inductive zero-sequence resistance of transformer
X_T inductive resistance of transformer.

The equivalent resistances and reactances of transformers can also be obtained from Figure 8.11.

8.3.2 Voltage-Regulating Transformers

For the compensation of voltage fluctuations in networks, the windings of transformers are provided with a tap so that the transformation ratios can be adjusted to keep the voltage for certain sections constant. Voltage-regulating transformers can be divided into two groups: controllable power transformers and series-regulating transformers. Controllable transformers have several taps on the voltage side to be regulated with which the transformation ratio can be increased or decreased in the same proportions, usually in steps of 1% or 2%.

This is also known as in-phase voltage control since only the magnitude of the voltage is regulated. Voltage regulation takes place stepwise with a switching device, which can be described as a stepping switch or stepping switch device. Switchover between the steps must take place under load, since load-dependent voltage fluctuations are regulated during operation.

The transformation ratio of the transformers is determined from the rated voltages. With stepping switches, it is possible to match the transformation ratio to

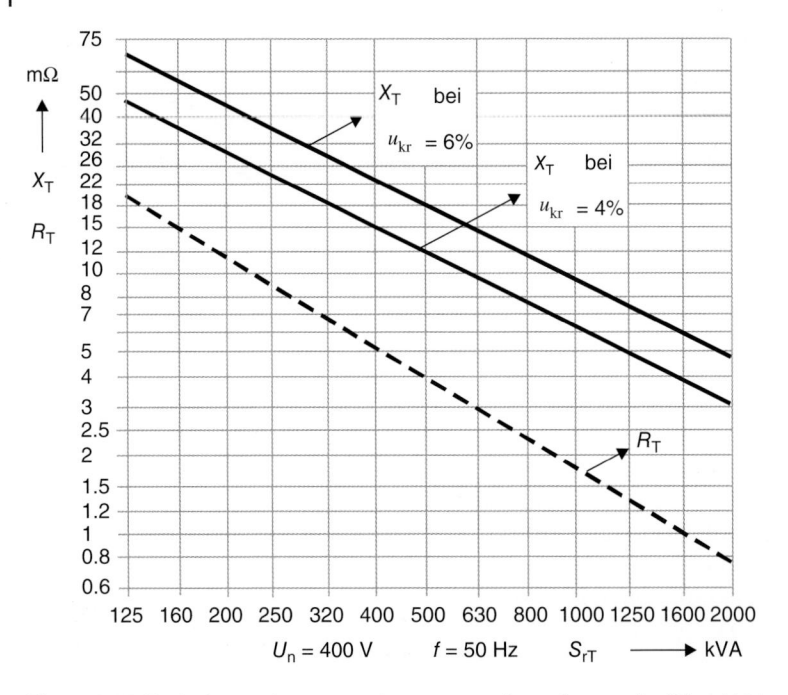

Figure 8.11 Equivalent resistances and reactances of transformers for LV and MV networks [7].

the load. The transformation ratio can be calculated by taking into account the step setting:

$$t = (1 + p_T) \cdot t_r$$

where, for t_r:

$$t_r = \frac{U_{rTHV}}{U_{rTLV}}$$

Table 8.2 Characteristic values of high-voltage (HV) transformers.

Rated voltage, U_r (kV)	Rated power, S_{rT} (MVA)	Short-circuit voltage, u_{kr} (%)	Impedance losses, P_{krT} (%)	No-load losses, P_{0rT} (%)	No-load current, i_{0rT} (%)
≤30	2–4	6	0.9–0.8	0.17–0.14	1.3–1.1
	5–10	7	0.8–0.7	0.13–0.11	1.0–0.8
	12.5–40	10	0.6–0.4	0.08–0.06	0.8–0.5
$30 < U_{rTHV} \leq 110$	6.3–10	10	0.9–0.8	0.18–0.14	0.9–0.8
	12.5–40	12	0.8–0.5	0.10–0.07	0.8–0.5
	50, 60	13	0.4	0.06	0.5–0.05
	80	14	0.5	0.05	0.45–0.05
110	100–350	12–16	0.31–0.19	0.05–0.03	0.45–0.05
$110 < U_{rTHV} \leq 220$	100–1000	10–20	0.32–0.19	0.065–0.035	0.47–0.04
$220 < U_{rTHV} \leq 380$	100–1000	11–20	0.4–0.2	0.07–0.04	0.48–0.04

In addition to maintaining constant voltage levels, regulating transformers are used for controlling the load flow. They too can be switched under load. These can be divided into quadrature control transformers and phase angle control transformers. For quadrature control transformers, an additional voltage is generated, which is phase-shifted by 90° from the voltage of a conductor. The additional voltage is added to the side on which the voltage is regulated. Phase angle control transformers are a combination of quadrature control transformers and in-phase control transformers. Stepping regulators are implemented via power electronic components, which are fast and require little maintenance. The most important features of HV transformers are tabulated in Table 8.2.

8.4 Cables and Overhead Lines

When the cross-section is known, the short-circuit impedances for LV networks can be obtained from the tables of IEC 60909-4.

For cables and lines (Figures 8.12 and 8.13), we can calculate the determining values as follows:

$$\underline{Z}_L = R_L + jX_L \tag{8.28}$$
$$R_L = l \cdot R'_L \tag{8.29}$$
$$X_L = l \cdot X'_L \tag{8.30}$$

Length-specific values for overhead lines:
Resistance in Ω/km:

$$R_L = \frac{l}{\kappa \cdot S} \tag{8.31}$$

The zero-sequence resistances of lines can be calculated from:

$$R_{0L} = \frac{R_{0L}}{R_L} \cdot R_L \tag{8.32}$$

$$X_{0L} = \frac{R_{0L}}{R_L} \cdot X_L \tag{8.33}$$

Inductive load reactance in Ω/km:

$$X_L = \omega \cdot L_b = \frac{\omega \cdot \mu_0}{2 \cdot \pi} \left(\ln \frac{d}{r} + \frac{l}{4 \cdot n} \right) \tag{8.34}$$

Double line:

$$X_L = \omega \cdot L_b = \frac{\omega \cdot \mu_0}{2 \cdot \pi} \left(\ln \frac{d \cdot d'}{r_e \cdot d''} + \frac{l}{4 \cdot n} \right) \tag{8.35}$$

Figure 8.12 Cables and lines in the positive-sequence system.

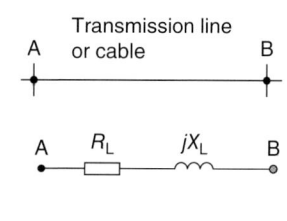

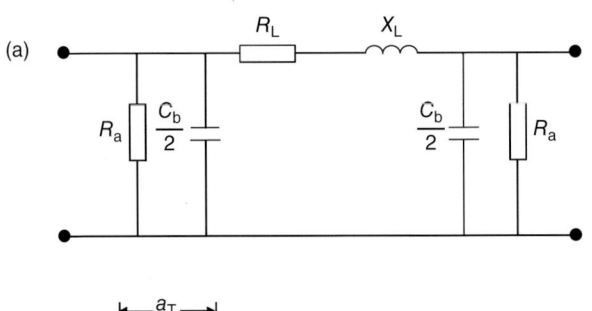

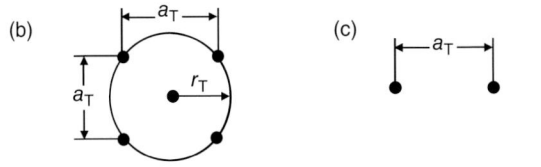

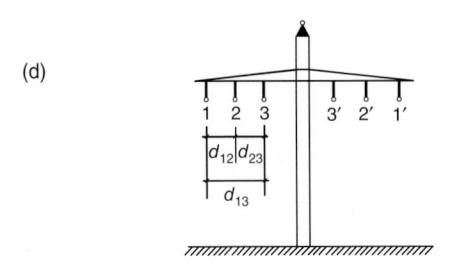

8 Impedances of Three-Phase Operational Equipment

Figure 8.13 The basis of calculation for overhead lines: (a) equivalent circuit of an overhead line; (b) 4× conductor bundle line; (c) 2× conductor bundle line; and (d) mast diagram.

Permeability:
$$\mu_0 = 4 \cdot \pi \cdot 10^{-4} \, \text{V s}/(\text{A m}) \tag{8.36}$$

Equivalent radius:
$$r_e = \sqrt[n]{n \cdot r \cdot R^{n-1}} \tag{8.37}$$

Average geometrical distance between conductors:
$$d = \sqrt[3]{d_{12} \cdot d_{23} \cdot d_{31}} \tag{8.38}$$

$$d' = \sqrt[3]{d'_{12} \cdot d'_{23} \cdot d'_{31}} \tag{8.39}$$

$$d'' = \sqrt[3]{d''_{11} \cdot d''_{22} \cdot d''_{33}} \tag{8.40}$$

Equivalent capacitive reactance in Ω/km:
$$C_b = \frac{2 \cdot \pi \cdot f_0}{\ln \frac{d \cdot d'}{r \cdot d''}} \tag{8.41}$$

$$X_b = \frac{1}{\omega \cdot C_B} \tag{8.42}$$

The meaning of the symbols is as follows:

a_T distance of conductor elements
C_b load capacitance
d average geometrical distance between three conductors or between center points of conductor bundles
L_b load inductance
n number of conductor elements
r conductor radius
r_e equivalent radius
R radius of conductor element
R_L effective resistance of a conductor
S cross-section
X_L inductive load reactance
μ_0 permeability.

The impedance values for overhead lines, cables, and conductors are generally available from the respective manufacturers. If no information is available, then the following Tables 8.3–8.19 can be followed.

Table 8.3 Impedances for polyvinyl chloride (PVC)-insulated three-phase NYY cables [8].

Cross-section of conductor, S (mm²)	3.5- and 4.5-conductive cable (mΩ/m)	Four- and five-conductive cable (mΩ/m)	One-conductive cable with PE, separated (mΩ/m)	One-conductive cable with PE, bundled (mΩ/m)
Cu conductor				
0.5	—	107.2	—	—
0.75	—	71.5	—	—
1	—	53.6	—	—
1.5	—	35.7	—	—
2.5	—	21.44	—	—
4	—	13.4	—	—
6	—	8.93	—	—
10	—	5.36	—	—
16	—	3.35	—	—
25	3.11	2.15	3.13	3.12
35	2.23	1.54	2.25	2.24
50	1.56	1.08	1.60	1.59
70	1.12	0.78	1.17	1.16
95	0.84	0.59	0.89	0.88
120	0.67	0.47	0.74	0.72
150	0.55	0.39	0.63	0.61
185	0.45	0.33	0.56	0.53
240	0.37	0.27	0.49	0.46
300	0.31	0.24	0.45	0.42

(Continued)

Table 8.3 (Continued)

Cross-section of conductor, S (mm²)	3.5- and 4.5-conductive cable (mΩ/m)	Four- and five-conductive cable (mΩ/m)	One-conductive cable with PE, separated (mΩ/m)	One-conductive cable with PE, bundled (mΩ/m)
Al conductor				
16	—	5.70	—	—
25	—	3.64	—	—
35	—	2.60	—	—
50	—	1.83	—	—
70	—	1.31	—	—
95	—	0.97	—	—
120	—	0.77	—	—
150	0.90	0.63	0.96	0.94
185	0.74	0.52	0.80	0.79
240	0.58	—	0.66	0.64
300	—	—	0.57	0.55

Table 8.4 Resistance values for PVC-insulated four- and five-conductor cables with copper conductors at 55 °C conductor temperature [8].

Cross-section of conductor, S (mm²)	Resistance, r (mΩ/m)	Reactance, x (mΩ/m)	Impedance, z (mΩ/m)
4 × 0.5	81.90	0.23	81.90
4 × 0.75	55.74	0.23	55.74
4 × 1	41.18	0.23	41.18
4 × 1.5	27.53	0.23	27.53
4 × 2.5	16.86	0.22	16.96
4 × 4	10.49	0.21	10.49
4 × 6	7.01	0.20	7.01
4 × 10	4.16	0.19	4.16
4 × 16	2.62	0.18	2.63
4 × 25	1.654	0.176	1.663
4 × 35	1.192	0.160	1.203
4 × 50	0.880	0.159	0.894
4 × 70	0.610	0.155	0.629
4 × 95	0.440	0.154	0.466
4 × 120	0.348	0.151	0.379
4 × 150	0.282	0.151	0.320
4 × 185	0.226	0.151	0.272
4 × 240	0.172	0.149	0.228
4 × 300	0.136	0.149	0.202

For PVC-insulated cables with aluminum conductors, multiply the resistance value R' by the factor 1.7.

Table 8.5 Resistance values for PVC insulated 3.5- and 4.5-conductor three-phase cables with copper conductors at 55 °C conductor temperature [8].

Cross-section of conductor, q (mm²)	Resistance, r (mΩ/m)	Reactance, x (mΩ/m)	Impedance, z (mΩ/m)
3 × 25/16	2.135	0.182	2.143
3 × 35/16	1.904	0.183	1.913
3 × 50/25	1.267	0.179	1.279
3 × 70/35	0.901	0.174	0.918
3 × 95/50	0.660	0.168	0.681
3 × 120/70	0.479	0.160	0.505
3 × 150/70	0.446	0.167	0.476
3 × 185/95	0.333	0.163	0.371
3 × 240/120	0.260	0.164	0.307
3 × 300/150	0.209	0.162	0.264

For PVC-insulated cables with aluminum conductors, multiply the resistance value R' by the factor 1.7.

Table 8.6 Resistance values for PVC-insulated single-conductor three-phase cables with PE or PEN next to each other, with copper conductors at 55 °C conductor temperature [8].

Cross-section of conductor, q (mm²)	Resistance, r (mΩ/m)	Reactance, x (mΩ/m)	Impedance, z (mΩ/m)
1 × 25/16	2.1350	0.390	2.170
1 × 35/16	1.904	0.377	1.941
1 × 50/25	1.267	0.373	1.321
1 × 70/35	0.901	0.366	0.973
1 × 95/50	0.660	0.333	0.739
1 × 120/70	0.479	0.330	0.582
1 × 150/70	0.446	0.327	0.553
1 × 185/95	0.333	0.326	0.466
1 × 240/120	0.260	0.320	0.412
1 × 300/150	0.209	0.328	0.381

For PVC-insulated cables with aluminum conductors, multiply the resistance value R' by the factor 1.7.

Table 8.7 Resistance values for PVC-insulated single-conductor three-phase cables with PE or PEN at a distance d from each other, with copper conductors at 55 °C conductor temperature [8].

Cross-section of conductor, q (mm²)	Resistance, r (mΩ/m)	Reactance, x (mΩ/m)	Impedance, z (mΩ/m)
1 × 25/16	2.135	0.390	2.170
1 × 35/16	1.904	0.377	1.941
1 × 50/25	1.267	0.373	1.321
1 × 70/35	0.901	0.366	0.973
1 × 95/50	0.660	0.362	0.753
1 × 120/70	0.479	0.359	0.599
1 × 150/70	0.446	0.356	0.571
1 × 185/95	0.333	0.354	0.486
1 × 240/120	0.260	0.349	0.435
1 × 300/150	0.209	0.347	0.405

For PVC-insulated cables with aluminum conductors, multiply the resistance value R' by the factor 1.7.

Table 8.8 Resistance values at 80 °C for copper cables and conductors [9].

Cross-section of conductor, q (mm²)	Copper			Aluminum		
	Resistance, r (mΩ/km)	Reactance, x (mΩ/km)	Impedance, z (mΩ/km)	Resistance, r (mΩ/km)	Reactance, x (mΩ/km)	Impedance, z (mΩ/km)
4 × 1.5	15	0.115	15	—	—	—
4 × 2.5	9.02	0.110	9.02	—	—	—
4 × 4	5.654	0.106	5.654	—	—	—
4 × 6	3.757	0.100	3.758	—	—	—
4 × 10	2.244	0.094	2.264	—	—	—
4 × 16	1.413	0.090	1.415	—	—	—
4 × 25	0.895	0.086	0.899	1.68	0.086	1.682
4 × 35	0.649	0.083	0.654	1.226	0.083	1.228
4 × 50	0.479	0.083	0.486	0.794	0.083	0.798
4 × 70	0.332	0.082	0.341	0.551	0.082	0.557
4 × 95	0.239	0.082	0.252	0.396	0.082	0.404
4 × 120	0.192	0.080	0.208	0.316	0.080	0.325
4 × 150	0.153	0.080	0.172	0.257	0.080	0.270
4 × 185	0.122	0.080	0.146	0.203	0.080	0.221
4 × 240	0.093	0.079	0.122	0.155	0.079	0.173
4 × 300	0.074	0.079	0.108	0.124	0.079	0.147

Table 8.9 Resistance values at 20 °C for copper cables and conductors [9].

	Copper			Aluminum		
Cross-section of conductor, q (mm²)	Resistance, r (mΩ/km)	Reactance, x (mΩ/km)	Impedance, z (mΩ/km)	Resistance, r (mΩ/km)	Reactance, x (mΩ/km)	Impedance, z (mΩ/km)
4 × 1.5	12.1	0.114	12.1	—	—	—
4 × 2.5	7.28	0.110	7.28	—	—	—
4 × 4	4.56	0.106	4.56	—	—	—
4 × 6	3.03	0.100	3.03	—	—	—
4 × 10	1.81	0.0945	1.812	—	—	—
4 × 16	1.14	0.0895	1.143	—	—	—
4 × 25	0.722	0.0879	0.729	1.20	0.088	1.203
4 × 35	0.524	0.0851	0.530	0.876	0.086	0.880
4 × 50	0.387	0.0848	0.396	0.641	0.084	0.646
4 × 70	0.268	0.0819	0.280	0.443	0.082	0.450
4 × 95	0.193	0.0819	0.209	0.320	0.082	0.330
4 × 120	0.155	0.0804	0.174	0.253	0.080	0.265
4 × 150	0.124	0.0804	0.147	0.206	0.080	0.220
4 × 185	0.0991	0.0804	0.127	0.164	0.080	0.182
4 × 240	0.0754	0.0797	0.109	0.125	0.079	0.147
4 × 300	0.0601	0.0797	0.998	0.100	0.079	0.127

Table 8.10 Resistances per unit length r in a positive-sequence system for overland line conductors manufactured in accordance with DIN 48 201 and $f = 50$ Hz at 20 °C [9].

Cross-section of conductor, q (mm²)	Nominal cross-section, q (mm²)	Copper, r (Ω/km)	Aluminum, r (Ω/km)
10	10	1.804	2.855
16	15.9	1.134	1.795
25	24.2	0.745	1.18
35	34.4	0.524	0.83
50	49.5	0.364	0.577
70	65.8	0.276	0.436
95	93.2	0.195	0.308
120	117	0.155	0.246

Table 8.11 Inductive reactances per unit length x in Ω/km in a positive-sequence system for overland line conductors at $f = 50$ Hz [9].

Cross-section of conductor, q (mm²)	Average distance d between conductors (cm)					
	50	60	70	80	90	100
10	0.37	0.38	0.40	0.40	0.41	0.42
16	0.36	0.37	0.38	0.38	0.40	0.40
25	0.34	0.35	0.37	0.37	0.38	0.39
35	0.33	0.33	0.35	0.36	0.37	0.38
50	0.32	0.32	0.34	0.35	0.36	0.37
70	0.31	0.32	0.33	0.34	0.35	0.35
95	0.29	0.31	0.32	0.33	0.34	0.34
120	0.29	0.30	0.31	0.32	0.33	0.34

Table 8.12 Quotients of effective resistances and inductive reactances in the zero-sequence and positive-sequence systems for NAYY and NYY cables as a function of the ground return system at $f = 50$ Hz [9].

q (mm²)	R_{OL}/R_L				X_{OL}/X_L			
	Copper		Aluminum		Copper		Aluminum	
	a	c	a	c	a	c	a	c
4 × 1.5	4.0	1.03	—	—	3.99	21.28	—	—
4 × 2.5	4.0	1.05	—	—	4.01	21.62	—	—
4 × 4	4.0	1.11	—	—	3.98	21.36	—	—
4 × 6	4.0	1.21	—	—	4.03	21.62	—	—
4 × 10	4.0	1.47	—	—	4.02	20.22	—	—
4 × 16	4.0	1.86	—	—	3.98	17.09	—	—
4 × 25	4.0	1.35	—	—	4.13	12.97	—	—
4 × 35	4.0	2.71	4.0	2.12	3.78	10.02	4.13	15.47
4 × 50	4.0	2.95	4.0	2.48	3.76	7.61	3.76	11.99
4 × 70	4.0	3.18	4.0	2.84	3.66	5.68	3.66	8.63
4 × 95	4.0	3.29	4.0	3.07	3.65	4.63	3.65	6.51
4 × 120	4.0	3.35	4.0	3.19	3.65	4.21	3.65	5.53
4 × 150	4.0	3.38	4.0	3.26	3.65	3.94	3.65	4.86
4 × 185	4.0	3.41	4.0	3.32	3.65	3.74	3.65	4.35
4 × 240	4.0	3.42	—	—	3.67	3.62	—	—
4 × 300	4.0	3.44	—	—	3.66	3.52	—	—

a, ground return system through fourth conductor.
c, ground return system through fourth conductor and ground.

Table 8.13 Resistances r per unit length of conductors for copper conductors.

Cross-section of conductor, q (mm²)	20 °C (Ω/km)	30 °C (Ω/km)
1.5	12.1	12.57
2.5	7.41	7.56
4	4.61	4.73
6	3.08	3.15
10	1.83	1.88
16	1.15	1.18
25	0.727	0.75
35	0.524	0.54
50	0.387	0.40
70	0.268	0.28
95	0.193	0.20
120	0.153	0.16
150	0.124	0.13
185	0.0991	0.10

Table 8.14 Impedance z for main outgoing and return lines of power supply companies.

Cross-section of conductor (NYM or NYY), q (mm²)	Impedance, z (mΩ/m)
1.5	0.03001
2.5	0.01838
4	0.01131
6	0.00752
10	0.00449
16	0.00284
25	0.00180
35	0.00131
50	0.00098
70	0.00069
95	0.00052

Table 8.15 Resistances of conductors in XLPE-insulated cables (6–30 kV) at 20 °C [10].

Cross-section of conductor, q (mm²)	Copper conductor (Ω/km)	Aluminum conductor (Ω/km)
25	0.727	1.20
35	0.524	0.868
50	0.387	0.641
70	0.268	0.443
95	0.193	0.320
120	0.153	0.253
150	0.124	0.206
185	0.0991	0.164
240	0.0754	0.125
300	0.0601	0.100
400	0.0470	0.0778
500	0.0366	0.0605

Table 8.16 Resistances per unit length of XLPE-insulated copper cables (6–20 kV) for $f = 50$ Hz [10].

Cross-section of conductor, q (mm²)	6/10 Ω/km		12/20 Ω/km	
	Single conductors on top of each other	Single conductors next to each other	Single conductors on top of each other	Single conductors next to each other
35	0.671	0.673	0.671	0.672
50	0.497	0.498	0.496	0.498
70	0.345	0.346	0.345	0.346
95	0.249	0.251	0.249	0.250
120	0.198	0.200	0.198	0.200
150	0.163	0.165	0.163	0.165
185	0.132	0.134	0.131	0.133
240	0.102	0.104	0.101	0.103
300	0.082	0.085	0.082	0.084
400	0.068	0.071	0.067	0.070
500	0.055	0.058	0.055	0.058

8.4 Cables and Overhead Lines

Table 8.17 Inductances of XLPE-insulated copper cables (6 30 kV) for $f = 50$ Hz [10].

Cross-section of conductor, q (mm²)	6/10 kV		12/20 kV		18/30 kV	
	mH/km	mH/km	mH/km	mH/km	mH/km	mH/km
35	0.45	0.76	0.48	0.76	—	—
50	0.42	0.73	0.45	0.74	0.48	0.75
70	0.39	0.70	0.43	0.70	0.45	0.71
95	0.38	0.67	0.41	0.68	0.43	0.68
120	0.36	0.65	0.39	0.65	0.42	0.66
150	0.35	0.63	0.38	0.63	0.41	0.64
185	0.34	0.61	0.36	0.62	0.39	0.63
240	0.32	0.59	0.35	0.59	0.37	0.60
300	0.31	0.57	0.33	0.58	0.36	0.59
400	0.30	0.55	0.33	0.55	0.34	0.56
500	0.29	0.53	0.31	0.53	0.33	0.54

Table 8.18 Effective capacitances of XLPE-insulated copper cables [10].

Nominal voltage	6/10 kV	12/20 kV	18/30 kV
Cross-section of conductor, q (mm²)	µF/km	µF/km	µF/km
35	0.22	0.16	—
50	0.25	0.18	0.14
70	0.28	0.20	0.15
95	0.31	0.22	0.17
120	0.34	0.23	0.18
150	0.37	0.25	0.19
185	0.40	0.27	0.20
240	0.44	0.30	0.22
300	0.48	0.32	0.24
400	0.55	0.36	0.27
500	0.60	0.40	0.29

8 Impedances of Three-Phase Operational Equipment

Table 8.19 Ground fault currents of XLPE-insulated copper cables [11].

Nominal voltage	6/10 kV	12/20 kV	18/30 kV
Cross-section of conductor, q (mm²)	A/km	A/km	A/km
35	1.2	1.7	—
50	1.4	1.9	2.3
70	1.5	2.1	2.5
95	1.7	2.4	2.7
120	1.9	2.6	2.9
150	2.0	2.7	3.1
185	2.2	3.0	3.3
240	2.4	3.3	3.7
300	2.6	3.5	4.0
400	3.0	4.0	4.4
500	3.3	4.3	4.8

8.5 Short-Circuit Current-Limiting Choke Coils

The short-circuit current-limiting choke coils (Figure 8.14) are used to limit the current flow as a result of a fault condition in series systems with insufficient stability against short circuits. They are used to reduce the breaking

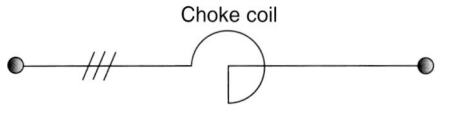

Figure 8.14 (a) Short-circuit current-limiting choke coil, equivalent circuit; and (b) application.

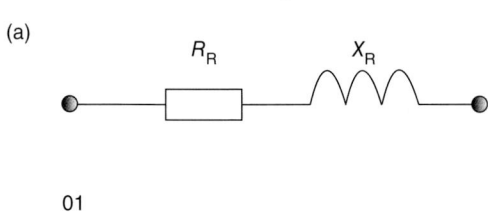

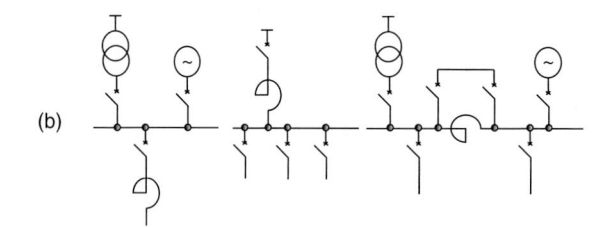

capacity of the circuit breakers to a permissible value. The following conditions apply:

$$X_R = \frac{u_{kR}}{100\%} \cdot \frac{U_n}{\sqrt{3} \cdot I_{rR}} \tag{8.43}$$

$$R_R \ll X_R \tag{8.44}$$

The meaning of the symbols is as follows:
X_R reactance of choke coil
U_n nominal power line voltage
R_R resistance of choke coil
u_{kR} rated voltage drop of choke coil (given on nameplate)
I_{rR} rated current of choke coil (given on nameplate).

8.6 Asynchronous Machines

Motors are energy converters, which convert the electric power into mechanical energy in the form of linear or rotary movement. The asynchronous motor is the most commonly used engine in plant engineering. It has a simple design and is robust and economical (Figure 8.15).

Asynchronous motors (Figure 8.16) have filed conditions similar to those of synchronous motors following a short circuit across the terminal.

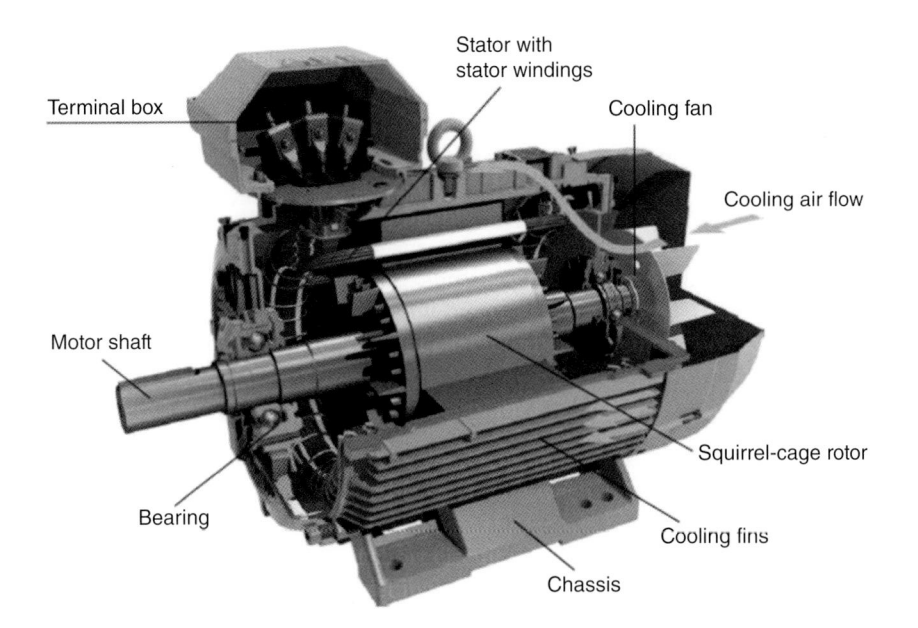

Figure 8.15 Overview of an asynchronous machine (ABB).

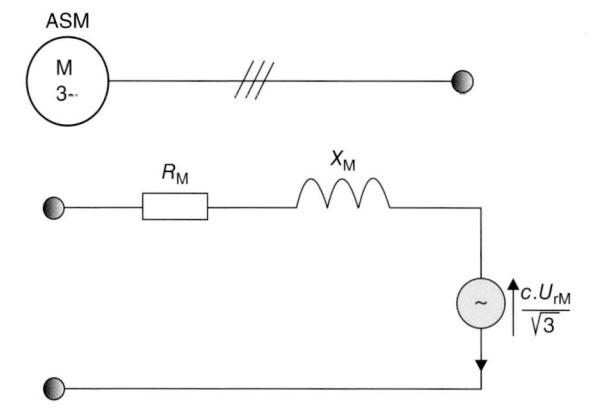

Figure 8.16 Asynchronous machine and equivalent circuit.

The equivalent circuit consists of an internal voltage source and an impedance. The value of this impedance can be calculated as follows:

$$Z_M = \frac{1}{I_{LR}/I_{rM}} \cdot \frac{U_{rM}}{\sqrt{3} \cdot I_{rM}} = \frac{1}{I_{LR}/I_{rM}} \cdot \frac{U_{rM}^2}{S_{rM}} \tag{8.45}$$

$$S_{rM} = \frac{P_{rM}}{\eta_r \cdot \cos \varphi_r} \tag{8.46}$$

The meaning of the symbols is as follows:

Z_M impedance of motor
U_{rM} rated voltage of motor
I_{rM} rated current of motor
S_{rM} rated apparent power of motor
P_{rM} rated effective power of motor
I_{LR}/I_{rM} ratio of locked rotor current to rated current of motor.

8.7 Consideration of Capacitors and Nonrotating Loads

The short-circuit currents are determined with the aid of the equivalent voltage source. The load flow before the occurrence of the short circuit, the capacitances of the conductors, the passive loads, the position of the step switch of transformers, and the state of the exciter of the generators are not considered. Independently of the point in time at which the short circuit occurs, the discharge current of the parallel capacitors can be neglected for the calculation of i_p. The influence of the series capacitors can also be neglected if these are provided with voltage-limiting systems connected in parallel, which respond in the event of a short circuit.

8.8 Static Converters

For the calculation of short-circuit currents, static converters are treated similarly to asynchronous motors. Reversing mechanisms supplied from a static converter contribute only to the initial symmetrical short-circuit current and the peak short-circuit current.

8.9 Wind Turbines

For mechanical to electrical energy conversion, SGs and AGs are used. Direct current link and inverter are in common use.

The AG needs an inductive reactive power for magnetization. The reactive power is drawn between different generator systems that reflect the excitation power from the grid and provide such inductive reactive power. Wind turbines can be connected individually or by multiple wind turbines in a wind farm via underground cables.

The electric power is usually fed at the HV side. Figure 8.17 shows a wind farm with three wind turbines.

For the calculation of short-circuit currents, the generators and transformers for wind turbines are combined into one unit. The short-circuit current is calculated on the HV side of the unit transformer.

The AG supplies in the event of a short circuit to contribute to the short-circuit current. This affects the thermal, electrical, and mechanical dimensioning of busbars and protection devices. These shorts are as follows:

1) Three-phase short circuit with and without earth
2) Two-phase short circuit with and without earth
3) Single-phase short circuit (ground fault).

Wind turbines with AGs or SGs up to 1 MW are directly coupled to the network. For higher ratings from 1.5 MW, particular SGs and AGs are used with inverters. They are able to adjust the variable torque of the rotor and the varying voltage to the requirements of the network.

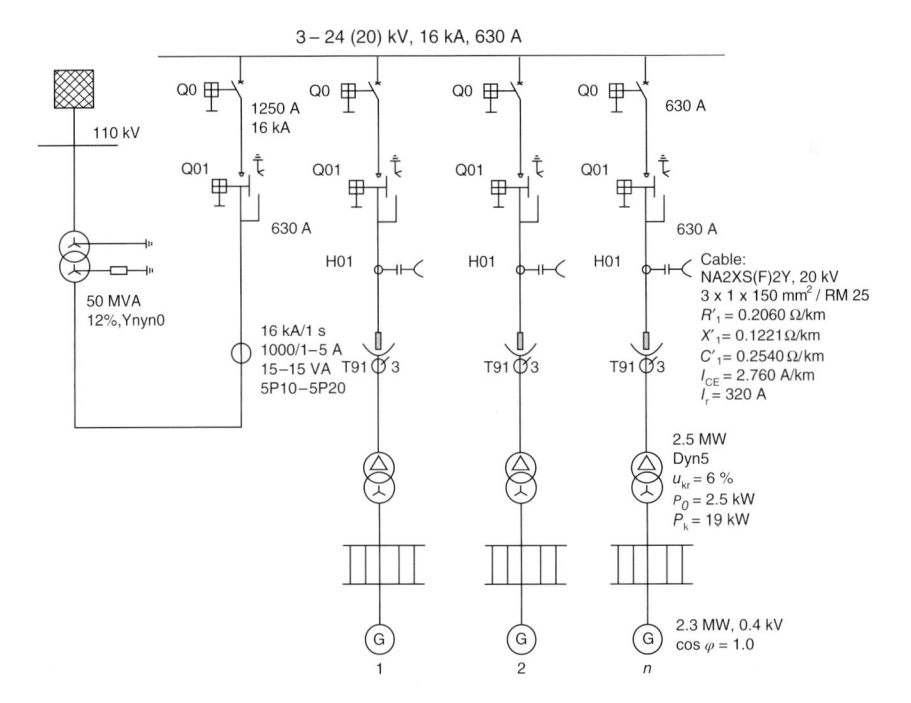

Figure 8.17 Connection to the HV power network.

The gearless SGs with a full-power converter and double-fed AGs have a strong presence in the market. However, the power converter feeds current harmonics into the grid. Wind power plants are usually divided into four types, which are used currently in the electricity networks. Using the IEEE definitions, the four types have the following properties.

8.9.1 Wind Power Plant with AG

A wind turbine with an AG (WA type one) is connected directly to the network without a power converter. WA has a soft starter, which is only active during startup. Figure 8.18 shows the main electrical and mechanical components of a WA, which is connected to the wind turbine rotor and the induction generator via a gearbox. The capacitor bank is used for reactive power compensation. This type of wind turbine with ride-through capabilities typically uses a variable capacitor bank that is dynamically controlled during and after errors. The power switch simultaneously disconnects the generator and the capacitors. The wind turbine terminals can be arranged on both sides of the transformer, as specified in IEC 61400-21.

Synchronous generators are modeled by an equivalent impedance Z_G, which can be calculated as follows:

$$Z_G = \frac{1}{\frac{I_{LR}}{I_{rG}}} \cdot \frac{U_{rG}}{\sqrt{3} \cdot I_{rG}} = \frac{1}{\frac{I_{LR}}{I_{rG}}} \cdot \frac{U_{rG}^2}{S_{rG}} \tag{8.47}$$

For calculating the short-circuit current on the HV power network, first the short-circuit positive-sequence impedance of a wind power plant with a SG is calculated:

$$\underline{Z}_W = t_r^2 \cdot \underline{Z}_G + \underline{Z}_{THV} \tag{8.48}$$

where U_{rG} is the rated voltage of the AG, I_{rG} is the rated current of the AG, S_{rG} is the rated apparent power of the AG, I_{LR}/I_{rG} is the ratio of the symmetrical locked rotor current to the rated current of the AG, Z_G is the impedance of the AG, Z_{THV} is the impedance of the unit transformer at the HV side, t_r is the rated transformation ratio of the unit transformer: $t_r = U_{rTHV}/U_{rTLV}$.

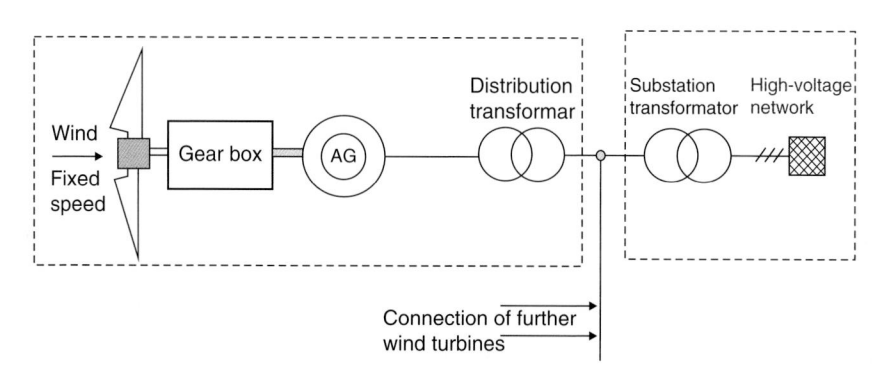

Figure 8.18 Wind power plant with asynchronous generator.

The type two wind turbine is constructed similarly to type one. The only difference is that the turbine is equipped with a variable rotor resistance and, therefore, uses a variable resistor and a rotor–blade–angle regulation.

8.9.2 Wind Power Plant with a Doubly Fed Asynchronous Generator

In a DFAG (type three), the stator is connected directly to the network. The rotor is connected by a back-to-back power converter. Figure 8.19 shows the main electrical and mechanical equipment.

The total positive-sequence short-circuit impedance $\underline{Z}_{WD}$ of a wind power station with a DFAG should be calculated as follows:

$$\underline{Z}_{WD} = \frac{\sqrt{2} \cdot \kappa_{WD} \cdot U_{rTHV}}{\sqrt{3} \cdot i_{WD\,max}} \tag{8.49}$$

where U_{rTHV} is the nominal system voltage, κ_{WD} is given by the manufacturer and referred to the HV side. If κ_{WD} is not known, then $\kappa_{WD} = 1.7$ can be taken, i_{WDmax} is the highest instantaneous short-circuit value in the case of a three-phase short circuit.

8.9.3 Wind Power with Full Converter

A wind turbine with a full converter, a full-scale power converter (type 4), is connected directly to the network via a converter. Figure 8.20 shows the essential electrical and mechanical components. Type 4 uses either SGs or AGs. Some of these types are used without a transmission (gear box) with direct drive.

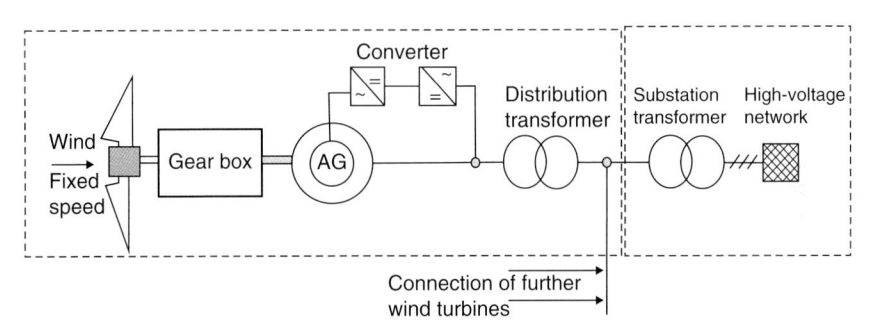

Figure 8.19 Wind power plant with a doubly fed asynchronous generator.

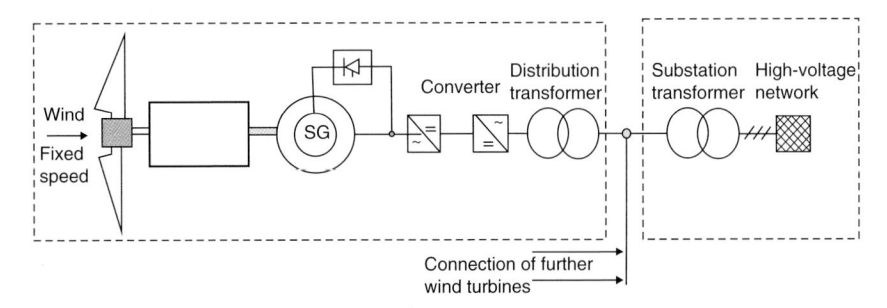

Figure 8.20 Wind power with full converter.

Directly related to the network-connected generators are modeled simply by their subtransient synchronous reactance. Short-circuit currents in generators with inverters and PV systems are modeled as voltage-controlled current sources. The currents injected in these investments are limited to the rated current and do not contribute, or only marginally, for a short-circuit current that must be specified by manufacturers.

8.10 Short-Circuit Calculation on Ship and Offshore Installations

The standard IEC 61363-1 is used for the short-circuit calculation on ship and offshore installations. The short-circuit current is considered to be close to the generator because the connections between the components in the entire network are very short.

The magnitude of the short-circuit current is caused by the reactance of generators, which are rapidly decayed by different preconditions. In the vicinity of the generator, very high short-circuit currents are to be expected, which strongly stresses the mechanical and thermal short-circuit strengths of the systems that are of great importance for the design of switching and protective devices.

Generators normally feed the three-phase system from 11 kV to 690 V. Other LV systems are supplied with 400 V levels, which have a radial-shaped design.

The MV main BB system provides essential loads of high power (e.g., the propellers or the engines for a transverse motion and large current motors for air conditioning or typical functions in the service areas, small loads, or lighting).

The electrical network on ships is made up of different plants and contains, for example, generators, synchronous and asynchronous motors, drives, and transformers.

For the calculation of short-circuit currents, the maximum number of generators that can be operated simultaneously and the total number of motors connected to the network at the same time should be taken into consideration. IEC 61363-1 describes the determination of the aperiodic component $I_{ac}(t)$, which is characterized by the subtransient, transient, and synchronous or steady state.

For the determination of the characteristic parameters are the consideration of the first 100 ms.

Therefore, there is a more stable time dependence of the short-circuit currents through the small distances to the fault location.

The synchronous machines used on marine/offshore electrical installations include SGs, motors, and capacitors. Knowledge about the short-circuit currents of these machines is fundamental for calculating the short-circuit current of an electrical system. During the first few cycles of a short circuit, all synchronous machines react in a similar manner and according to the short-circuit current have the same basic characteristics. For shunt energized machines, the excitation current during the short-circuit condition can drop to close to zero by discharging the short-circuit current. On connected machines, the short-circuit current is controlled and the excitation current maintained. Compound excited machines produce higher values of short-circuit currents after the subtransients effects have subsided.

In the case of a short circuit of generators, three different reactances are produced that increase in the order indicated. The subtransient reactance is smaller than the transient reactance and this is in turn smaller than the stationary reactance. Thus, the short-circuit currents decrease. The reactance of the generator decreases faster than the direct current component. This can lead to switch off and saturation problems of the magnetic circuits. In this case, the zero crossing of the current takes place after several periods.

The calculation of the short-circuit current of a synchronous machine is based on the curve of the maximum values of the machine with the actual, time-dependent short-circuit current.

The resultant envelope is a function of the basic machine parameters (current, impedance, etc.) and the active voltages. The impedances are dependent on the machine-operating conditions immediately before the occurrence of the short-circuit condition.

The calculation of the short-circuit current is the highest value of the current, which varies as a function of time along the upper envelope curve of the complex time-dependent function. The current defined by this upper envelope curve is calculated by the following equation:

$$i_k(t) = \sqrt{2} \cdot I_{a.c.}(t) + i_{d.c.}(t) \tag{8.50}$$

With the subtransient and transient periods, the ac component of the time function is described by subtransient, transient, and continuous currents. These periods are controlled by the subtransient time constant T_d'' and the transient time constant T_d'.

$$I_{ac}(t) = (I_{kd}'' - I_{kd}') \cdot e^{-(t/T_d'')} + (I_{kd}' - I_{kd}) \cdot e^{-(t/T_d')} + I_{kd} \tag{8.51}$$

The subtransient and transient initial values of the three-phase short-circuit currents, I_{kd}'' and I_{kd}', can be evaluated with the active voltages behind the respective impedance using the following equations:

$$I_{kd}'' = \frac{E_q''}{Z_d''} = \frac{E_q''}{\sqrt{R_a^2 + X_a''^2}}$$

$$I_{kd}' = \frac{E_q'}{Z_d'} = \frac{E_q'}{\sqrt{R_a^2 + X_a'^2}} \tag{8.52}$$

The continuous short-circuit current $I_{kd} = I_k$ can be obtained from the data sheet of the manufacturer. The active voltage currents E_q'' and E_q' are dependent on the current and current can be evaluated using the following equations that are obtained from the vector equations.

The subtransient voltage of the generator in the q-axis is

$$E_q'' = \sqrt{\left(\frac{U_{rG}}{\sqrt{3}} \cdot \cos\varphi + R_a \cdot I_{rG}\right)^2 + \left(\frac{U_{rG}}{\sqrt{3}} \cdot \sin\varphi + X_d'' \cdot I_{rG}\right)^2} \tag{8.53}$$

Transient voltage of the generator in the q-axis is

$$E'_q = \sqrt{\left(\frac{U_{rG}}{\sqrt{3}} \cdot \cos\varphi + R_a \cdot I_{rG}\right)^2 + \left(\frac{U_{rG}}{\sqrt{3}} \cdot \sin\varphi + X'_d \cdot I_{rG}\right)^2}$$

$$E''_q = \frac{U_{rG}}{\sqrt{3}}\varphi + I_{rG} \cdot \underline{Z}''_d$$

$$E'_q = \frac{U_{nG}}{\sqrt{3}}\varphi + I_{rG} \cdot \underline{Z}'_d \tag{8.54}$$

In which

$$\underline{Z}''_d = R_a + jX''_d$$
$$\underline{Z}'_d = R_a + jX'_d \tag{8.55}$$

The d.c. component can be expressed with the equation

$$i_{d.c.}(t) = \sqrt{2} \cdot (I''_{kd} - I_{rG} \cdot \sin\varphi) \cdot e^{-(t/T_{dc})} \tag{8.56}$$

The peak short-circuit current occurs between the time $t=0$ and $t=T/2$ of the short-circuit condition. The exact time depends on the preload conditions, the generator impedance, and the time constant. However, it is acceptable to calculate ip at time $T/2$ in the first half-wave of the short-circuit state using the equation:

$$i_p(t) = \sqrt{2} \cdot I_{a.c.}(t) + i_{d.c.}(t) \tag{8.57}$$

8.11 Examples

8.11.1 Example 1: Calculate the Impedance

Given: $U_{nQ} = 20\,\text{kV}$, $I''_{kQ} = 10\,\text{kA}$, $c_{max} = 1.1$, $R_Q = 0.1 \cdot X_Q$, $X_Q = 0.995 \cdot Z_Q$

$$Z_{Qt} = \frac{c \cdot U_{nQ}}{I''_{kQ}} \cdot \frac{1}{t_r^2} = \frac{1.1 \cdot 110\,\text{kV}}{10\,\text{kA}} \left(\frac{20\,\text{kV}}{110\,\text{kV}}\right)^2 = 0.23\,\text{m}\Omega$$

$X_{Qt} = 0.995 \cdot Z_{Qt} = 0.228\,\text{m}\Omega$

$R_{Qt} = 0.1 \cdot X_{Qt} = 0.0228\,\text{m}\Omega$

$Z_{Qt} = (0.0228 + j0.228)\,\text{m}\Omega$

8.11.2 Example 2: Calculation of a Transformer

Given: the nameplate of a transformer
$S_{rT1} = 630\,\text{kVA}$, $U_{rTHV} = 20\,\text{kV}$, $U_{rTLV} = 420\,\text{V}$ Dyn5, $u_{kr} = 4\%$, $P_{krT1} = 6.4\,\text{kW}$, $R_{(0)T}/R_T = 1.0$, $X_{(0)T}/X_T = 0.95$

a) Calculate the impedance:

$$Z_T = \frac{u_{krT}}{100\%} \cdot \frac{U^2_{rTLV}}{S_{rT}} = \frac{4\%}{100\%} \cdot \frac{(420\,\text{V})^2}{630\,\text{kVA}} = 11.2\,\text{m}\Omega$$

$$R_T = \frac{P_{krT}}{3 \cdot I_{rTLV}^2} = \frac{P_{krT} \cdot U_{rTLV}^2}{S_{rT}^2} = \frac{6.4\,\text{kW} \cdot (420\,\text{V})^2}{(630\,\text{kVA})^2} = 2.84\,\text{m}\Omega$$

$$u_{Rr} = \frac{P_{krT}}{S_{rTLV}} \cdot 100\% = 1.015\%$$

$$u_{xr} = \sqrt{(u_{kr}^2 - u_{Rr}^2)} = 3.869\%$$

$$X_T = \sqrt{(Z_T^2 - R_T^2)} = 10.83\,\text{m}\Omega$$

b) Calculate the correction factor:

$$K_T = 0.95 \cdot \frac{c_{max}}{1 + 0.6 \cdot x_T} = 0.95 \cdot \frac{1.05}{1 + 0.6 \cdot 0.03869} = 0.974$$

$$Z_{TK} = Z_T \cdot K_T = (2.76 + j10.54)\,\text{m}\Omega$$

8.11.3 Example 3: Calculation of a Cable

For a distribution panel, the following data are given for a cable:

$l_1 = 55\,\text{m}$, $S = 2 \times 4 \times 185\,\text{mm}^2$, Cu, $\underline{Z}'_L = (0.101 + j0.080)\,\Omega/\text{km}$,
$R_{(0)L}/R_L = 4$, $X_{(0)L}/X_L = 3.65$

a) Calculate the positive sequence:

$$\underline{Z}_L = 0.5 \cdot (0.101 + j0.080)\frac{\Omega}{\text{km}} \cdot 0.055\,\text{km} = (2.77 + j2.2)\,\text{m}\Omega$$

b) Calculate the zero sequence:

$$R_{(0)L} = 4.0 \cdot R_L = 4.0 \cdot 2.77 = 11.08\,\text{m}\Omega$$
$$X_{(0)L} = 3.61 \cdot X_L = 3.61 \cdot 2.2 = 7.492\,\text{m}\Omega$$
$$Z_{(0)L} = (11.08 + j7.492)\,\text{m}\Omega$$

8.11.4 Example 4: Calculation of a Generator

The following data are given for a generator:

$$X''_{Gen} = x''_d \cdot \frac{U_{rG}^2}{S_{rG}} = 0.2 \cdot \frac{10.5\,\text{kV}}{20\,\text{MVA}} = 1.1\,\Omega$$
$$R_G = 0.07 \cdot X''_d = 0.0772\,\Omega$$

a) Calculate the impedance of the generator:

$$Z_G = \sqrt{R_G^2 + X_G^2} = \sqrt{0.0772\,\Omega^2 + 1.1\,\Omega^2} = 1.1\,\Omega$$

Correction factor:

$$K_G = \frac{c}{1 + x_d'' \cdot \sin \varphi_{rG}} = \frac{1.1}{1 + 0.17 \cdot 0.63} = 0.994$$

Impedance of the generator:

$$Z_G = K_G \cdot Z_G = 0.994 \cdot 1.1\,\Omega = 1.093\,\Omega$$

8.11.5 Example 5: Calculation of a Motor

The following data are given for a motor:

$P_{rM} = 2.3\,\text{MW}, U_{rM} = 6\,\text{kV}, \cos \varphi_{rM} = 0.86$
$p = 2, I_a/I_{rM} = 5, \eta = 0.97$.

a) Calculate the impedance of the motor:

$$Z_M = \frac{\eta \cdot \cos \varphi}{I_{an}/I_{rM}} \cdot \frac{U_{rM}^2}{P_{rM}} = \frac{1}{2} \cdot \frac{0.86 \cdot 0.97}{5} \cdot \frac{(6\,\text{kV})^2}{2.3\,\text{MW}} = 2.611\,\Omega$$

8.11.6 Example 6: Calculation of an LV motor

$$S_{rM} = \frac{P_{rM}}{\eta_{rM} \cdot \cos_{rM}}$$

$$Z_M = \frac{1}{I_{an}/I_{rM}} \cdot \frac{U_{rM}^2}{P_{rM}} = \frac{1}{5.5} \cdot \frac{(20\,\text{kV})^2 \cdot 0.9 \cdot 0.973}{6\,\text{MW}} = 10.61\,\Omega$$

$$X_M = \frac{Z_{rM}}{\sqrt{1 + (R_M/X_M)^2}} = \frac{10.61\,\Omega}{\sqrt{1 + (0.1)^2}} = 10.55\,\Omega$$

$$R_M = X_M \cdot \left(\frac{R_M}{X_M}\right) = 10.55\,\Omega \cdot 0.1 = 1.055\,\Omega$$

8.11.7 Example 7: Design and Calculation of a Wind Farm

8.11.7.1 Description of the Wind Farm

The wind farm will be connected to the 154-kV network via a subdistribution power network (Figure 8.21). During normal operation, the wind farm network is designed as a radial distribution system with five branches.

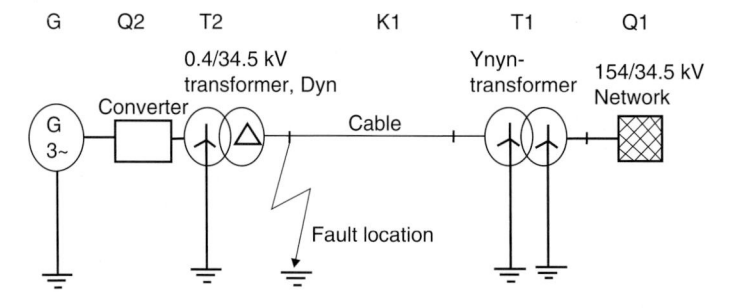

Figure 8.21 Wind power with full converter.

The following data are given for the design and calculation:
Network:

$$S''_{kQ} = 280 \, \text{MVA}$$
$$X_Q = 0.995 \cdot Z_Q$$
$$R_Q = 0.1 \cdot X_Q$$
$$Z_{0Q} = 3 \cdot Z_Q$$
$$X_{0Q} = 0.995 \cdot Z_{0Q}$$
$$R_{0Q} = 0.1 \cdot X_{0Q}$$

WP:

Number of WP: 15-voltage level
0.4 kV/34.5 kV
Rated power: $S_{rG} = 600 \, \text{kVA}$.

TR 1:

YNyn0 154/34.5 kV → $t = 4.46 : 1$

$$S_{nT1} = 50 \, \text{MVA}$$
$$u_{k1} = 12\%$$
$$X_{T1} = Z_{VT}$$
$$R_{T1} = 0$$
$$\frac{X_0}{X_1} = 3, \ldots, 10$$

Grounding: star point
MV-site $R_E = 20 \, \Omega$
HV-site direct.

TR 2:

Dyn11
$S_{rT2} = 700 \, \text{kVA}$
$u_{k2} = 6\%$
$X_{T2} = Z_{T2}$
$R_{T2} = 0$
$t = 1 : 86.25$.
Grounding: LV-site direct

Cable:

Resistance (at 20 °C):
196 mΩ/km
Reactance with:

$$L = 0.447 \, \text{mH/km} \quad 2 \cdot \pi \cdot 50 \, \text{Hz} \cdot 0.447 \frac{\text{mH}}{\text{km}} = 140 \, \text{m}\Omega$$

Zero sequence:
Resistance: 1.039 Ω/km
Reactance: 0.563 Ω/km
Length: $l = 0.75 \, \text{km}$

The positive-sequence and negative-sequence impedances are as follows:

$$Z_{1Q} = 1.1 \cdot \frac{(34.5\,\text{kV})^2}{280\,\text{MVA}} = 4.68\,\Omega$$

$$X_{1Q} = 0.995 \cdot 4.68\,\Omega = 4.66\,\Omega$$

$$R_{1Q} = 0.1 \cdot 4.66\,\Omega = 0.466\,\Omega$$

The negative-sequence impedance is given by:

$$Z_{0Q} = 14.04\,\Omega$$

$$X_{0Q} \approx 14\,\Omega$$

$$R_{0Q} = 1.4\,\Omega$$

Power transformer:
YNyn0 154/34.5 kV → $t = 4.46 : 1$

$$S_{rT1} = 50\,\text{MVA}$$

$$u_{k1} = 12\%$$

$$X_{T1} = Z_{PT}$$

$$R_{T1} = 0$$

Neutral-point connection:
MV $R_E = 20\,\Omega$, HV direct connected.

The resistance of the power transformer is negligible. The impedance Z is calculated from the reactance and the resistance, which are geometrically added.

Grounding arrangement:

The grounding arrangement in the transformer station is to be measured after being installed. Its grounding resistance is expected with $< 2\,\Omega$.

For three-phase, three-legged core transformers with star-to-star (Y–Y) winding [6]:

$$\frac{X_0}{X_1} = 3, \ldots, 10$$

Thus, when the correct values are not known in every single calculation, the worst-case values are taken. For the minimal short circuit, the biggest value and for the maximum short circuit, the smallest value are taken.

With

$$Z_T = X_T = \frac{u_{kr}}{100\%} \cdot \frac{U_{rT}^2}{S_{rT}}$$

The positive and negative impedances are as follows:

$$X_{1T1} = X_{2T1} = 0.12 \cdot \frac{(34.5\,\text{kV})^2}{50\,\text{MVA}} = 2.86\,\Omega$$

The negative impedance is as follows:

$$X_{0T1} = (3, \ldots, 10) \cdot 2.8566\,\Omega = 8.58, \ldots, 28.6\,\Omega$$

According to the IEC 60909-0 for short-circuit calculations, a correction factor for the transformers is necessary. Therefore

$$K_{T1} = 0.95 \cdot \frac{c_{max}}{1 + 0.6 \cdot x_{T1}} \tag{8.58}$$

$$x_{T1} = \frac{X_{T1}}{\left(\frac{U_n^2}{S_N}\right)}$$

And $Z_{T1} = X_{T1}$ follows that $x_{T1} = u_{k1} = 0.12$.
K_{T1} is calculated as:
$K_{T1} = 0.975$
With

$$X_{T1K} = K_{T1} \cdot X_{T1} \tag{8.59}$$

It follows that

$$X_{T1K} = 0.975 \cdot 2.86\,\Omega = 2.79\,\Omega$$

The equivalent circuit diagram of a Y–Y transformer within the zero sequence is as shown in Figure 8.22.

Transformer:

Dyn11 0.4/34.5 kV → $t = 1 : 86.25$
$S_{rT2} = 700\,\text{kVA}$
$u_{k2} = 6\%$
$X_{T2} = Z_{T2}$
$R_{T2} = 0$.

Neutral point connection:
Low voltage: direct
For three-phase, three-legged core transformers with delta-to-star (Δ–Y) winding, it is obtained as:

$$\frac{X_0}{X_1} \approx 1$$

The positive and negative impedances are calculated as follows:

$$X_{1T2} = X_{2T2} = 0.06 \cdot \frac{(34.5\,\text{kV})^2}{700\,\text{kVA}} = 102\,\Omega$$

Then, the negative impedance is

$X_{0T2} = X_{1/2T2} = 102\,\Omega$ and $Z_{T2} = X_{T2}$, follows that $x_{T2} = u_{k2} = 0.06$.
K_{T2} is calculated as: $K_{T2} \approx 1$.

Figure 8.22 Equivalent circuit diagram of a star-to-star (Y–Y) transformer in zero sequence.

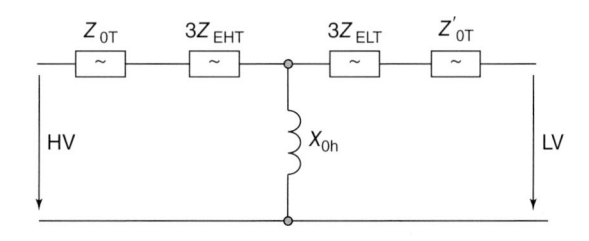

Each WEP consists of a permanently excited SG that produces two three-phase systems. The energy is inverted and supplied in the network after being transmitted via an intermediate direct currency link. The value of the supplied current comes up to the power that is available. A special attribute of this principle is that the generator increases the short-circuit current in the system by its rated current. The increase in the short-circuit power of the system is negligible. While normal generators would increase the current during a short circuit, this system just supplies a current equal to the rated current. Consequently, the WEP acts as a constant current generator, and it is essential that $\underline{Z}_2 = \underline{Z}_1$ is taken for the whole wind farm. The evaluation of such a generator is difficult. On this account, simplifications are made.

Wind energy plant data:

Number of plants: 70
Rated power: $S_{rG} = 600\,\text{kVA}$.

With

$$I_{rG} = \frac{S_{rG}}{\sqrt{3}\cdot U_{rG}} \qquad (8.60)$$

The maximal rated current is as follows:

$$I_{rG} = \frac{600\,\text{kVA}}{\sqrt{3}\cdot 34.5\,\text{kV}} = 10\,\text{A}$$

Cable:
XLPE-insulated cables are installed. The indication of these cables is N2XS2Y.
Cable in the branches:
The cables are installed in earth as a bundle of three cables. The data of the cables are as follows:

Cross-section: 95 mm²
Cross-section of shield: 16 mm²
Maximum current: 328 A
Resistance (20 °C): 196 mΩ/km.

Reactance 0.447 mH/km → $2\cdot\pi\cdot 50\,\text{Hz}\cdot 0.447\,\text{mH/km} = 140\,\text{m}\Omega$
Zero sequence:

Resistance: 1.039 Ω/km
Reactance: 0.563 Ω/km.

Cable connected from 34.5-kV BB to power transformer:

Cross-section: 240 mm²
Cross-section of shield: 25 mm²
Maximum current: 540 A
Resistance (20 °C): 0.0808 mΩ/km.

Reactance: 0.384 mH/km → $2\cdot\pi\cdot 50\,\text{Hz}\cdot 0.384\,\text{mH/km} = 120.6\,\text{m}\Omega$
Zero sequence:

Resistance: 0.712 Ω/km
Reactance: 0.302 Ω/km
Length: $l = 25\,\text{m} = 0.025\,\text{km}$

Two parallel, three-phase bundles are installed:

$$R_{1K1} = R_{2K1} = R'_{1K1} \times l = \frac{80.8\,\text{m}\Omega \times 0.025\,\text{km}}{2} = 1.01\,\text{m}\Omega$$

$$X_{1K1} = X_{2K1} = X'_{1K1} \cdot l = \frac{120.6\,\text{m}\Omega \cdot 0.025\,\text{km}}{2} = 1.5\,\text{m}\Omega$$

$$R_{0K1} = R'_{0K1} \cdot l = \frac{712\,\text{m}\Omega \cdot 0.025\,\text{km}}{2} = 8.9\,\text{m}\Omega$$

$$X_{0K1} = X'_{0K1} \cdot l = \frac{302\,\text{m}\Omega \cdot 0.025\,\text{km}}{2} = 3.8\,\text{m}\Omega$$

8.11.7.2 Calculations of Impedances

The installed WEPs are to be considered as though there would not be rotating elements because of their attitude. This means that $\underline{Z}_2 = \underline{Z}_1$ and

$$\frac{\underline{Z}_2}{\underline{Z}_1} = 1$$

The calculated impedances prove that $\underline{Z}_0 \geq \underline{Z}_1$, which means

$$\frac{\underline{Z}_0}{\underline{Z}_1} > 1 \quad \text{and} \quad \frac{\underline{Z}_1}{\underline{Z}_0} < 1$$

From Table 8.20 and Figure 8.23, the maximal current results from a three-phase short circuit and the minimal current results from a one-phase short circuit may be inferred.

During a one-phase short circuit, the WEPs and their transformer affect the impedances of the positive and negative sequences. The calculation of their impedances shows that these values are very large when compared with values of the network and power transformer. Also, the impedance of the inverter is expected to be very large. As a result, the impedances are negligible, so that the positive and negative impedances could be increased to downsize the short-circuit current. It also would be possible, because of calm, that every single WEP would not supply power. To calculate even the minimal short circuit, these simplifications are acceptable.

The coupling impedance of the power transformer is also negligible. Furthermore, the real value of this impedance must be measured. All capacitances are neglected because of the resulting big resistance.

In case of a maximal short circuit, the current of the WEP is just the rated current and is to be added to the calculated current. All negligences are considered

Table 8.20 Rating of the ground short-circuit current I''_{k2E}.

Z_2/Z_1	Z_2/Z_0	I''_{k2E}
1	0–1	$<I''_{k3} > I''_{k1}$
1	1	$=I''_{k3} = I''_{k1}$
1	1–4.4	$>I''_{k3} < I''_{k1}$
1	>4.4	$>I''_{k3} > I''_{k1}$

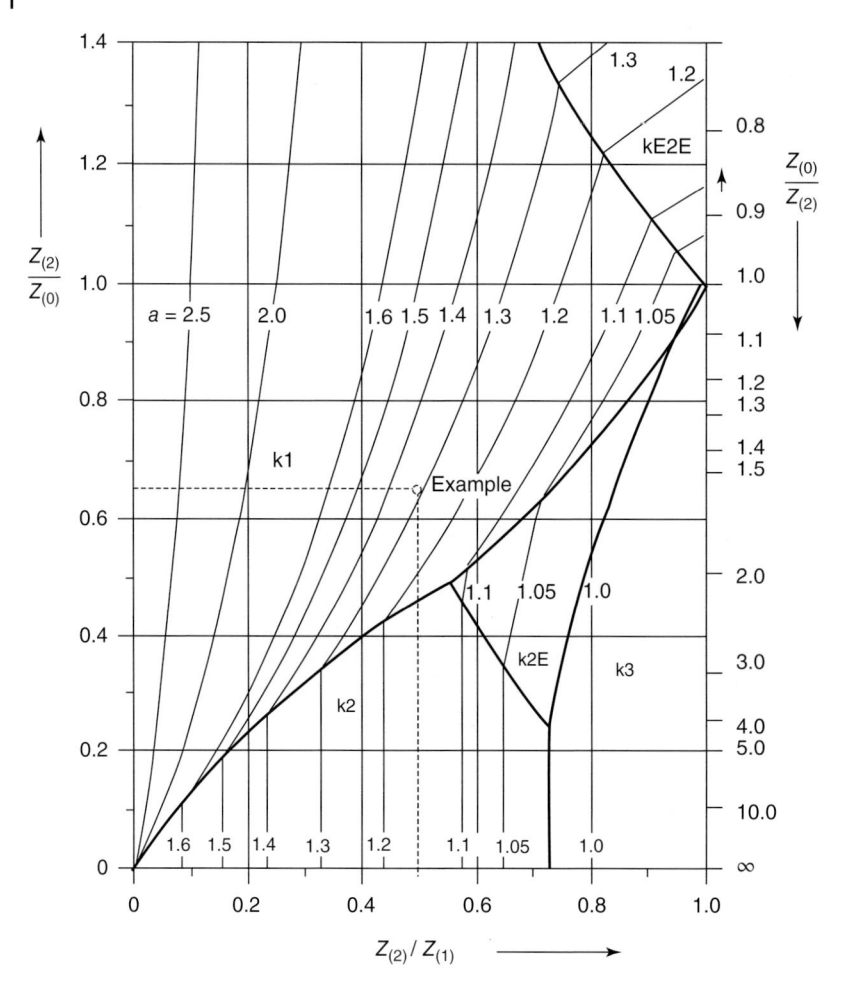

Figure 8.23 Diagram to assign failure with maximal current [12].

within the computer-added calculations. Within NEPLAN, the value of the abovementioned coupling impedance is equal to the impedance of the positive sequence. In the following example (Figure 8.24), a short circuit at the end of a 750-m cable is calculated (short circuit at the 34.5-kV BB is connected to WEP 44).

Symmetrical components of a partial network for one-phase short-circuit are shown in Figure 8.25.

According to the standards for short-circuit calculations, for the minimal short-circuit calculations, an increased cable temperature and, therefore, an increased resistance is necessary. The following results are based on an increased cable temperature of 80 and 160 °C.

Temperature coefficient of copper:

$$\alpha = 39 \cdot 10^{-4} \frac{1}{K}$$

8.11 Examples

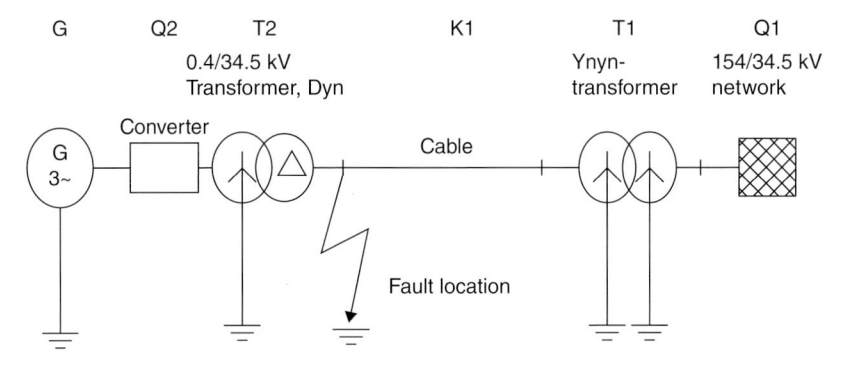

Figure 8.24 Diagram of a one-phase short circuit in partial network.

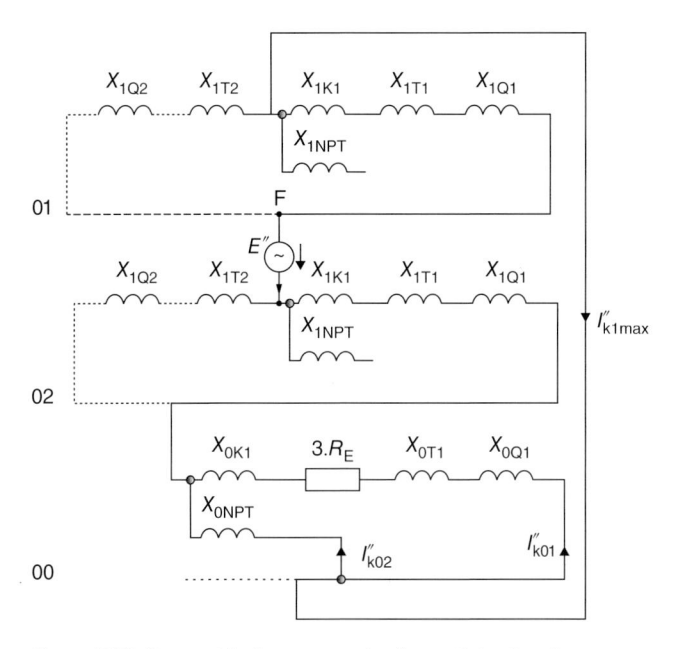

Figure 8.25 Symmetrical components of a partial network.

With

$$R_{80} = R_{20} \cdot (1 + \alpha \cdot \Delta \vartheta) \tag{8.61}$$

as follows:

Cable 95 mm²:
Positive and negative sequence:

$$R_{1K2/20} = R' \cdot 196 \frac{m\Omega}{km} \cdot 0.75 \, km = 147 \, m\Omega$$

$$R_{1K2/80} = 147 \, m\Omega \cdot \left(1 + 39 \cdot 10^{-4} \frac{1}{K} \cdot 60 \, K\right) = 181 \, m\Omega$$

$$R_{1K2/160} = 147 \, m\Omega \cdot \left(1 + 39 \cdot 10^{-4} \frac{1}{K} \cdot 140 \, K\right) = 227 \, m\Omega$$

Zero sequence:

$$R_{0K2/20} = R' \cdot 1039 \frac{m\Omega}{km} \cdot 0.75 \, km = 779 \, m\Omega$$

$$R_{0K2/80} = 779 \, m\Omega \cdot \left(1 + 39 \cdot 10^{-4} \frac{1}{K} \cdot 60 \, K\right) = 962 \, m\Omega$$

$$R_{0K2/160} = 779 \, m\Omega \cdot \left(1 + 39 \cdot 10^{-4} \frac{1}{K} \cdot 140 \, K\right) = 1.2 \, \Omega$$

Cable 240 mm²:
Positive and negative sequence:

$$R_{1K1/20} = 1.01 \, m\Omega$$

$$R_{1K1/80} = 1.01 \, m\Omega \cdot \left(1 + 39 \cdot 10^{-4} \frac{1}{K} \cdot 60 \, K\right) = 1.25 \, m\Omega$$

$$R_{1K1/160} = 1.01 \, m\Omega \cdot \left(1 + 39 \cdot 10^{-4} \frac{1}{K} \cdot 140 \, K\right) = 1.55 \, m\Omega$$

Zero sequence:

$$R_{0K1/20} = 8.9 \, m\Omega$$

$$R_{0K2/80} = 8.9 \, m\Omega \cdot \left(1 + 39 \cdot 10^{-4} \frac{1}{K} \cdot 60 \, K\right) = 11 \, m\Omega$$

$$R_{0K2/160} = 8.9 \, m\Omega \cdot \left(1 + 39 \cdot 10^{-4} \frac{1}{K} \cdot 140 \, K\right) = 13.75 \, m\Omega$$

Note: These values are only given for the completeness. The impedance of the 240-mm² cable is too small that it can be neglected. The minimal one-phase short-circuit current is calculated as follows:

$$I''_{k1min} = \frac{c_{min} \cdot \sqrt{3} \cdot U_n}{|2 \cdot \underline{Z}_1 + \underline{Z}_0|} \tag{8.62}$$

Analog to the minimal, the maximal one-phase short-circuit current is calculated as follows:

$$I''_{k1max} = \frac{c_{max} \cdot \sqrt{3} \cdot U_n}{|2 \cdot \underline{Z}_1 + \underline{Z}_0|} \tag{8.63}$$

The total impedance of the positive sequence, $\underline{Z}_1$, is calculated with:

$$Z_1 = Z_{1Q1} + X_{1T1} + Z_{1K1}$$

as follows:

$$\underline{Z}_1 = 0.466 \, \Omega + j4.66 \, \Omega + j2.86 \, \Omega + 0.181 \, \Omega + j0.14 \, \Omega = 0.65 \, \Omega + j7.63 \, \Omega$$

The total impedance of the zero sequence, $\underline{Z}_0$, is calculated with:

$$Z_0 = Z_{0Q1} + X_{0T1} + Z_{0K1}$$

as follows:

$$\underline{Z}_0 = 1.4 \, \Omega + j13.97 \, \Omega + j28.6 \, \Omega + 0.962 \, \Omega + j0.563 \, \Omega = 2.36 \, \Omega + j43 \, \Omega$$

8.11 Examples

The one-phase short-circuit is calculated with equation as follows:

$$I''_{k1min} = \frac{0.95 \cdot \sqrt{3} \cdot 34.5\,kV}{|2 \cdot (0.65\,\Omega + j7.63\,\Omega) + 2.36\,\Omega + j43\,\Omega + 3 \cdot 20\,\Omega|}$$

$$= \frac{56.76\,kV}{|63.66\,\Omega + j58.48\,\Omega|} = 656.7\,A$$

With an increase in cable temperature to 160 °C, the following occurs

$$\underline{Z}_1 = 0.466\,\Omega + j4.66\,\Omega + j2.86\,\Omega + 0.227\,\Omega + j0.14\,\Omega = 0.7\,\Omega + j7.63\,\Omega$$
$$\underline{Z}_0 = 1.4\,\Omega + j13.97\,\Omega + j28.6\,\Omega + 1.2\,\Omega + j0.563\,\Omega = 2.6\,\Omega + j43\,\Omega$$

The minimal one-phase short-circuit current is

$$I''_{k1min} = \frac{0.95 \cdot \sqrt{3} \cdot 34.5\,kV}{|2 \cdot (0.65\,\Omega + j7.63\,\Omega) + 2.36\,\Omega + j43\,\Omega + 3 \cdot 20\,\Omega|}$$

$$= \frac{56.76\,kV}{|63.95\,\Omega + j58.48\,\Omega|} = 655\,A$$

The maximal three-phase short-circuit current is calculated with (Figure 8.26)

$$I''_{k3max} = \frac{c_{max} \cdot U_n}{\sqrt{3} \cdot |\underline{Z}_1|} \tag{8.64}$$

According to IEC 60909-0, a resistance with a smaller value of $0.3 \cdot X_k$ is negligible.

The impedances are calculated as follows:

$$\underline{Z}_1 = j4.66\,\Omega + j2.86\,\Omega + 0.147\,\Omega + j0.14\,\Omega = 0.147\,\Omega + j7.63\,\Omega$$
$$|\underline{Z}_1| = 7.63\,\Omega$$

The short circuit is given as follows:

$$I''_{k3max} = \frac{1.1 \cdot 34.5\,kV}{\sqrt{3} \cdot 7.63\,\Omega} = \underline{2871\,A}$$

In this case, the influence of all WEPs is not included. With a total number of 70 WEPs with a maximum particular power of 600 kW, the additional current is given as

$$I_{k+} = 70 \cdot 10\,A = 700\,A$$

The complete current is

$$I''_{k3max} = 2871\,A + 700\,A = 3571\,A$$

Figure 8.26 Impedances for calculating a three-phase short-circuit.

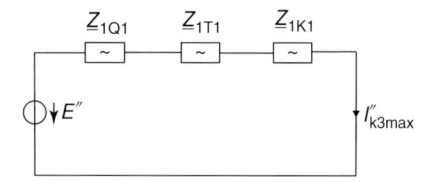

Table 8.21 Results of the minimal one-phase short-circuit calculation in NEPLAN with a cable temperature of 80 °C.

Location	$I''_{k1\,min}$ (kA)	Location	$I''_{k1\,min}$ (kA)	Location	$I''_{k1\,min}$ (kA)
BB 28	0.714	BB 01	0.668	BB 14	0.677
BB 32	0.719	BB 13	0.693	BB 25	0.704
BB 42	0.705				
BB 44	0.706	BB 43	0.707		
BB 70	0.681	BB 65	0.692		

Table 8.22 Results of the minimal one-phase short-circuit calculation in NEPLAN with a cable temperature of 160 °C.

Location	$I''_{k1\,min}$ (kA)	Location	$I''_{k1\,min}$/(kA)	Location	$I''_{k1\,min}$ (kA)
BB 28	0.713	BB 01	0.660	BB 14	0.671
BB 32	0.719	BB 13	0.689	BB 25	0.702
BB 42	0.703				
BB 44	0.704	BB 43	0.705		
BB 70	0.674	BB 65	0.688		

The results of the computer-added calculations of the minimal one-phase short circuit are shown as follows (the numbers of the BB are relative to the numbers of the WEP) (Tables 8.21 and 8.22). It is not necessary to look on every node in the wind farm – just the beginning and end of a branch line is considered.

8.11.7.3 Backup Protection and Protection Equipment

If a one-phase short circuit occurs between the transformer of a WEP and the fuse next in line, a big current flows from the 154-kV network toward the failure location. In this case, the fuse (40 A) is the protection system that needs to be activated before the circuit breaker. The time–current characteristic gives us the needed activation time (Figure 8.27). Concerning the worst case, the smallest and, therefore, the longest activation times are to be observed.

For the minimal short-circuit current:

Branch of WEP 28–42: 705 A on BB 42
Branch of WEP 1–13: 668 A on BB 42
Branch of WEP 14–253: 691 A on BB 42
Branch of WEP 44–70: 681 A on BB 42

Branch of WEP 43–65: 692 A on BB 42
Time before occurrence of the arc t (s)
Prospected current I (A).

The time–current characteristics for a 40-A fuse shows that the activation time for all currents shown above is $t_s < 0.03$ s. To have a full backup protection, the activation time of the independent-time-measuring relay is to be set up with a delay of $t_v = 100$ ms (Figure 8.28). The final activation time is $t_v \geq 30 + 100$ ms $= 130$ ms.

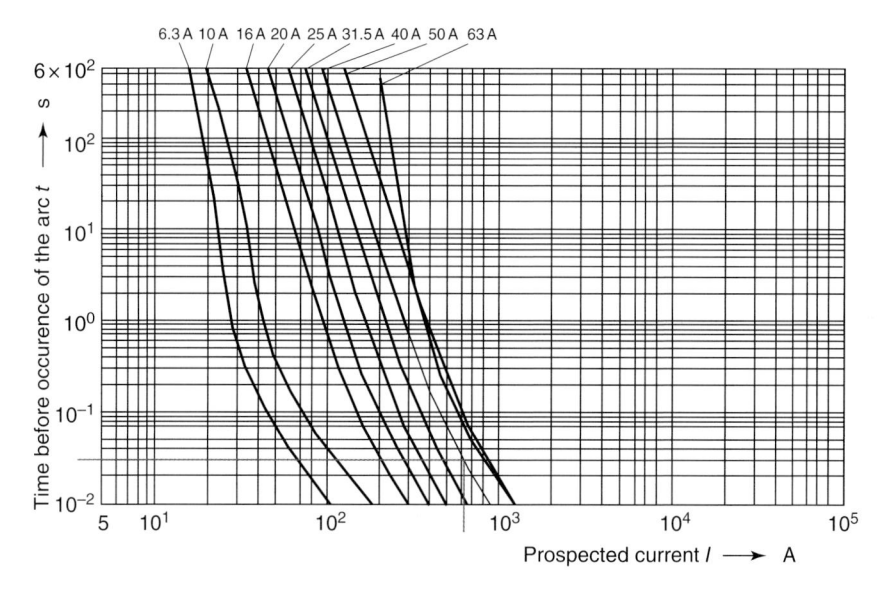

Figure 8.27 Characteristic diagram of high-voltage fuses.

Figure 8.28 Characterization diagram of high-voltage fuse flow operating to a circuit breaker.

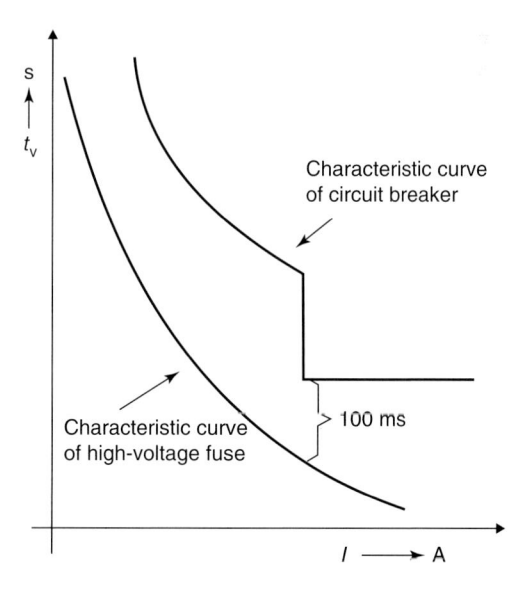

8.11.7.4 Thermal Stress of Cables

It shows that a short circuit on BB 32 affects the biggest current. This branch with 15 WEPs is connected via a 120-m cable to the main BB in the transformer station. If this short circuit occurs, the whole current flow excluding the rated current of 15 WEPs is on that cable. The current is $I_{diff} = 3661\,A - 150\,A = 3511\,A$.

To collect the thermal stress of cables, the thermal equivalent short-time current is to be calculated first as follows:

$$I_{th} = I_k'' \cdot \sqrt{(m+n)}$$

The coefficient m is reproducing the thermal effect of the direct current portion of a three-phase and single-phase current. The coefficient n is reproducing the thermal effect of the a.c. portion of a three-phase system. Both values are explained in Chapter 11.

For electronic devices like cables and switches, the rated short-time current I_{thr} for the rated short-circuit time T_{kr} (mostly ones) is given by the manufacturer. I_{th} is the root mean square (RMS) current to be carried by the electrical device for the duration T_k. Electrical devices are thermally stable if:

$$I_{th} \leq I_{thr} \quad \text{if } T_k \leq T_{kr}$$

And

$$I_{th} \leq I_{thr} \cdot \sqrt{\frac{T_{kr}}{T_k}} \quad \text{if } T_k \geq T_{kr}$$

According to Ref. [1], n is a function of I''_k/I_k. In this case, $I''_k = I_k$ means that $n = 1.2$, and m is a function of R_1/X_1 in failure location. $\underline{Z}_1$ is as follows:

$$\underline{Z}_1 = j4.66\,\Omega + j2.86\,\Omega + 0.196\frac{m\Omega}{km} \cdot 0.12\,km + j0.14\frac{m\Omega}{km} \cdot 0.12\,km$$
$$= 0.023\,\Omega + j7.54\,\Omega$$

From this, the following occurs

$$\frac{R_1}{X_1} = \frac{0.023\,\Omega}{7.54\,\Omega} = 0.03$$

The factor $\kappa \approx 1.95$ can be read from the curves in Figure 8.29. Given the factor $\kappa \approx 1.95$ and the permissible duration of the short circuit $t_{Ls} = 130$ ms, the thermal effect of the d.c. component $m = 1.05$ can also be obtained from Figure 8.30. With the equation $I_{th} = I_k'' \cdot \sqrt{(m+n)}\,I_{th}$, the following occurs

$$I_{th} = 3511\,A \cdot \sqrt{(1.05 + 1)} = 3511\,A \cdot 1.43 = 5026\,A$$

The rated short-time current for cables with a cross-section of $q = 95\,mm^2$ is as follows:

$$I_{thr} = 13.6\,kA$$

This is the specific value for a rated short-circuit time $T_{kr} = 1$ s. In this case, the short-circuit time is $T_k = 0.13$ s. Thus, the condition $I_{th} \leq I_{thr}$ is fulfilled.

In the case of a short circuit, a high thermal load is not applied.

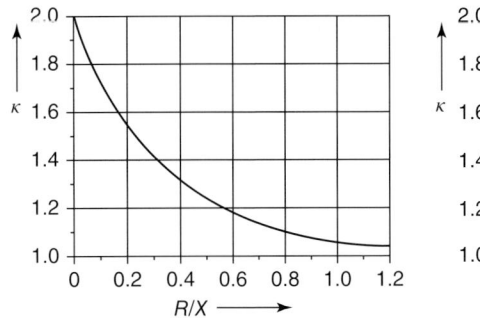

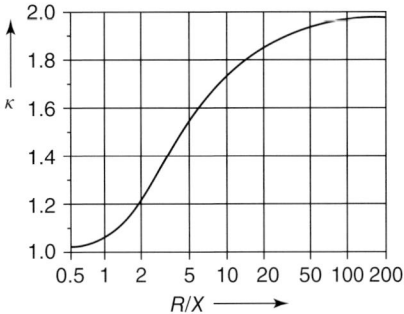

Figure 8.29 Factor k.

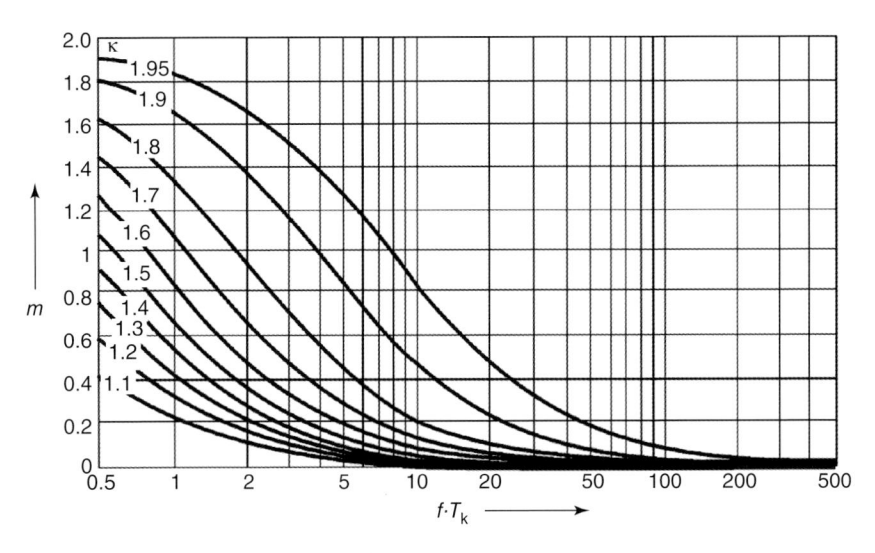

Figure 8.30 Factor m and parameter k.

8.11.7.5 Neutral Point Connection

Star-to-star (Y–Y) transformers are installed. The delivery of a WEP unit provides a star-to-delta (Y–Δ) transformer, which can be grounded on the LV side but not on the MV side. Thus, the conditions of a service company envision a neutral point transformer (NPT) for the purpose of grounding the Y–Δ transformer. An NPT is analyzed and its possible applications are discussed as follows.

8.11.7.6 Neutral Point Transformer (NPT)

In practice, two types of NPTs are used.

1) NPT with Y–Δ winding
2) NPT with zigzag–zigzag (Z–Z) winding.

NPTs with Y–Δ winding are to be handled as though they would be normal Y–Δ transformers (Figure 8.31).

With equal conditions, an *NPT with Z–Z winding (Figure 8.32)* gives a minimized need of material and a reduction in the nominal power by $\sqrt{3}$. However,

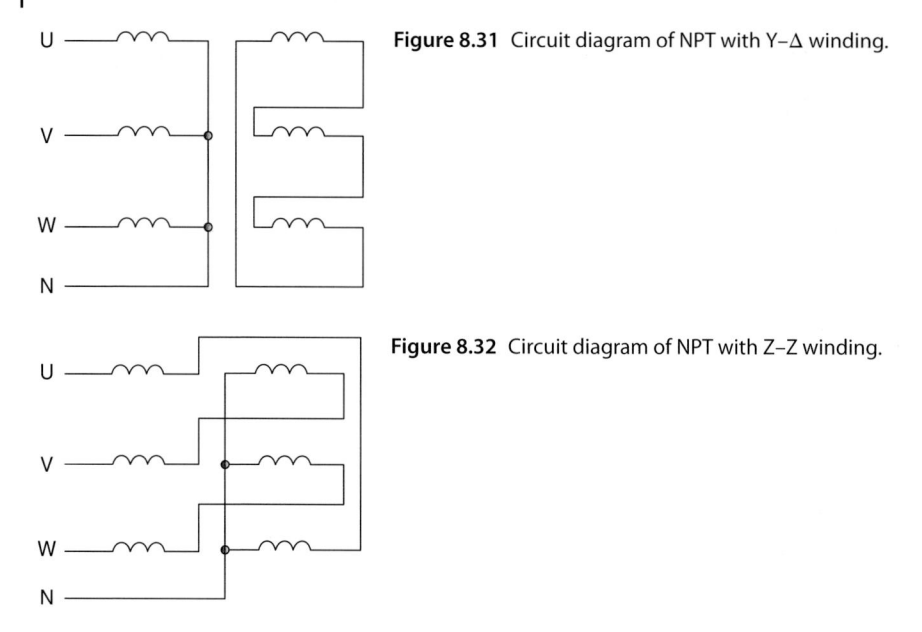

Figure 8.31 Circuit diagram of NPT with Y–Δ winding.

Figure 8.32 Circuit diagram of NPT with Z–Z winding.

these types require a galvanic connection of both windings, and if the rated voltage is >1.15 kV, a large gap between the windings and the core is also necessary.

Due to earlier cited reasons, only an NPT with Y–Δ winding is considered.

Furthermore, the different operational areas of the NPT are shown.

8.11.7.7 Network with Current-Limiting Resistor

In this example, the first cable with a length of 750 m of a branch is considered and a one-phase short circuit is calculated. The following figures show the simple diagram without (Figure 8.33) and with a connected NPT (Figure 8.34). Also, the equivalent circuit diagrams are shown as follows (Figures 8.35 and 8.36). The impedances of the WEP and its transformer are neglected.

First of all, the zero sequence is observed, because modifications only occur in this sequence when connecting an NPT. As shown, the impedance of the NPT is connected in parallel to one of the power transformers and the network. At first, imaginary data are adopted to calculate the changed impedance as well as the current. With this current, it is possible to find out the current I''_{k02} in the

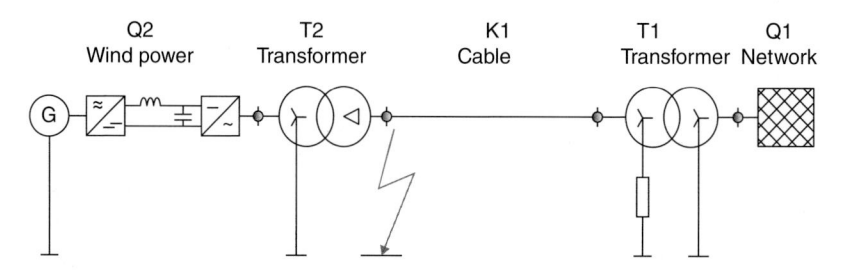

Figure 8.33 Diagram of a branch without an NPT.

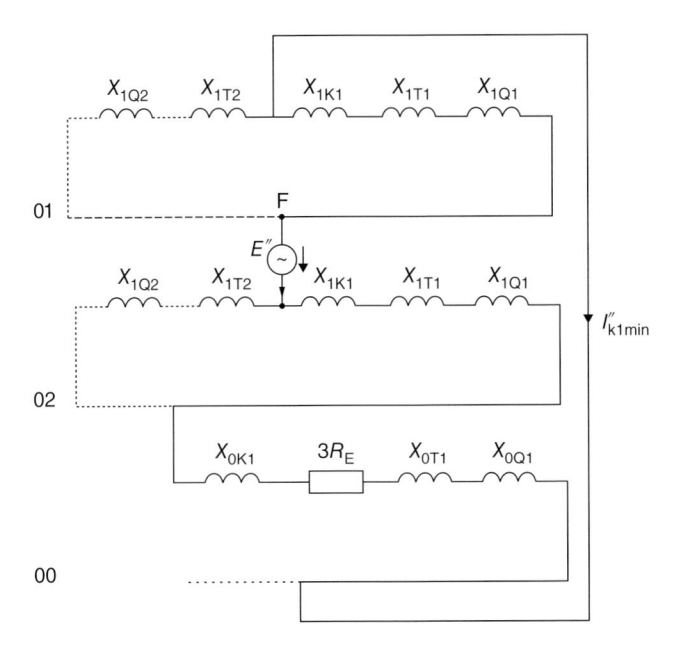

Figure 8.34 Symmetrical components of branch without an NPT.

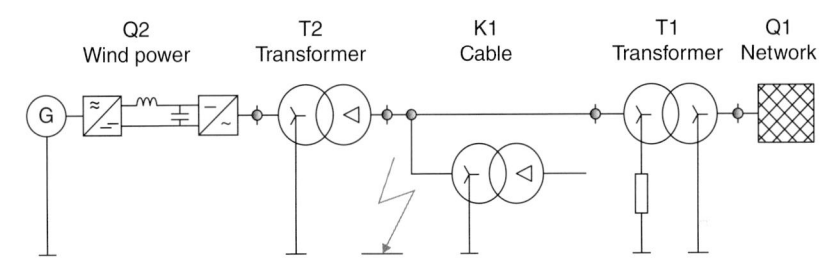

Figure 8.35 Diagram of a branch with an NPT.

neutral point of the NPT. This, again, is a dimension for the rated power of the NPT and vice versa.

The example shows the cycle of the calculation program by computing the maximal one-phase short circuit with an NPT connected to the network. In addition to the named simplification, the reactances of all resources are neglected. The data for the resources are as follows:

Network data:

$$X_{1Q} = 4.66\,\Omega$$

$$X_{0Q} \approx 14\,\Omega$$

Data of power transformer:

$$X_{1T} = 2.79\,\Omega$$

With $X_0/X_1 = 3$ (minimal impedance for the biggest current) follows:

$$X_{0Q} = 8.57\,\Omega$$

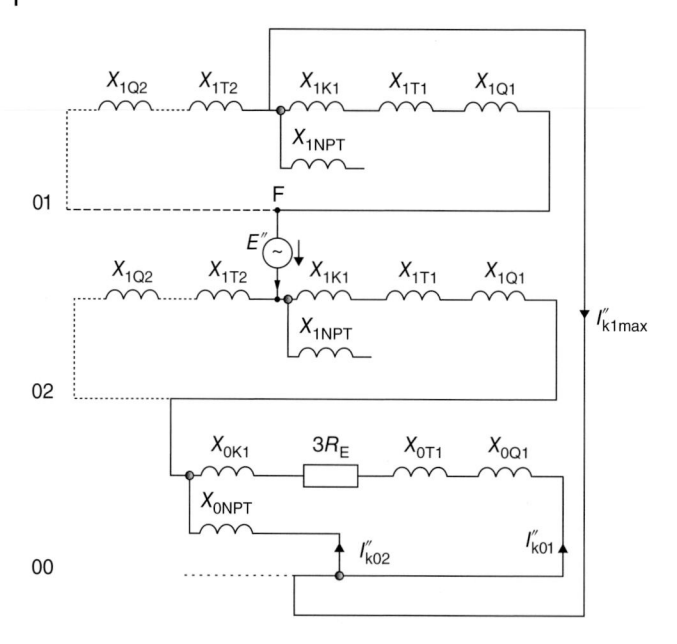

8 Impedances of Three-Phase Operational Equipment

Figure 8.36 Symmetrical components of branch with an NPT.

Data of the 95-mm² cable:

$$l = 0.75 \, \text{km}$$

$$X_{1K2} = 0.75 \, \text{km} \cdot 0.14 \frac{\Omega}{\text{km}} = 0.105 \, \Omega$$

$$X_{0K2} = 0.75 \, \text{km} \cdot 0.563 \frac{\Omega}{\text{km}} = 0.422 \, \Omega$$

The 240-mm² cable is neglected.
The NPT is a common Y–Δ transformer with

$S_{nPT} = 300 \, \text{kW}$
$U_n = 34.5 \, \text{kV}$
$u_{kr} = 6\%$
$X_0/X_1 = 1$ (reference value, see Ref. [1])
$R_{1NPT} = R_{0NPT} = 0$.

And follows:

$$X_{1NPT} = X_{0NPT} = \frac{6\%}{100\%} \cdot \frac{(34.5 \, \text{kV})^2}{300 \, \text{kVA}} = 238.1 \, \Omega$$

The impedance of the zero sequence is as follows:

$$\underline{Z}_0 = (X_{0Q1} + X_{0T1} + 3 \cdot R_E + X_{0K2}) // X_{St0}$$
$$= \frac{(j14 \, \Omega + j8.57 \, \Omega + 3 \cdot 20 \, \Omega + j0.422 \, \Omega) \cdot j238.1 \, \Omega}{j14 \, \Omega + j8.57 \, \Omega + 3 \cdot 20 \, \Omega + j0.422 \, \Omega + j238.1 \, \Omega}$$
$$= 47.36 \, \Omega + j31.86 \, \Omega$$

The maximal one-phase short circuit is calculated as follows:

$$I''_{k1max} = \frac{1.1 \times \sqrt{3} \times 34.5\,\text{kV}}{|2 \times (j4.66\,\Omega + j2.86\,\Omega + j0.105\,\Omega) + 47.36\,\Omega + j31.86\,\Omega|}$$

$$= \frac{65.73\,\text{kV}}{67\,\Omega} = 981\,\text{A}$$

This current is divided between the impedances $X_{0Q} + X_{0T1} + 3 \cdot R_E + X_{0K2}$ and X_{St0}:

$$|X_{0Q} + X_{0T} + 3 \cdot R_E + X_{0K2}| = |j14\,\Omega + j8.57\,\Omega + 3 \cdot 20\,\Omega + j0.422\,\Omega|$$

$$= |60\,\Omega + j23\,\Omega| = 64.3\,\Omega$$

$$|X_{0St}| = 238.1\,\Omega$$

With $\frac{I_{k01}}{I_{k02}} = \frac{238.1\,\Omega}{64.3\,\Omega}$ follows $I_{k01} \approx 3.7 \cdot I_{k02}$, $I_{k02} \approx 0.27 \cdot I_{k01}$.

We know that $I_{k01} + I_{k02} = I''_{k1max}$.

The currents are divided as shown in Figure 8.37. After dividing the currents, the values are $I_{k02} = 209\,\text{A}$ and $I_{k01} = 772\,\text{A}$.

The current I_{k02} is the dimension for the required power of the NPT. If a failure like this occurs, the disconnecting time from the wind farm will be $t_k < 10\,\text{s}$. Thus, the power of the NPT could be reduced to about 8%.

$$S_{NPT} = r \cdot \frac{I_{rNPT} \cdot U_r}{\sqrt{3}} \qquad (8.65)$$

The rated power is as follows:

$$S_{NPT} = 0.08 \cdot \frac{208\,\text{A} \cdot 34.5\,\text{kV}}{\sqrt{3}} = 331\,\text{kVA}$$

As shown, the adopted power of the NPT increases. A downsized impedance follows from an increased power, which does not make sense. The only way to adapt the NPT is to increase the short-circuit voltage. With Microsoft Excel, these calculations are very efficient and easy. Acceptable values for an NPT are as follows:

$S_{NPT} = 300\,\text{kVA}$
$U_r = 34.5\,\text{kV}$
$u_{kr} = 7\%$
$X_0/X_1 = 1$
$R_{1St} = R_{0St} = 0$.

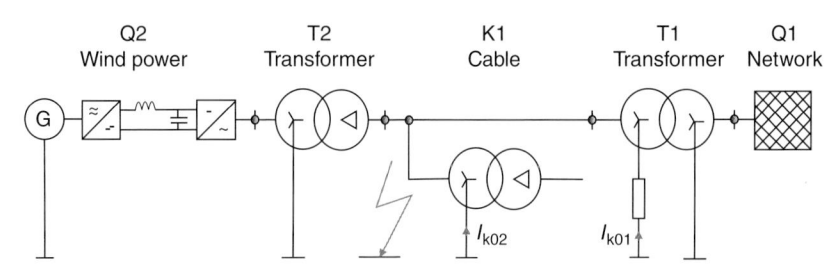

Figure 8.37 Diagram of a short circuit divided.

Table 8.23 Maximal one-phase short-circuit currents with and without a connected NPT.

$I''_{k1\,max}$ with NPT connected (A)	I''_{k1max} without NPT connected (A)
981	821

As a result, the installation as previously shown would not give the desired effects. The results are a parallel impedance and as a result of that an increased short-circuit current.

The current with and without an NPT is compared and tabulated in Table 8.23.

8.11.7.8 Compensated Network

A compensated neutral point is another possibility to operate the network, which means that a Peterson coil is to be installed into the neutral point. Note that a Y–Y transformer is unserviceable for connecting a coil. The coupling impedance transforms zero potential in case of an earth fault. That affects unsymmetrical voltage measures even on the faultless side of the transformer. Thus, an NPT is necessary.

If a one-phase earth fault occurs, the capacities of the cables provoke a capacitive current. It is possible to compensate that current with an additional current in opposite phase, which is supplied by a Peterson coil.

If the installed cables and their capacities are known, then it is possible to generate the awaiting current as follows:

$$I_E = 3 \cdot \omega \cdot C_E \cdot \frac{U_r}{\sqrt{3}} \tag{8.66}$$

C_E is the capacity of all cables installed:

Total length of 95-mm² cables: $l = 11.47\,\text{km}$
Total length of 240-mm² cables: $l = 2 \times 0.025\,\text{km}$

The capacity is

$$C_E = 11.47\,\text{km} \cdot 0.165 \frac{\mu F}{\text{km}} + 2 \cdot 0.025\,\text{km} \cdot 0.227 \frac{\mu F}{\text{km}} = 1.9\,\mu F$$

Earth current can be given as:

$$I_E = 3 \cdot 2 \cdot \pi \cdot 50\,\text{Hz} \cdot 1.9\,\mu F \cdot \frac{34.5\,\text{kV}}{\sqrt{3}} \approx 35.7\,\text{A}$$

The needed inductivity of the coil is

$$X_C \approx \frac{1}{3 \cdot \omega \cdot C_E} \tag{8.67}$$

The nominal power of the NPT infers from the calculated earth fault current. When compensating a network, one-phase earth faults can occur without activating the circuit breaker. The outcome of this is a better reliability. Without having a compensated network, every single connection between the earth and a single phase would disconnect the wind farm from the network.

The needed nominal power of the NPT is to be calculated by referring to the awaiting earth fault current:

$$S_{St} = \frac{35.7\,A \cdot 34.5\,kV}{\sqrt{3}} = 711\,kVA \approx 750\,kVA$$

When installing a coil, the current-limiting resistor is to be released.

In addition to a Peterson coil and an NPT, special instruments and the understanding of these instruments are necessary to permit operation while one-phase earth faults.

Among the installation of a compensating coil, additional devices are to be fitted. These are necessary to report an occurred earth fault or even to localize it. Due to the occurrence of voltage rise during an earth fault, it is simple to realize it. In compensated networks, not only the error recognition, but also the location is desired. To that, the following devices are to be fitted:

- cable-type current transformer
- earth fault relay (i.e., 7SN93, 7TG207SA511, or 7SN73 of Siemens)
- voltage converter for the recording of the displacement voltage.

If the mentioned resources are fitted, it is possible to locate an earth fault exactly. Once localized, the corresponding part of the cable can be disconnected without disturbing the normal operation.

People working in the control room and people checking the wind farm must be specially trained and developed to be able to act correctly in case of an earth fault. Basic knowledge about the behavior during an earth fault is inevitable.

Among the additional installation of protection devices and earth fault relays, the voltage jump is to be considered. All devices fitted in the network are to be designed for such an increased voltage.

8.11.7.9 Insulated Network

Along with the previously mentioned possibilities, there is also the possibility of operating the network with an insulated neutral point. In this case, the prescriptions [13] are to be observed.

It shows that the permissible value for an earth fault current with a nominal voltage of $U_n = 34.5\,kV$ is about the value of $I_E = 35.7\,A$.

Resulting from that, the current-limiting resistor could be removed. However, the risk of the intermittent earth fault is considered. From this point of view, this kind of neutral point connection is not recommended.

8.11.7.10 Grounding System

Among the neutral point connection, the efficiency of the grounding system is of primary interest according to the protection of human beings and animals. The EN 50522 is recommended (e.g., the voltage rise while failure is dangerous). The dimensioning and the judgment of the grounding systems quality is divided in four attributes:

- specification of the form and connection of the devices
- specification of the tolerable grounding voltage
- specification of the tolerable step and touch voltage
- specification of the maximal current permissible body current.

The regulation EN 50522 is based on the first three attributes.

Insulated and compensated networks are expressly attended because an earth fault within these networks does not involve a fast disconnection, but a continuous operation, as [2] the maximal tolerable touch voltage is $U_T = 75$ V a.c. In addition, Ref. [2] provides the conditions for the maximal resistance of the grounding system.

The following conditions are to be fulfilled:

$$Z_E < 2 \cdot \frac{U_T}{I_E}$$

For the whole network, the resistance of the grounding system is assumed with $Z_E < 2\,\Omega$.

The following is calculated:

$2 \cdot \frac{75\,\text{V}}{35.7\,\text{A}} = 4.2\,\Omega > Z_E$. The conditions are then fulfilled.

9

Impedance Corrections

The magnitude of the short-circuit currents in a network depends primarily on the design of the network, the generators or power station blocks, and the motors operating and secondarily on the operating state of the network before the occurrence of the short circuit. It is therefore difficult to find the loading state that leads to either the greatest or the smallest short-circuit current at the different fault locations of the network.

IEC 60909 thus recommends the method of calculation with the equivalent voltage source $c \cdot U_n/\sqrt{3}$ at the fault location. Investigations have shown that the voltage factor c is no longer sufficient for calculating the maximum short-circuit current when the subtransient behavior of the generators, power station blocks with or without step switches, and network transformers is considered.

The factor c in Table 1.1 shows that on the average, the highest voltage in a normal network does not deviate by more than about +5% for low voltage and +10% for high voltage from the network voltage U_n. In a common network with 50 or 60 Hz, the highest and lowest voltages do not differ by more than ±10% from the network voltage. In the North American countries, the highest voltage differs by not more than +5% and the lowest voltage −10% from the network voltage [1].

The conditions of Section 1.4.2 apply only when the voltage differences in a network are <10%. Special conditions can occur for generators and power station blocks with high values of x''_d and u_k, causing voltage drops of more than +10%. For this reason, the introduction of impedance correction factors is necessary, above all in order to obtain reliable values for the determination of the greatest short-circuit current.

For the dimensioning of electrical operational equipment, the calculation of the greatest short-circuit current and the greatest partial short-circuit current is necessary. In this section, we will discuss the required correction factors.

For the calculation of the largest short-circuit current and the transferred short-circuit current, it is necessary to make corrections to the generator impedances (K_G) and the power plant block impedances (K_{KW}) in addition to the factor c_{max}, especially when the subtransient reactances x''_d of the generators are large and the transformation ratio of the block transformers differs from the network voltages during the operation on both sides of the transformer [1].

Short Circuits in Power Systems: A Practical Guide to IEC 60909-0, Second Edition. Ismail Kasikci.
© 2018 Wiley-VCH Verlag GmbH & Co. KGaA. Published 2018 by Wiley-VCH Verlag GmbH & Co. KGaA.

9 Impedance Corrections

The calculation of the smallest short-circuit current requires special considerations, such as

- the smallest power supplied for thermal power stations
- the largest reactive power of machine units for pumping power stations
- special equipment for limiting the load angle
- loading state of power station units during low-load periods.

9.1 Correction Factor K_G for Generators

The impedance correction factor K_G is applied to the impedance of generators connected directly to the network (Figure 9.1).

K_G is derived from the overexcited generator, by taking account of the subtransient reactance X_d'' and subtransient internal voltage $\underline{E}''$ [1]. The impedance of the generator in the positive-phase system is as follows:

$$\underline{Z}_G = R_G + jX_d'' \tag{9.1}$$

$$\underline{Z}_{(GK)} = K_G \cdot \underline{Z}_G = K_G(R_G + jX_d'') \tag{9.2}$$

With the correction factor:

$$K_G = \frac{U_n}{U_{rG}} \cdot \frac{c_{max}}{1 + x_d'' \cdot \sqrt{1 - \cos^2 \varphi_{rG}}} \tag{9.3}$$

In accordance with IEC 60909, for three-phase short-circuit currents with direct connection to the network:

$$I_k'' = \frac{c \cdot U_n}{\sqrt{3} \cdot (R_G + jX_d'') \cdot K_G} \tag{9.4}$$

Substituting the correction factor in the above equation yields

$$I_k'' = \frac{c}{c_{max}} \cdot \frac{U_{rG}}{\sqrt{3} \cdot R_G + jX_d''} \cdot (1 + x_d'' \cdot \sin \varphi_{rG}) \tag{9.5}$$

Figure 9.1 Connection and equivalent circuit of a generator.

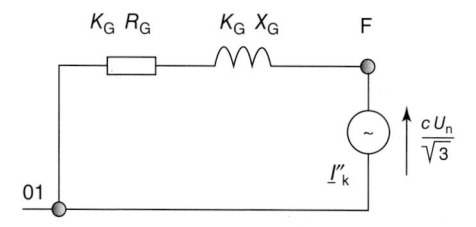

For salient-phase machines with different values for X_d'' and X_q'', we introduce

$$X_{(2)G} = \frac{1}{2}(X_d'' + X_q'') \tag{9.6}$$

$$\underline{Z}_{(0)G} = K_G(R_{(0)G} + jX_{(0)G}) \tag{9.7}$$

The meaning of the symbols is as follows:

c_{max} voltage factor
U_n nominal voltage of the network
U_{rG} rated voltage of the generator
$\underline{Z}_{GK}$ corrected impedance of the generator
$\underline{Z}_G$ impedance of the generator
x_d'' subtransient reactance of the generator
φ_{rG} phase angle between $U_{rG}/\sqrt{3}$ and I_{rG}.

9.2 Correction Factor K_{KW} for Power Plant Block

For the determination of the impedance correction factor K_{KW} as in Figure 9.2, the following considerations are necessary [1].

- Whether the block transformer is equipped with a step switch or with a higher transformation ratio.
- Whether the rated voltages of the generator and the low-voltage side of the block transformer are different.
- Whether the rated apparent powers of the generator and transformer are also different.

The voltage control is implemented either on the high-voltage side of the transformer or via the inherent voltage regulation of the generator. Here, it is necessary to distinguish between a transformer with and without tap changer.

Power station units with on-load tap changer:

$$\underline{Z}_{SK} = K_S \cdot (t_r^2 \cdot \underline{Z}_G + \underline{Z}_{THV}) \tag{9.8}$$

$$K_S = \frac{U_{nQ}^2}{U_{rG}^2} \cdot \frac{U_{rTLV}^2}{U_{rTHV}^2} \cdot \frac{c_{max}}{1 + |x_d'' - x_T| \cdot \sqrt{1 - \cos^2\varphi_{rG}}} \tag{9.9}$$

Power station units without on-load tap changer:

$$\underline{Z}_{SOK} = K_{SO} \cdot (t_r^2 \cdot \underline{Z}_G + \underline{Z}_{THV}) \tag{9.10}$$

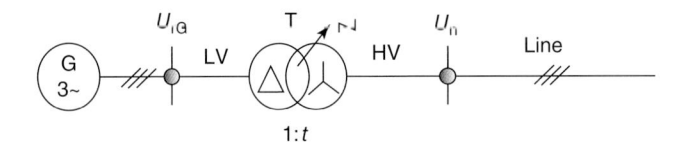

$$K_{SO} = \frac{U_{nQ}}{U_{rG} \cdot (1 + p_G)} \cdot \frac{U_{rTLV}}{U_{rTHV}} \cdot (1 \pm p_T) \cdot \frac{c_{max}}{1 + x_d'' \cdot \sqrt{1 - \cos^2\varphi_{rG}}} \tag{9.11}$$

Figure 9.2 Impedance correction for power station.

Where
Z_G is the subtransient impedance of the generator,
Z_{THV} is the impedance of the unit transformer related to the high-voltage side (without correction factor K_T),
U_{nQ} is the nominal system voltage at the feeder connection point Q of the power station unit,
U_{rG} is the rated voltage of the generator; $U_{Gmax} = U_{rG}(1+p_G)$, with, for instance, $p_G = 0.05$–0.10,
$\cos \varphi_{rG}$ is the power factor of the generator under rated conditions,
x''_d is the relative saturated subtransient reactance of the generator related to the rated impedance; where
t_r is the rated transformation ratio of the unit transformer
$1 \pm p_T$ is to be introduced if the unit transformer has off-load taps and if one of these taps is permanently used, if not choose $1 \pm p_T = 1$. If the highest partial short-circuit current of the power station unit at the high-voltage side of the unit transformer with off-load taps is searched for, then choose $1 - p_T$.

9.3 Correction Factor K_T for Transformers with Two and Three Windings

The correction factor K_T for transformers with two and three windings in accordance with IEC 60909-0 can be calculated as follows.

For transformers with two windings, with or without step switching:

$$\underline{Z}_T = R_T + jX_T \tag{9.12}$$

$$\underline{Z}_{TK} = K_T \cdot \underline{Z}_T \tag{9.13}$$

$$K_T = 0.95 \frac{c_{max}}{1 + 0.6 x_T} \tag{9.14}$$

$$x_T = \frac{X_T}{U_{rT}^2 / S_{rT}} \tag{9.15}$$

For transformers with three windings, with or without step switching:

$$K_{TAB} = 0.95 \cdot \frac{c_{max}}{1 + 0.6 x_{TAB}} \tag{9.16}$$

$$K_{TAC} = 0.95 \cdot \frac{c_{max}}{1 + 0.6 x_{TAC}} \tag{9.17}$$

$$K_{TBC} = 0.95 \cdot \frac{c_{max}}{1 + 0.6 x_{TBC}} \tag{9.18}$$

The meaning of the symbols is as follows:
Z_{KW} corrected impedance of power plant block for high-voltage side
Z_G impedance of the generator
$\underline{Z}_{THV}$ impedance of block transformer for high-voltage side
t_r rated value of transformation ratio for transformer with step switch set to principal tapping

9.3 Correction Factor K_T for Transformers with Two and Three Windings

Table 9.1 Impedance corrections.

Appliances	Impedance	Corrections
Generator	$\underline{Z}_{GK} = \underline{Z}_G \cdot K_G$	$K_G = \dfrac{U_n}{U_{rG}} \cdot \dfrac{c_{max}}{1 + x_d'' \sin \varphi_{rG}}$
Transformer	$\underline{Z}_{TK} = \underline{Z}_T \cdot K_T$	$K_T = \dfrac{U_n}{U_{rT}} \cdot \dfrac{c_{max}}{1 + x_T(I_{rT}^b/I_{rT}) \sin \varphi_T^b}$
		$K_T = 0.95 \cdot \dfrac{c_{max}}{1 + 0.6 x_T}$
Power plant with tapping change (TC)	$\underline{Z}_S = K_{STC} \cdot (t_r^2 \underline{Z}_G + \underline{Z}_{THV})$	$K_{STC} = \dfrac{U_{nQ}^2}{U_{rG}^2} \cdot \dfrac{U_{rTLV}^2}{U_{rTHV}^2}$ $\cdot \dfrac{c_{max}}{1 + |x_d'' - x_T| \sin \varphi_{rG}}$
Power plant without tapping change (WTC)	$\underline{Z}_S = K_{SWTC}(t_r^2 \underline{Z}_G + \underline{Z}_{THV})$	$K_{SWTC} = \dfrac{U_{nQ}}{U_{rG}(1 + p_G)}$ $\cdot \dfrac{U_{rTLV}}{U_{rTHV}} \dfrac{(1 \pm p_T)c_{max}}{1 + x_d'' \sin \varphi_{rG}}$

$\underline{Z}_{T,KW}$ corrected impedance of the block transformer
$\underline{Z}_{G,KW}$ corrected impedance of the generator
$1 + p_T$ this value is introduced when the block transformer has tappings and is used continuously. Otherwise $1 + p_T = 1$
K_{with} correction factor with tap changer
$K_{without}$ correction factor without tap changer
x_T relative reactance of transformer.

Table 9.1 summarizes impedance corrections.

10

Power System Analysis

This chapter describes the mathematical basic circuit theory of the three-phase system to be repeated. Figure 10.1 shows a delta and star connection with three-phase voltages. Star point is connected to a neutral point (N).

In the single-phase load, the neutral conductor current I_N is flowing back to the source. The beginning and the end of the delta circuit are connected to one another. The neutral conductor is not applicable for medium-voltage and high-voltage networks. The symbols of the live conductors are given according to the IEC 60027 or International System of Units (SI system). In the literature and in some countries, the letters a, b, and c are also used.

The currents of the three conductors L1–L2–L3 (before R, S, and T) in three-phase systems are of symmetrical construction (i.e., they have the same size and each phase shifted by 120° and applies to impedances $\underline{Z}_R = \underline{Z}_S = \underline{Z}_T = Z$).

Therefore, voltages, currents, and powers are drawn as a single phase calculated and simulated by a single-phase equivalent circuit diagram.

The symmetry of a three-phase system can be disturbed by ground fault, short circuit, and multiple faults. In this case, the currents are not equal in size and have different phase angles to each other. The single-phase diagram is no longer sufficient.

Here and in other chapters, the calculation equations for three-phase systems and the method of symmetrical components will be described.

The individual-phase voltages can be represented as follows:

$$u_{1(t)} = \hat{u} \cdot \sin(\omega t)$$
$$u_{2(t)} = \hat{u} \cdot \sin(\omega t - 120°) \quad (10.1)$$
$$u_{3(t)} = \hat{u} \cdot \sin(\omega t - 240°)$$

Another way to describe the phase voltages is

$$\underline{U}_1 = U_1 \cdot e^{j0°}$$
$$\underline{U}_2 = U_1 \cdot e^{-j120°} \quad (10.2)$$
$$\underline{U}_3 = U_1 \cdot e^{-j240°}$$

The sum of the three symmetrical voltages is always zero.

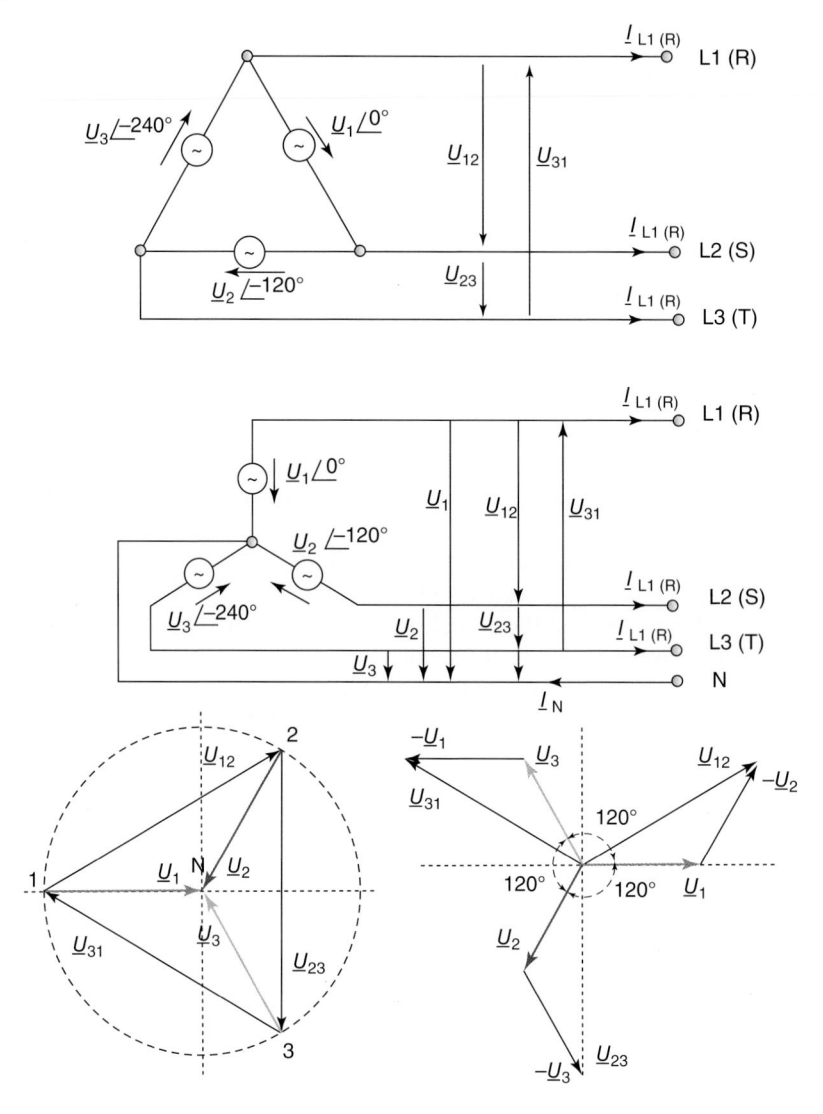

Figure 10.1 Delta and star (wye) connection with neutral point (N).

The line–neutral voltages are
$$\underline{U}_1 + \underline{U}_2 + \underline{U}_3 = 0$$
and the line–line voltages are
$$\underline{U}_{12} + \underline{U}_{23} + \underline{U}_{31} = 0$$

$$\underline{U}_{12} = \underline{U}_1 - \underline{U}_2 = \sqrt{3} \cdot U \angle -30°$$
$$\underline{U}_{23} = \underline{U}_2 - \underline{U}_3 = \sqrt{3} \cdot U \angle -90° \qquad (10.3)$$
$$\underline{U}_{31} = \underline{U}_3 - \underline{U}_1 = \sqrt{3} \cdot U \angle -210°$$

The relationship between phase and line voltages is

$$\underline{U}_L = \underline{U}_{ph} \cdot \sqrt{3} \tag{10.4}$$

Ohm's law gives the currents for each phase:

$$\underline{I}_1 = \frac{\underline{U}_1}{\underline{Z}_1}, \quad \underline{I}_2 = \frac{\underline{U}_2}{\underline{Z}_2}, \quad \underline{I}_3 = \frac{\underline{U}_3}{\underline{Z}_3}$$

The neutral current for a symmetric system can be given as:

$$\underline{I}_N = \underline{I}_1 + \underline{I}_2 + \underline{I}_3 = \underline{I}_1 \cdot [1 + e^{-j120°} + e^{-j240°}] = 0 \tag{10.5}$$

For completing the proceeding equations for currents, we can summarize as:
Ohm's law gives the currents for each phase:

$$\underline{I}_1 = \frac{\underline{U}_1}{\underline{Z}}, \quad \underline{I}_2 = \frac{\underline{U}_2}{\underline{Z}} = \frac{\underline{U}_1 \cdot e^{-j120°}}{\underline{Z}} = \underline{I}_1 \cdot e^{-j120°},$$

$$\underline{I}_3 = \frac{\underline{U}_3}{\underline{Z}} = \frac{\underline{U}_1 \cdot e^{-j240°}}{\underline{Z}} = \underline{I}_1 \cdot e^{-j240°}$$

$$\underline{I}_1 = I_1 \cdot e^{j0°}$$
$$\underline{I}_2 = I_1 \cdot e^{-j120°} \tag{10.6}$$
$$\underline{I}_3 = I_1 \cdot e^{-j240°}$$

For unbalanced circuits and for the delta connection, the following equations can be given:

Line currents: $\underline{I}_1 + \underline{I}_2 + \underline{I}_3 = 0$ and the phase currents between the two nodes are $\underline{I}_{12} + \underline{I}_{23} + \underline{I}_{31} = 0$.

$$\underline{I}_{12} = \frac{\underline{U}_{12}}{\underline{Z}_{12}}, \quad \underline{I}_{23} = \frac{\underline{U}_{12}}{\underline{Z}_{23}} = \frac{\underline{U}_{12} \cdot e^{-j120°}}{\underline{Z}_{23}}, \quad \underline{I}_3 = \frac{\underline{U}_{12}}{\underline{Z}_{12}} = \frac{\underline{U}_{12} \cdot e^{-j240°}}{\underline{Z}_{12}}$$

$$\underline{I}_{12} = \underline{I}_1 - \underline{I}_2 = \sqrt{3} \cdot I \angle -30°$$
$$\underline{I}_{23} = \underline{I}_2 - \underline{I}_3 = \sqrt{3} \cdot I \angle -90° \tag{10.7}$$
$$\underline{I}_{31} = \underline{I}_3 - \underline{I}_1 = \sqrt{3} \cdot I \angle -210°$$

The relationship between the phase and line currents is

$$\underline{I}_L = \underline{I}_{ph} \cdot \sqrt{3} \tag{10.8}$$

Electrical power systems operate normally in a balanced three-phase sinusoidal steady-state mode. In the case of an unbalanced operation, it is difficult to describe the system. In 1918, C.L. Fortescue explained how unbalanced three-phase voltages and currents could be transformed into balanced three-phase components. He called it *symmetrical components* of the original phasors. The theory of symmetrical components is of great importance to the electrical power systems and is the subject of the next topic.

10.1 The Method of Symmetrical Components

The method of symmetrical components is used for the calculation of asymmetrical faults. This section discusses the fundamentals of this method. A characteristic rotational operator is a complex number with the magnitude 1. Multiplication with a rotational operator, therefore, describes the rotation of an arbitrary phasor without changing its magnitude. From the set of complex numbers, we know the rotational operator $j = \sqrt{-1}$, which gives rise to a rotation by 90°. Accordingly, $j^2 = -1$ then gives rise to a rotation by 180°, and j^4 brings the phasor back to its original position. In three-phase systems, the phase angles $\varphi = 120°$ and 240° are of special importance.

Figure 10.2 shows a symmetrical system, which consists of three-unit phasors separated by angles of 120°.

The rotational operator for $\varphi = 120°$ is designated and for 240°, it is designated $\underline{a}^2$ so that:

$$\underline{a} = e^{j120} = e^{j(2\pi/3)} = \frac{1}{2}(1 - j\sqrt{3}) \tag{10.9}$$

$$\underline{a}^2 = e^{j240} = e^{j(4\pi/3)} = \frac{1}{2}(1 - j\sqrt{3}) \tag{10.10}$$

$$\underline{a}^3 = 1 \tag{10.11}$$

As can be easily seen from the above equations, their sum is equal to zero:

$$1 + \underline{a} + \underline{a}^2 = 0 \tag{10.12}$$

With the rotational operators $\underline{a}$ and $\underline{a}^2$, the symmetrical three-phase system belongs to the set of complex numbers.

$$\underline{U}_R = U_R \tag{10.13}$$
$$\underline{U}_S = \underline{a}^2 \cdot U_R \tag{10.14}$$
$$\underline{U}_T = \underline{a} \cdot U_R \tag{10.15}$$

For the voltages in the external conductors:

$$\underline{U}_{RS} = \underline{U}_R - \underline{U}_S = \sqrt{3} \cdot U_R \cdot e^{j30} \tag{10.16}$$
$$\underline{U}_{ST} = \underline{U}_S - \underline{U}_T = \sqrt{3} \cdot U_R \cdot e^{j270} \tag{10.17}$$
$$\underline{U}_{TR} = \underline{U}_T - \underline{U}_R = \sqrt{3} \cdot U_R \cdot e^{j150} \tag{10.18}$$

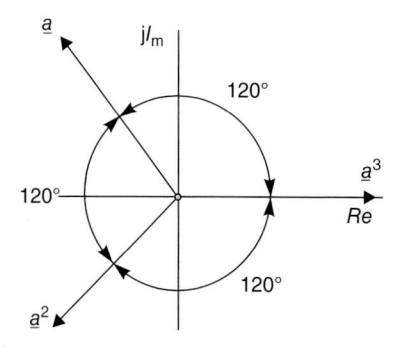

Figure 10.2 Phasor diagram for the positive-sequence, negative-sequence, and zero-sequence systems.

10.2 Fundamentals of Symmetrical Components

Symmetrical faults are calculated from the equivalent circuit in the positive-sequence system. The network is thereby reduced to a single conductor and drawn as a single-phase system. Three-phase short circuits load the network symmetrically. For the other types of short circuits, it is no longer possible to use the positive-sequence system as the network is loaded asymmetrically. It is here that the method of symmetrical components is well suited.

For each conductor of a three-phase system, the corresponding equations (currents or voltages in the conductors) are written. Asymmetrical operation, a short circuit to ground, line interruptions, or switching mechanisms can be the result of network loading. The voltages and currents at the position of the short circuit are determined by the geometrical addition of the symmetrical component voltages and currents.

With this method, the three-phase network is resolved into three independent single-phase systems, namely, the positive-sequence, negative-sequence, and zero-sequence systems.

The impedances of these three systems can then be given for individual operational systems at a fault position. According to the position of the fault, unequal currents can arise in the conductors. Then, the equivalent single-phase circuit can no longer be used. The transformation of the original R (L1), S (L2), T (L3) to a symmetrical image space with the coordinates 1, 2, 0 is, therefore, necessary (method of symmetrical components).

The procedure that the method entails will now be discussed in detail [14, 15, 16]. In the case of an unsymmetrical three-phase system, the corresponding

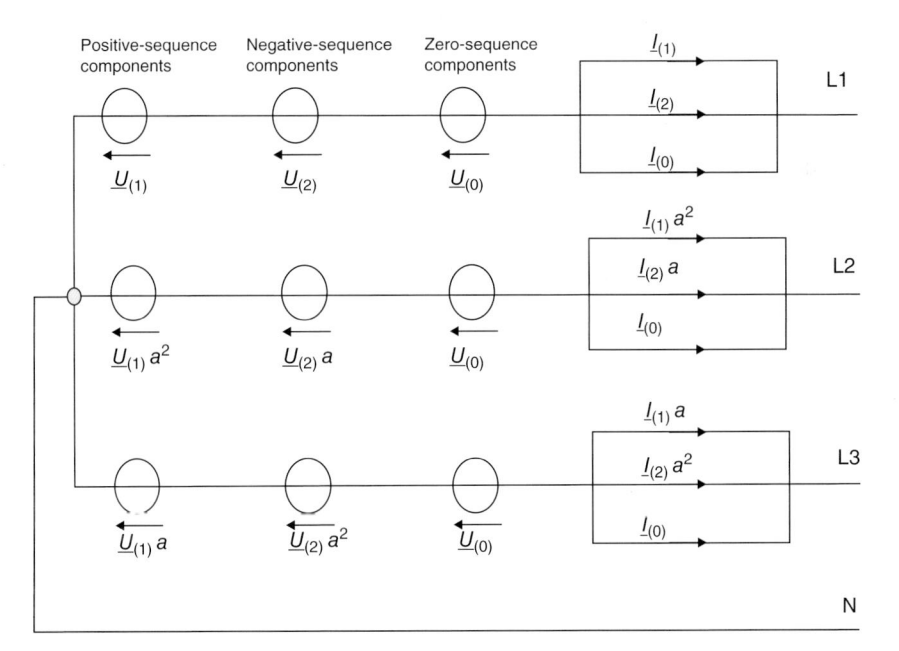

Figure 10.3 Power transmission, three-phase system.

equations (conductor currents or voltages) are set up for each conductor of the three-phase system. Therefore, each unsymmetrical system can be clearly shown in a partial three-symmetrical system (Figure 10.3).

A three-phase fault affects the three-phase network symmetrically. All three conductors are equally involved and carry the same amount of short-circuit current. Therefore, the calculation is performed for only one conductor. On the other hand, all other short-circuit conditions incur asymmetrical loadings. A suitable method for investigating such events is to split the asymmetrical system into its symmetrical components (Figure 10.4). With a symmetrical voltage

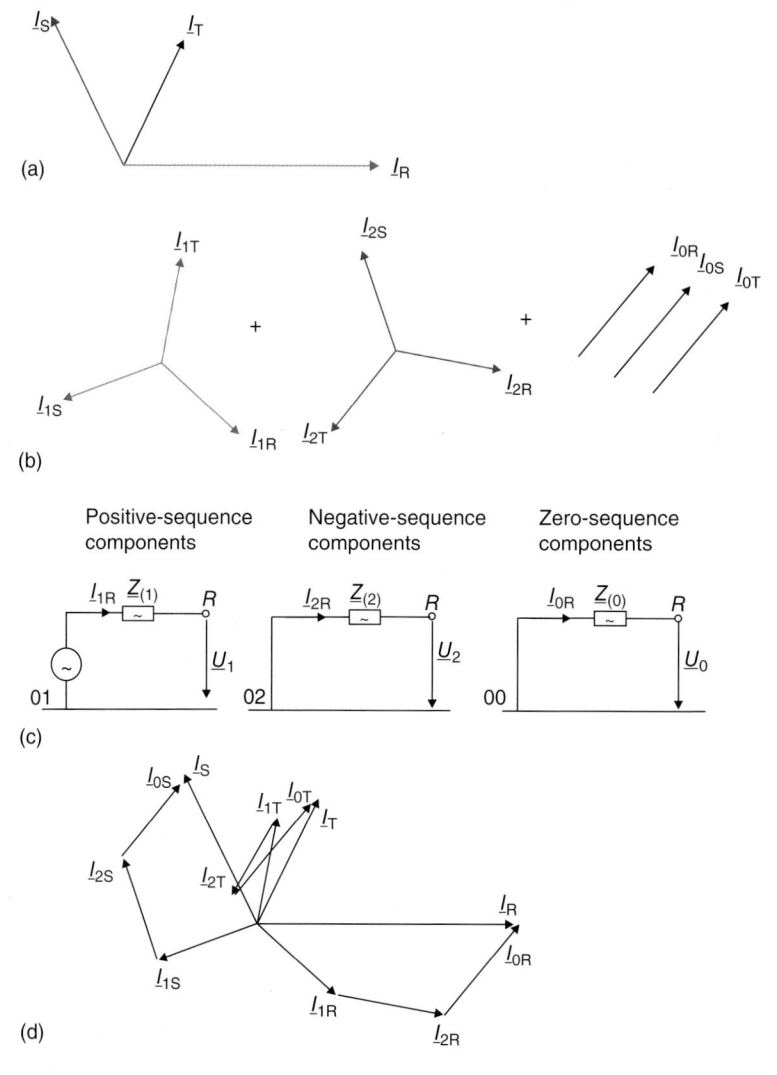

Figure 10.4 Schematic of a three-phase network and relationship between components of positive-sequence, negative-sequence, and zero-sequence systems. (a) Unsymmetrical system, (b) Symmetrical system, (c) equivalent one-line diagram of the components, and (d) graphical addition of three components.

system, the currents produced by an asymmetrical loading (I_1, I_2, and I_3) can be determined with the aid of the symmetrical components (positive-sequence, negative-sequence, and zero-sequence systems).

The symmetrical components can be found with the aid of complex calculation or by graphical means. If the current vector leading the current in the reference conductor (L1) is rotated 120° backward, and the lagging current vector 120° forward, the resultant is equal to three times the vector I_m in the reference conductor. The negative-sequence components are apparent. If one turns in the other direction, the positive-sequence system is evident, and the resultant is three times the vector I_2 in the reference conductor. Geometrical addition of all three current vectors (I_1, I_2, and I_3) yields three times the vector I_0 in the reference conductor. If the neutral conductor is unaffected, there is no zero-sequence system. Unsymmetrical faults at one point in the network can be caused by unbalanced loads or unbalanced short circuits through impedances or open conductors. Basically, a procedure is followed that describes the unsymmetry in the three-phase mains, and this description then transfers to the symmetrical components so as to determine in what way the positive-sequence, negative-sequence, and zero-sequence systems must be connected together at the point of unsymmetry.

10.2.1 Derivation of the Transformation Equations

The decomposition presented in positive-sequence, negative-sequence, and zero-sequence systems can be described by the following equations (using the example of the currents; applies equivalently for voltages):

$$\underline{I}_R = \underline{I}_{1R} + \underline{I}_{2R} + \underline{I}_{0R} \tag{10.19}$$

$$\underline{I}_S = \underline{I}_{1S} + \underline{I}_{2S} + \underline{I}_{0S}$$

$$\underline{I}_T = \underline{I}_{1T} + \underline{I}_{2T} + \underline{I}_{0T} \tag{10.20}$$

$$\underline{I}_{0R} = \underline{I}_{0S} + \underline{I}_{0T} + \underline{I}_{0}$$

The system of equations can be further simplified by taking advantage of the fact that the three phasors of the positive-sequence and negative-sequence systems are out of phase with each other with the same amount to 120°, respectively, and that the three phasors of the zero-sequence system are identical to each other. These relationships can be taken into account when considering each case, a phasor in the three systems as a reference. Usually, the components pointer $\underline{I}_{1R}$, $\underline{I}_{2R}$, and $\underline{I}_{0R}$ of the conductor R are chosen as a reference. The phase position can thereby be expressed as follows:

$$1 = e^{j0°} \quad \underline{a} = e^{j120} = e^{j(2\pi/3)} \quad \underline{a}^2 = e^{-j120} = e^{-j(2\pi/3)}$$

Calculations in 120° space:

$$\underline{I}_{1S} = \underline{I}_{1R} \cdot \underline{a}^2$$

$$\underline{I}_{1T} = \underline{I}_{1R} \cdot \underline{a}$$

$$\underline{I}_{2S} = \underline{I}_{2R} \cdot \underline{a} \tag{10.21}$$

$$\underline{I}_{2T} = \underline{I}_{2T} \cdot \underline{a}^2$$

$$\underline{I}_{0R} = \underline{I}_{0S} \cdot \underline{I}_{0T}$$

Fault currents in the original space:

$$\underline{I}_R = \underline{I}_{1R} + \underline{I}_{2R} + \underline{I}_{0R} \tag{10.22}$$

$$\underline{I}_S = \underline{I}_{1R} \cdot \underline{a}^2 + \underline{I}_{2R} \cdot \underline{a} + \underline{I}_{0R}$$

$$\underline{I}_T = \underline{I}_{1T} \cdot \underline{a} + \underline{I}_{2R} \cdot \underline{a}^2 + \underline{I}_{0R} \tag{10.23}$$

Then, in matrix notation, we have

$$\begin{bmatrix} I_R \\ I_S \\ I_T \end{bmatrix} = \frac{1}{3} \begin{bmatrix} 1 & 1 & 1 \\ \underline{a}^2 & \underline{a} & 1 \\ \underline{a} & \underline{a}^2 & 1 \end{bmatrix} \cdot \begin{bmatrix} \underline{I}_{1R} \\ \underline{I}_{2R} \\ \underline{I}_{0R} \end{bmatrix} \tag{10.24}$$

Inverse transformation from the image space to the original space:

$$[I_{RST}] = [T] \cdot [I_{120}] \tag{10.25}$$

With this transformation, we can transform the unknown current components with the matrix T to the actual currents in the components:

$$\underline{I}_{1R} = \frac{1}{3} \cdot (\underline{I}_R + \underline{I}_S \cdot \underline{a} + \underline{I}_T \cdot \underline{a}^2) \tag{10.26}$$

$$\underline{I}_{2R} = \frac{1}{3} \cdot (\underline{I}_R + \underline{I}_S \cdot \underline{a}^2 + \underline{I}_T \cdot \underline{a}) \tag{10.27}$$

In matrix notation, this yields

$$\begin{bmatrix} I_{1R} \\ I_{2R} \\ I_{0R} \end{bmatrix} = \frac{1}{3} \begin{bmatrix} 1 & \underline{a} & \underline{a}^2 \\ 1 & \underline{a}^2 & \underline{a} \\ 1 & 1 & 1 \end{bmatrix} \cdot \begin{bmatrix} \underline{I}_R \\ \underline{I}_S \\ \underline{I}_T \end{bmatrix} \tag{10.28}$$

$$[I_{120}] = [S] \cdot [I_{RST}] \tag{10.29}$$

We are now in a position to calculate the asymmetrical currents and voltages, where

$$[S] = [T^{-1}] \tag{10.30}$$

10.3 General Description of the Calculation Method

The method for calculating the short-circuit currents can be applied according to the following general scheme to the different types of errors. The procedure of symmetrical components will also be shown using the example of a single-phase short-circuit current (Figure 10.5). The short circuits are for reference conductors R symmetrically.

The following proceeding are considered in the calculation:

1) Set up the fault conditions in RST components.
2) Transformation of the fault conditions in symmetrical components (012).
3) Draw the circuit diagram in symmetrical components.
4) Set of the equations and calculation of short-circuit currents and voltages.
5) Transformation of the short-circuit currents and voltages in RST components.

10.3 General Description of the Calculation Method

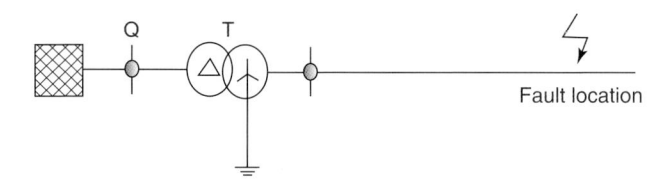

Figure 10.5 Circuit diagram of a one-phase short circuit.

Step 1: The fault conditions in RST components arising on the short-circuit point as follows:

A short circuit with earth contact of the conductor R has the consequence that the voltage of the conductor assumes the value zero.

$$\underline{U}_R = 0$$
$$\underline{I}_S = \underline{I}_R = 0$$

Step 2: With the transformation, the error conditions are detected in symmetrical components.

$$I_{120} = S \cdot I_{RST}$$

$$\begin{bmatrix} T \end{bmatrix} = \frac{1}{3} \begin{bmatrix} 1 & 1 & 1 \\ a^2 & a & 1 \\ a & a^2 & 1 \end{bmatrix}$$

$$\underline{U}_1 + \underline{U}_2 + \underline{U}_0 = \underline{U}_R = 0$$

$$\underline{I}_1 + \underline{I}_2 + \underline{I}_0 = \underline{I}_R = \frac{1}{3} \cdot \underline{I}_R$$

Step 3: The fault conditions can be realized only if all positive-sequence, negative-sequence, and zero-sequence components are connected in series. In this case, the zero system into the power grid is not included in the calculation of the single-phase short-circuit current because a decoupling of zero systems enters through the transformer. The impedance of the network must be converted with the square of the transmission ratio to the short side.

By combining the individual impedances, the relationships of equations can be given as follows:

$$\underline{Z}_1 = \underline{Z}_{1Qt} + \underline{Z}_{1T} + \underline{Z}_{1L}$$
$$\underline{Z}_2 = \underline{Z}_{2Qt} + \underline{Z}_{2T} + \underline{Z}_{2L}$$
$$\underline{Z}_0 = \underline{Z}_{0T} + \underline{Z}_{0L}$$

Step 4: For the currents and voltages, the relationship will be

$$\underline{I}_1 = \underline{I}_2 = \underline{I}_0 = \frac{\underline{E}_1}{\underline{Z}_1 + \underline{Z}_2 + \underline{Z}_0}$$

$$\underline{U}_1 = \underline{E} - \underline{Z}_1 \cdot \underline{I}_1$$
$$\underline{U}_2 = -\underline{Z}_2 \cdot \underline{I}_2$$
$$\underline{U}_0 = -\underline{Z}_0 \cdot \underline{I}_0$$

Step 5: The transformation into the RST components is performed.

$$T^{-1} = \begin{bmatrix} 1 & 1 & 1 \\ 1 & a^2 & a \\ 1 & a^2 & a \end{bmatrix}$$

$$\underline{I}_R = \underline{I}''_{k1} = \underline{I}_0 + \underline{I}_1 + \underline{I}_2 = 3\underline{I}_0 = \frac{3\underline{E}_1}{\underline{Z}_1 + \underline{Z}_2 + \underline{Z}_0}$$

Finally, we get with

$$\underline{E}_1 = \frac{c \cdot U_n}{\sqrt{3}}$$

One-phase short-circuit current.

$$\underline{I}''_{k1} = \frac{\sqrt{3} \cdot c \cdot U_n}{\underline{Z}_1 + \underline{Z}_2 + \underline{Z}_0}$$

10.4 Impedances of Symmetrical Components

Only three-phase short circuits load the network symmetrically. It is sufficient to calculate with the "positive-sequence system." In all other cases, it is also necessary to use the method of symmetrical components to consider the negative-sequence and zero-sequence systems.

The method of symmetrical components is based on the principle of superposition. The determination of the voltage and current components requires equivalent single-phase circuits, which, during symmetrical operation of the network, are fully decoupled from each other.

The three impedances of the component systems are summarized briefly (Figure 10.6) [1].

1) *Positive-phase impedance Z_1 (index m or 1)*: A symmetrical positive-sequence system with a normal phase angle is present. The equivalent circuit and the data for the operational equipment are identical with the data for the equivalent single-phase circuit for the calculation of the three-phase short circuit.
2) *Negative-sequence impedance Z_2 (index i or 2)*: A symmetrical negative-sequence system is present. The negative-sequence impedance is the same as the positive-sequence impedance for operational equipment without load. The impedances are different when the machines are operating.

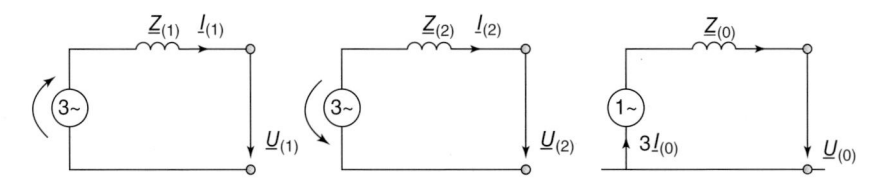

Figure 10.6 Equivalent circuits for positive-sequence, negative-sequence, and zero-sequence systems.

3) *Negative-sequence impedance Z_0 (index 0)*: A system consisting of three currents of equal value and having the same phase angle is present if we take the three main conductors connected in parallel for the outgoing line and a fourth conductor as a common return line and apply an a.c. voltage. Three times the zero-sequence current flows in this return line.

The circuitry of the N is considered in the zero-sequence impedance as follows:

- not grounded
- grounded through a ground fault neutralizer coil
- grounded through resistances or reactance
- direct grounding.

Under the assumption that the symmetrical components for the current and voltage are physically real values, they must be related through general physical laws. Then, it must be possible to assign an impedance to each of the three component systems according to Ohm's law.

Positive-sequence impedance:

$$\underline{Z}_{(1)} = \frac{\underline{U}_{(1)}}{\underline{I}_{(1)}} \tag{10.31}$$

Negative-sequence impedance:

$$\underline{Z}_{(2)} = \frac{\underline{U}_{(2)}}{\underline{I}_{(2)}} \tag{10.32}$$

Zero-sequence impedance:

$$\underline{Z}_{(0)} = \frac{\underline{U}_{(0)}}{\underline{I}_{(0)}} \tag{10.33}$$

Furthermore, it is possible to define equivalent circuits for the three component systems. The positive-sequence, negative-sequence, and zero-sequence impedances can be determined by applying the positive-sequence, negative-sequence, and zero-sequence voltages to the circuit to be measured, measuring the current and then calculating the impedance from Ohm's law. The positive-sequence impedance is identical with the impedance of a conductor in the three-phase system and is, therefore, the same as the impedance of an equivalent single-phase circuit in symmetrical operation. It is the sum total of the impedances in the conductor from the overland lines, machines, and other components. The negative-sequence impedance is determined as for the positive-sequence impedance but with a negative-sequence system voltage. As can be seen from the measurement circuit in Figure 10.7, the negative-sequence impedance must be the same as the positive-sequence impedance for all passive operational equipment such as overhead lines, cables, and transformers. For rotating machines, on the other hand, the negative-sequence impedance can be smaller than the positive-sequence impedance.

The zero-sequence impedance is, by definition (no phase shift of the individual components), measured single phase, and the three conductors are connected in

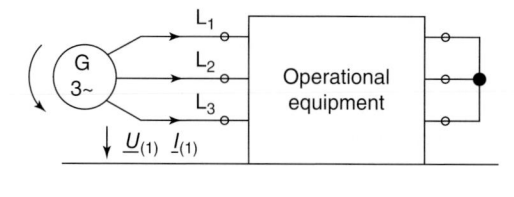

Figure 10.7 Measuring circuits for determining the positive-sequence, negative-sequence, and zero-sequence impedances (see also Ref. [1]).

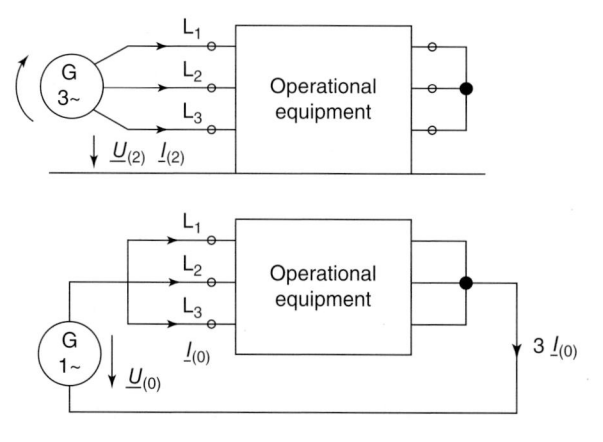

Figure 10.8 Schematic of the equivalent circuit components.

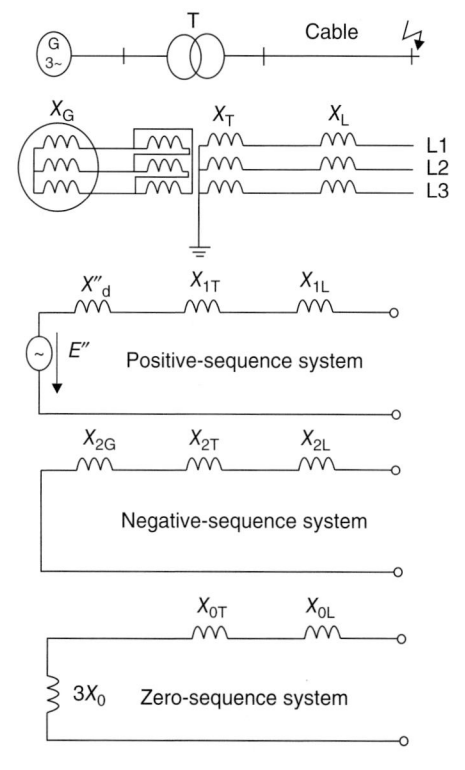

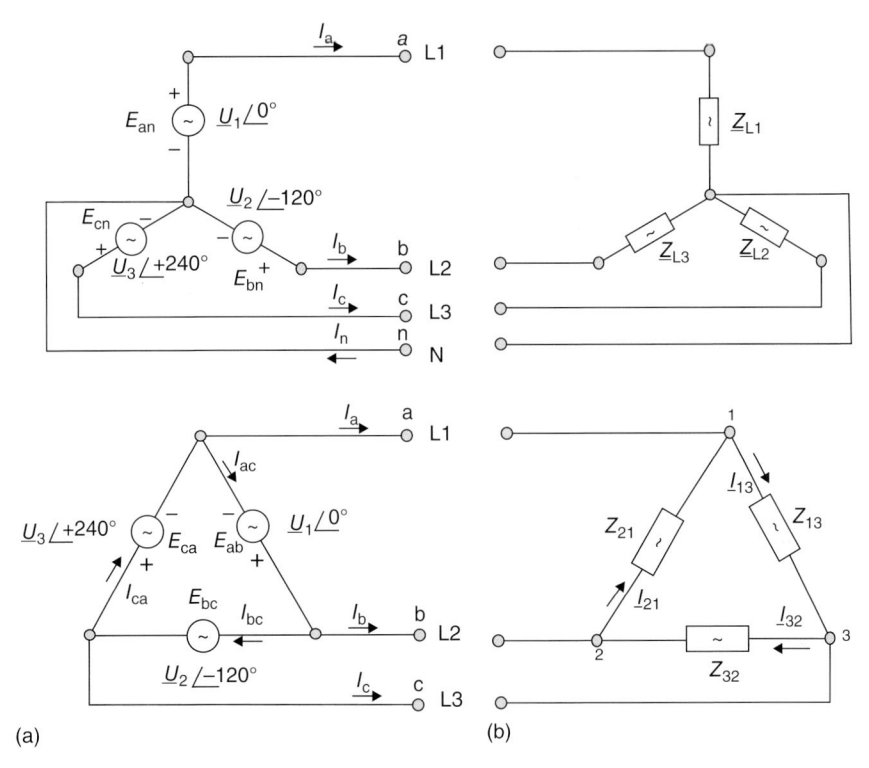

Figure 10.9 Schematic diagram of three-phase delta, star source, and loads. (a) Source and (b) load.

parallel. For imaging a symmetrical three-phase network, the positive-sequence, negative-sequence, and zero-sequence systems must be linked, and linking requires knowledge of the source voltage. This procedure is described in detail in the following sections.

In three-phase networks, the source voltage is generated symmetrically with synchronous generators. For this reason, the source voltage appears only in the positive-sequence system and is set to the value $\frac{c \cdot U_n}{\sqrt{3}}$ for the calculations.

All three equivalent circuit components are connected to the fault position, and for each type of fault, there is a different connection (Figure 10.8).

Figure 10.9 shows a schematic diagram of three-phase delta and star source, and loads explaining SI and other symbols. The letters a, b, c, R, Y, B are not internationally standardized and accepted.

11

Calculation of Short-Circuit Currents

In IEC 60909-0, the different types of short circuits are clearly defined. This chapter deals with the short-circuit currents and sets up the equations required to determine these currents. For the calculation, RST components are used instead of L1–L2–L3 for reasons of simplification.

11.1 Three-Phase Short Circuits

For the dimensioning of electrical systems, it is necessary to consider three-phase short circuits in order to guarantee the mechanical and thermal stabilities of the systems and the rated making and breaking capabilities of the overcurrent protection equipment.

The requirements for calculating the largest three-phase short-circuit current are as follows:

- The temperature of the conductor is 20 °C.
- The network circuitry is mostly responsible for this current.
- The network feeder delivers the maximum short-circuit power.
- The voltage factor is chosen in accordance with IEC 60909-0.

The three-phase short circuit is a symmetrical fault. The following fault conditions apply for the equivalent circuit (Figure 11.1):

$$U_R = U_S = U_T = 0 \tag{11.1}$$

$$I_R + I_S + I_T = 0 \tag{11.2}$$

It then follows that

$$\begin{bmatrix} \underline{U}_0 \\ \underline{U}_1 \\ \underline{U}_2 \end{bmatrix} = \frac{1}{3} \cdot \begin{bmatrix} 1 & 1 & 1 \\ 1 & a & a^2 \\ 1 & a^2 & a \end{bmatrix} \cdot \begin{bmatrix} \underline{U}_R \\ \underline{U}_S \\ \underline{U}_T \end{bmatrix} \tag{11.3}$$

$$\underline{U}_0 = \underline{U}_1 = \underline{U}_2 = 0 \tag{11.4}$$

$$\begin{bmatrix} \underline{I}_0 \\ \underline{I}_1 \\ \underline{I}_2 \end{bmatrix} = \frac{1}{3} \cdot \begin{bmatrix} 1 & 1 & 1 \\ 1 & a & a^2 \\ 1 & a^2 & a \end{bmatrix} \cdot \begin{bmatrix} \underline{I}_R \\ \underline{I}_S \\ \underline{I}_T \end{bmatrix} \tag{11.5}$$

Short Circuits in Power Systems: A Practical Guide to IEC 60909-0, Second Edition. Ismail Kasikci.
© 2018 Wiley-VCH Verlag GmbH & Co. KGaA. Published 2018 by Wiley-VCH Verlag GmbH & Co. KGaA.

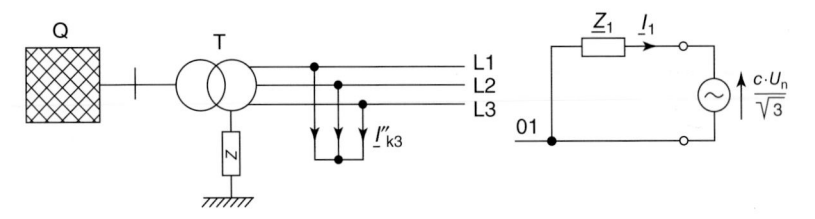

Figure 11.1 Equivalent circuit for a three-phase short circuit with equivalent voltage source at the position of the fault.

For three-phase short circuits:

$$I''_{k3} = \frac{c \cdot U_n}{\sqrt{3} \cdot \underline{Z}_1} \tag{11.6}$$

whereby for $\underline{Z}_1$:

$$Z_1 = \sqrt{(R_{1Q} + R_{1T} + R_{1L})^2 + (X_{1Q} + X_{1T} + X_{1L})^2} \tag{11.7}$$

Or with the impedances of the individual operational equipment:

$$\underline{Z}_1 = \underline{Z}_{1Q} + \underline{Z}_{1T} + \underline{Z}_{1L} \tag{11.8}$$

11.2 Two-Phase Short Circuits with Contact to Ground

This represents the general case of a two-phase short circuit. As can be seen from Figure 11.2, for the two-phase short circuit the following boundary conditions apply:

$$\underline{I}_R = 0, \quad \underline{I}_S = \underline{I}_T, \quad \underline{I}_{kE2E} = \underline{I}_S + \underline{I}_T, \quad \underline{U}_S = \underline{U}_T = 0$$

$$I''_{kE2E} = \frac{\sqrt{3} \cdot c \cdot U_n}{|\underline{Z}_1 + 2\underline{Z}_0|} \tag{11.9}$$

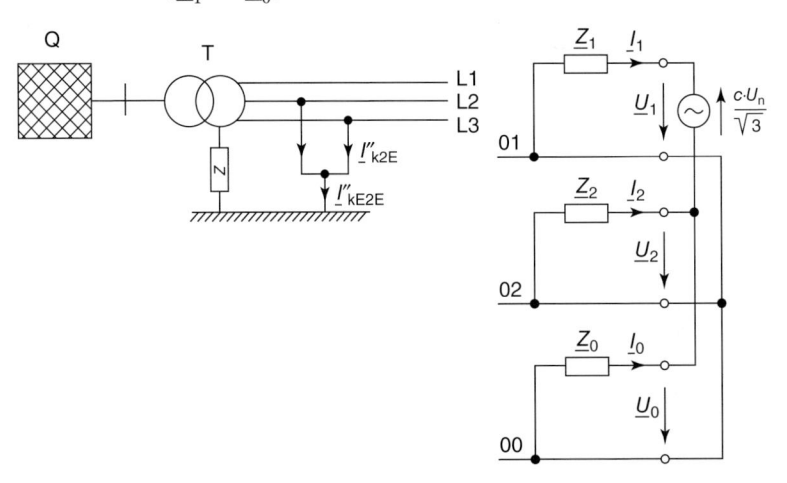

Figure 11.2 Equivalent circuit of a two-phase short circuit with contact to ground.

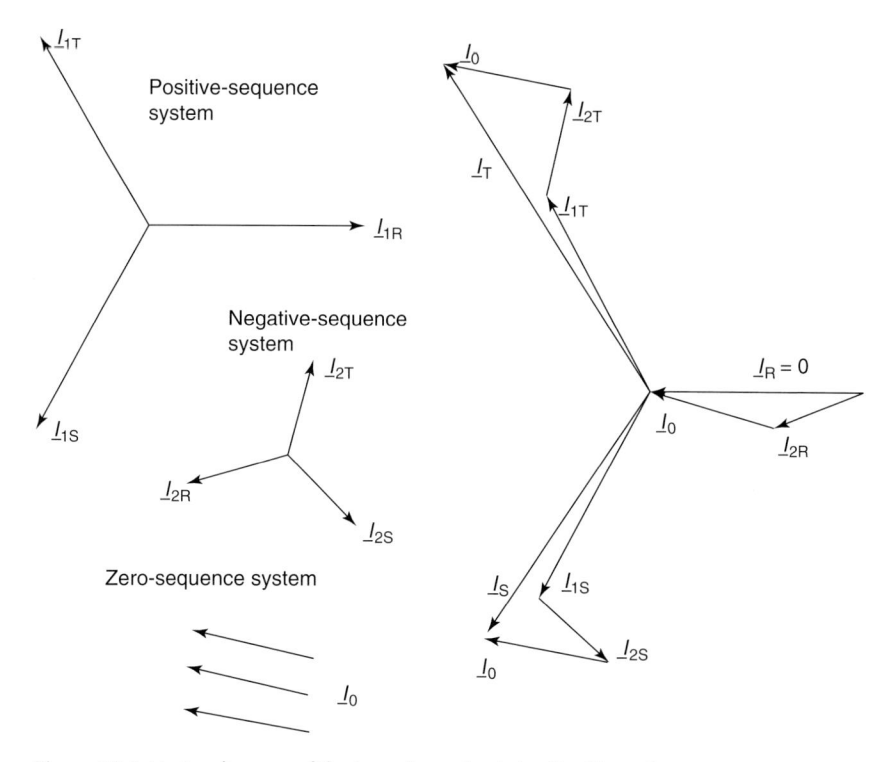

Figure 11.3 Vector diagram of the two-phase short circuit with earth contact.

Figure 11.3 shows the phasor diagrams of the corresponding positive, negative, and zero systems and the resulting components.

11.3 Two-Phase Short Circuit Without Contact to Ground

According to Figure 11.4, a two-phase fault without contact to ground should occur between the two conductors.

For the equations giving the currents:

$$\underline{I}_S = -\underline{I}_T, \quad \underline{I}_R = 0$$

The zero-sequence system current is zero, because no current flows through ground, that is, $I_0 = 0$, $U_0 = 0$.

For a two-phase short-circuit current, this results in

$$\underline{I}''_{k2} = \frac{c \cdot U_n}{|\underline{Z}_1 + \underline{Z}_2|} \tag{11.10}$$

$$I''_{k2} = \frac{\sqrt{3}}{2} I''_{k3} \tag{11.11}$$

The voltage system for a two-phase short circuit shifts in such a way that the voltage on the third, fault-free conductor, in this case U_R, remains unchanged.

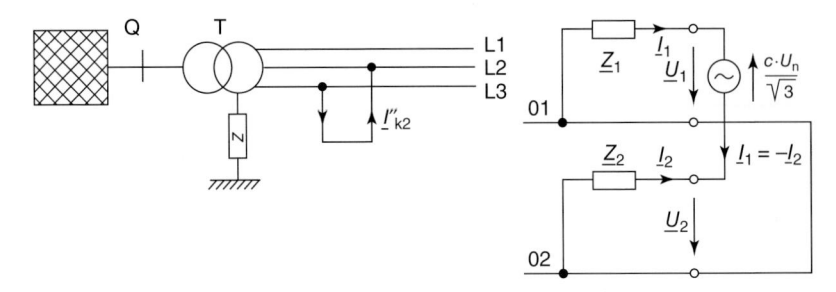

Figure 11.4 Equivalent circuit of a two-phase short circuit without contact to ground.

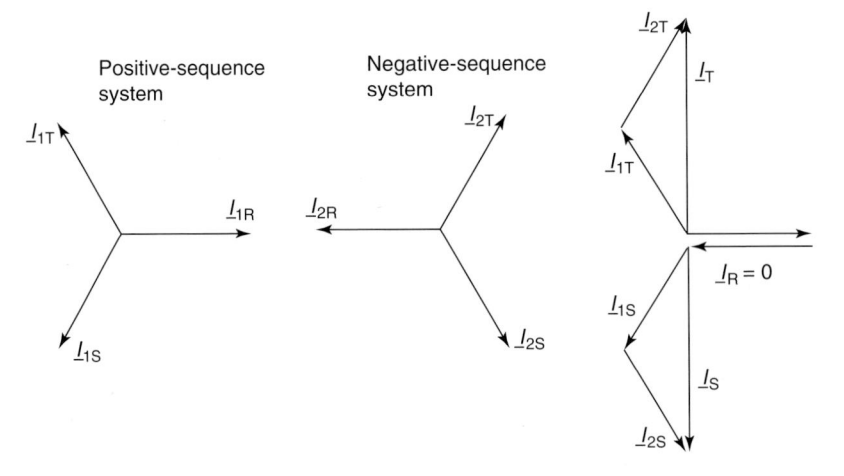

Figure 11.5 Vector diagram of the two-phase short circuit without earth contact.

Two-phase short-circuit currents without contact to ground can be larger with powerful asynchronous motors than for three-phase short circuits.

Figure 11.5 shows the phasor diagrams of the corresponding positive and negative systems and the resulting components.

11.4 Single-Phase Short Circuits to Ground

The single-phase short-circuit current occurs frequently in electrical networks (Figure 11.6). Its calculation is necessary in order to ensure

- The maximum conductor lengths (IEC 60364 part 52)
- Protection against indirect contact (IEC 60364 part 41)
- Protection against thermal stress (IEC 60364 part 43). Calculating the smallest short-circuit current requires that
- The voltage factor used is taken from IEC 60909-0
- Motors can be neglected
- In low-voltage networks, the temperature of the conductors is set to 80 °C
- Setting up the network so that the smallest $I''_{k1\min}$ flows. For the component systems shown in Figure 11.7, we can then use the values: $\underline{I}_S = \underline{I}_T = 0$, $\underline{I}''_{k1} = \underline{I}_R$, $\underline{U}_R = 0$.

11.4 Single-Phase Short Circuits to Ground

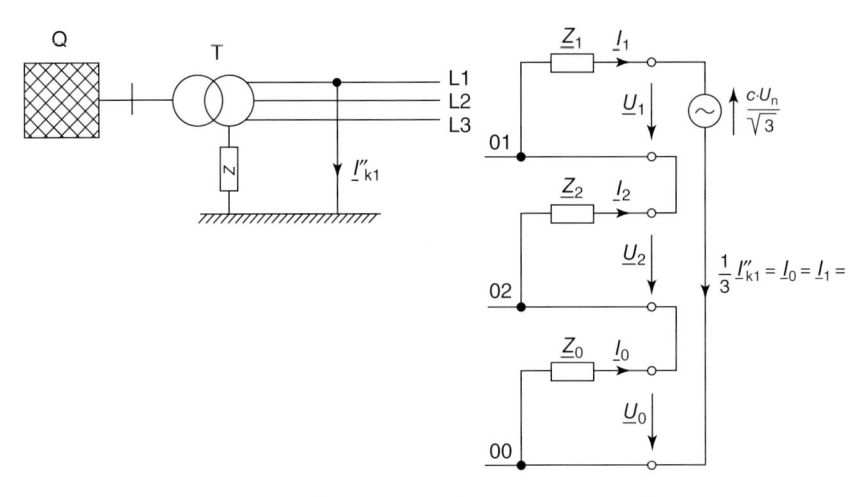

Figure 11.6 Equivalent circuit of single-phase short circuit to ground.

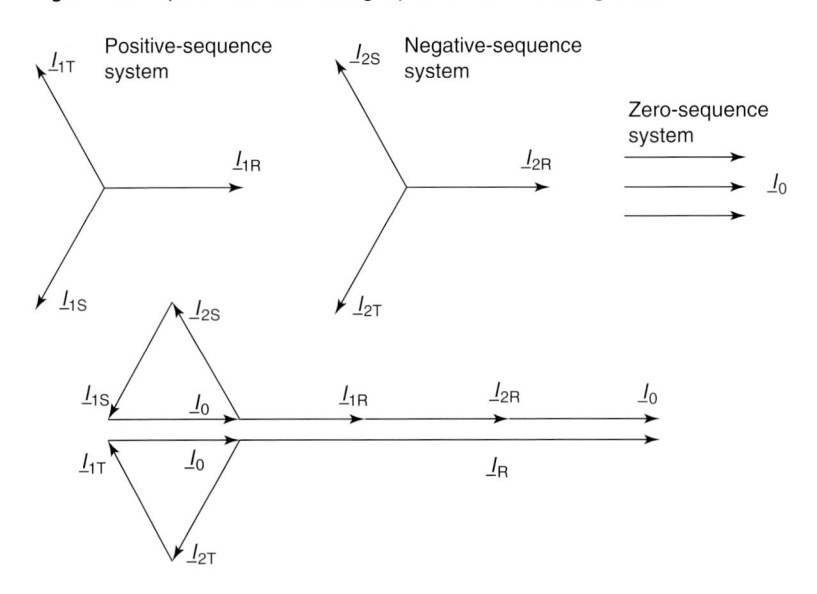

Figure 11.7 Vector diagram of the one-phase short circuit.

Since the currents in the positive-sequence, negative-sequence, and zero-sequence systems are identical, this means that the three systems must be connected in series. For the current, then

$$3 \cdot \underline{I}_0 = \underline{I}_R + \underline{I}_S + \underline{I}_T \tag{11.12}$$

$$\underline{I}_{1R} = \underline{I}_{2R} = \underline{I}_0 \tag{11.13}$$

$$\underline{I}_R = \underline{I}_{1R} + \underline{I}_{2R} + \underline{I}_{0R} = 3 \cdot \underline{I}_{1R} \tag{11.14}$$

$$\underline{I}_{1R} = \frac{E''}{\underline{Z}_1 + \underline{Z}_2 + \underline{Z}_0} \tag{11.15}$$

$$\underline{I}_R = \frac{3 \cdot E''}{\underline{Z}_1 + \underline{Z}_2 + \underline{Z}_0} \tag{11.16}$$

Using the relationship

$$E'' = \frac{3 \cdot U_n}{\sqrt{3}} \tag{11.17}$$

It follows for a single-phase short circuit under the condition $Z_1 = Z_2$ that

$$I''_{k1min} = \frac{\sqrt{3} \cdot c_{min} \cdot U_n}{|2\underline{Z}_1 + \underline{Z}_0|} \tag{11.18}$$

For the loop impedance of the short circuit:

$$I''_{k1min} = \frac{c_{min} \cdot U_n}{\sqrt{3} \cdot Z_s} \tag{11.19}$$

Equating the right sides of Equations (11.18) and (11.19) yields

$$Z_s = \frac{2Z_1 + Z_0}{3} \tag{11.20}$$

According to IEC 60909-0,

$$I''_{k1min} = \frac{\sqrt{3} \cdot c_{min} \cdot U_n}{\sqrt{(2R_{1Q} + 2R_{1T} + 2R_{1L} + R_{0T} + R_{0L})^2 + (2X_{1Q} + 2X_{1T} + 2X_{1L} + X_{0T} + X_{0L})^2}} \tag{11.21}$$

Equations (11.18), (11.19), and (11.21) are identical and give the same result for the calculation of I''_{k1min}.

Figure 11.7 shows the phasor diagrams of the corresponding positive, negative, and zero systems and the resulting components.

For asymmetrical short circuits, the largest short-circuit current can be determined with the aid of Figure 11.8 and depends on the network design. The double ground fault, I''_{kEE}, is not included in this figure, because it leads to smaller short circuits than the two-phase short circuit. The ranges of the different types of short circuits according to the neutral point treatment are indicated in this diagram. The phase angles of the impedances $\underline{Z}_1, \underline{Z}_2,$ and $\underline{Z}_0$ in this figure must not differ by more than 15°.

The symbols in Figure 11.8 have the meanings:

k2　two-phase short-circuit current
k3　three-phase short-circuit current
k2E　two-phase short-circuit current without contact to ground
k1　single-phase ground fault current;
δ　ground fault factor

$$a = \frac{\text{Short circuit current for asymmetrical short circuit}}{\text{Short circuit current for three-pole short circuit}}$$

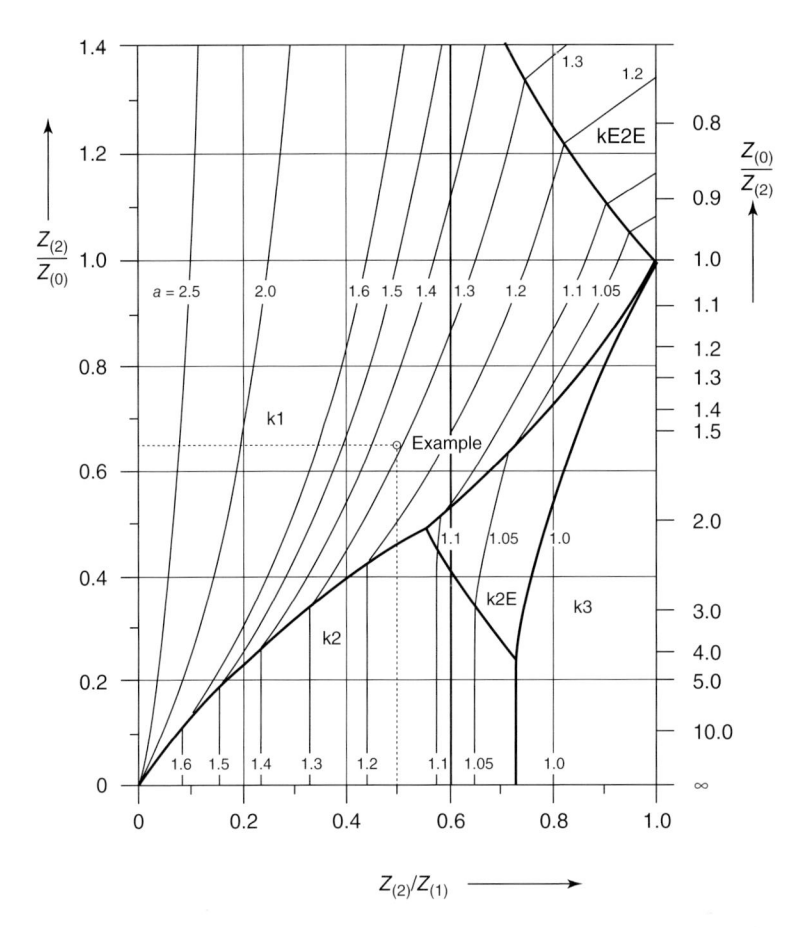

Figure 11.8 Largest short-circuit currents for asymmetrical short circuits (see also Ref. [1]).

11.5 Peak Short-Circuit Current, i_p

The initial short-circuit current, I_k'', and the withstand ratio, κ, determine the peak short-circuit current, i_p. The factor κ depends on the ratio R/X of the short-circuit path and takes account of the decay of the d.c. aperiodic component in the short circuit. The peak value of i_p occurs during the period immediately following the occurrence of the short circuit (transient period). If the ratio R/X is known, the factor κ can be read from the curves in Figure 11.9.

The peak short-circuit current calculated determines the dynamic loading of electrical systems.

The peak short-circuit current can be calculated in unmeshed networks from the equation:

$$i_p = \kappa \cdot \sqrt{2} I_k'' \qquad (11.22)$$

11 Calculation of Short-Circuit Currents

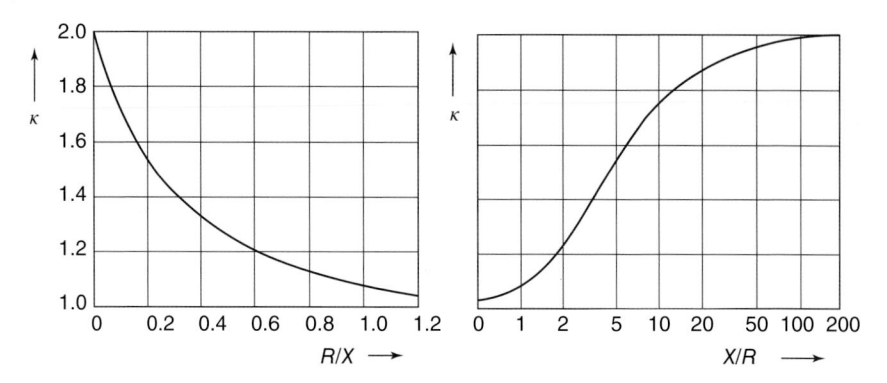

Figure 11.9 Factor κ for calculating the peak short-circuit current i_p (see also Ref. [1]).

Standard values:

$\kappa < 1.4$: in public networks
$\kappa \leq 1.8$–2.04: immediately downstream from transformer feeder.

κ can also be calculated from the following equation:

$$\kappa = 1.02 + 0.98 \cdot e^{-3(R/X)} \tag{11.23}$$

The peak short-circuit current, i_p, can be calculated in all the networks using the basic equation $i_p = \kappa \cdot \sqrt{2} \cdot I_k''$. With the three following procedures, it is possible to determine the factor κ in meshed networks [12].

1) *Procedure A* ($\kappa = \kappa_a$): κ is determined from the smallest R/X ratio of all the branches in the network. In low-voltage networks, $\kappa \leq 1.8$.
2) *Procedure B* ($\kappa = 1.15\kappa_b$): κ is determined from the R/X ratio of the short-circuit impedance at the position F of the short circuit and multiplied by a safety factor of 1.15 in order to take account of different R/X ratios in parallel branches.
    - For low-voltage networks: $\kappa \leq 1.8$.
    - For medium-voltage and high-voltage networks: $\kappa \leq 2.0$.
3) *Procedure C* ($\kappa = \kappa_c$): with procedure C, κ is determined with an equivalent frequency, as below:
    - Calculation of reactances for all network branches i for the equivalent frequency f_c in the positive-sequence system:

$$X_{ic} = \frac{f_c}{f} X_i$$

    f: nominal frequency – 50, 60 Hz.
    f_c: equivalent frequency – 20, 24 Hz.
    - Calculation of equivalent impedance at the position of the short circuit from the resistances R_i and the reactances X_i of the network branches in the positive-sequence system: $\underline{Z}_c = R_c + jX_c$
    - Determination of the factor κ_c from the ratio:

$$\frac{R}{X} = \frac{f_c}{f} \frac{R_c}{X_c}$$

11.6 Symmetrical Breaking Current, I_a

The symmetrical breaking current is the effective value of the short-circuit current $I_k''(t)$, which flows through the switch at the time of the first contact separation and is used for near-to-generator short-circuit feeder. For far-from-generator short circuits, the breaking currents are identical to the initial short-circuit currents:

$$I_a = \mu \cdot I_k'' \tag{11.24}$$

Synchronous machines:

$$I_a = \mu I_{kG}'' \tag{11.25}$$

I_a depends on the duration of the short circuit and the installation position of the switchgear at the position of the short circuit. μ characterizes the decay behavior of the short-circuit current and is a function of the variables I_{kG}''/I_{rG}'' and t_{min} (Figure 11.10).

The factor μ can be taken from Figure 11.10 or from the following equations:

$$\mu = 0.84 + 0.26 e^{-0.26 I_{kG}''/I_{rG}} \quad \text{for } t_{min} = 0.02\,\text{s}$$
$$\mu = 0.71 + 0.51 e^{-0.30 I_{kG}''/I_{rG}} \quad \text{for } t_{min} = 0.05\,\text{s}$$
$$\mu = 0.62 + 0.72 e^{-0.32 I_{kG}''/I_{rG}} \quad \text{for } t_{min} = 0.10\,\text{s}$$
$$\mu = 0.56 + 0.94 e^{-0.38 I_{kG}''/I_{rG}} \quad \text{for } t_{min} = 0.25\,\text{s}$$
$$\mu_{max} = 1$$

When $I_a = I_k''$, then $\mu = 1$, that is, a far-from-generator short circuit is present, if for each synchronous machine the following condition is satisfied, then:

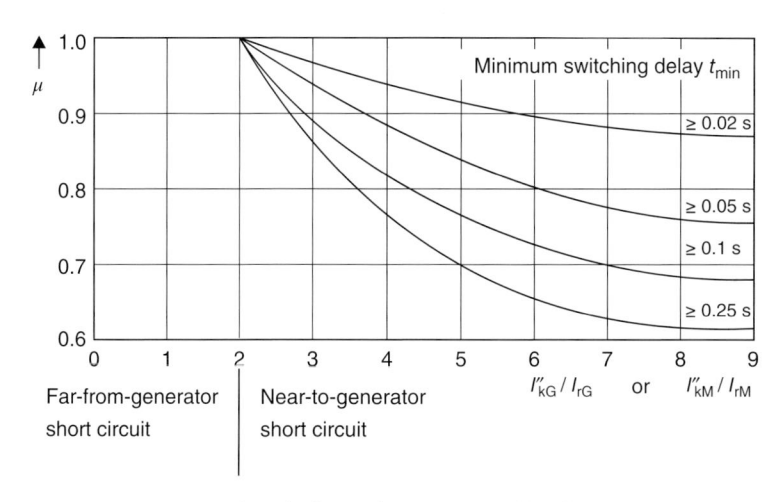

$$\frac{I_{k3}''}{I_{rG}} \leq 2 \tag{11.26}$$

Figure 11.10 Factor μ for calculating the symmetrical breaking current I_a (see also Ref. [1]).

For $I_a < I_k''$, that is, a near-to-generator short circuit:

$$\frac{I_{k3}''}{I_{rG}} \geq 2 \tag{11.27}$$

In practice: The minimum switching delay is 0.1 s.
Asynchronous machines:

$$I_a = \mu \cdot q \cdot I_{kM}'' \tag{11.28}$$

The factor q depends on the power per phase pair.
Networks:

$$I_{aQ} = I_{kQ}'' \tag{11.29}$$

More exact procedure for calculation of symmetrical breaking current in meshed networks [1]:

$$\underline{I}_a = \underline{I}_k'' - \sum_i \frac{\Delta \underline{U}_{Gi}''}{\frac{c \cdot U_n}{\sqrt{3}}}(1 - \mu_i) \cdot \underline{I}_{kGi}'' - \sum_i \frac{\Delta \underline{U}_{Mj}''}{\frac{c \cdot U_n}{\sqrt{3}}}(1 - \mu_j \cdot q_j) \cdot \underline{I}_{kMj}'' \tag{11.30}$$

With

$$\Delta \underline{U}_{Gi}'' = jX_{di}'' \cdot \underline{I}_{kGi}'' \tag{11.31}$$
$$\Delta \underline{U}_{Mj}'' = jX_{Mj}'' \cdot \underline{I}_{kMj}'' \tag{11.32}$$

Figure 11.11 shows the dependence of the factor q on the effective power per phase pair of the motor and the minimum switching delay t_{min}. For the equations used in the calculation of q, see IEC 60909-0.

The factor q applies to the induction motors and takes account of the rapid decay of the motor short circuit owing to the absence of an excitation field. It can be obtained from Figure 11.11 or from the following equations:

$$q = 1.03 + 0.12 \ln m \quad \text{for } t_{min} = 0.02 \text{ s}$$
$$q = 0.79 + 0.12 \ln m \quad \text{for } t_{min} = 0.05 \text{ s}$$
$$q = 0.57 + 0.12 \ln m \quad \text{for } t_{min} = 0.10 \text{ s}$$
$$q = 0.26 + 0.12 \ln m \quad \text{for } t_{min} = 0.25 \text{ s}$$
$$q_{max} = 1$$

The meaning of the symbols is as follows:

i	generator
j	motor
$\Delta \underline{U}_{Gi}''$	initial voltage difference at connection to synchronous machine i
$\Delta \underline{U}_{Mj}''$	initial voltage difference at connection to asynchronous machine j
$\frac{c \cdot U_n}{\sqrt{3}}$	equivalent voltage source at the position of short circuit
$\underline{I}_a, \underline{I}_k''$	symmetrical breaking current, initial symmetrical short-circuit current considering all network inputs, synchronous machines, and asynchronous machines
$\underline{I}_{kGi}''$	initial symmetrical short-circuit current of synchronous machine

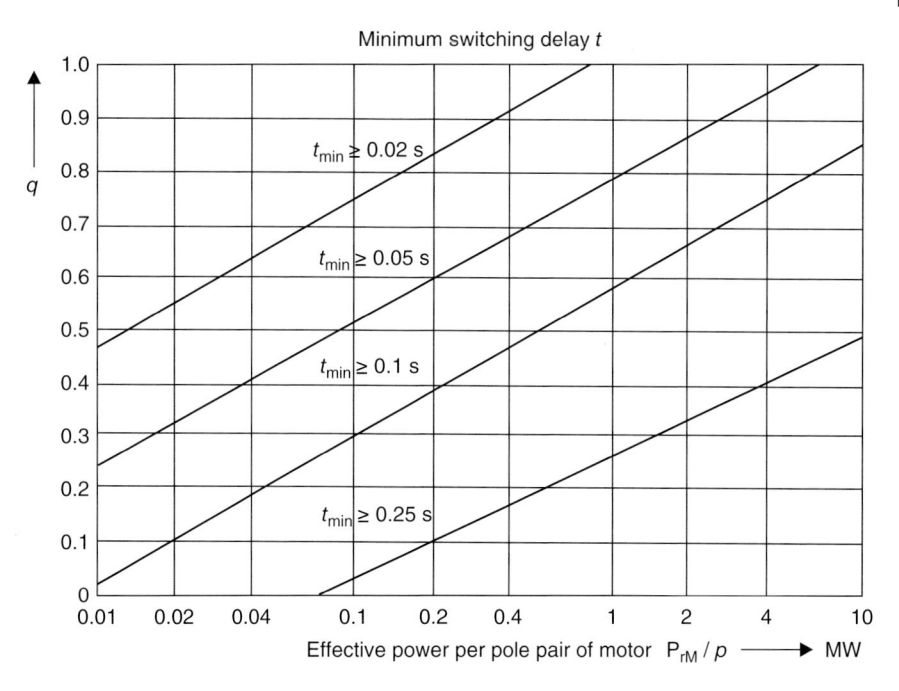

Figure 11.11 Factor q for calculation of the symmetrical breaking current for asynchronous machines (see also Ref. [1]).

$\underline{I}''_{kMj}$ initial symmetrical short-circuit current of asynchronous machine
μ_j factor j for asynchronous machines
μ_i factor i for synchronous machines
q_i factor j for asynchronous machines.

11.7 Steady-State Short-Circuit Current, I_k

The steady-state short-circuit current is the effective value of the short-circuit current I''_k remaining after the decay of all transient processes. It depends strongly on the excitation current, excitation system, and saturation of the synchronous machine:

For near-to-generator short circuits: $I_k < I''_k$.
For far-from-generator short circuits: $I_k = I''_k = I_a$.

The following relationships show a dependence on the fault position:

$$I_k = \lambda \cdot I_{rG} \tag{11.33}$$

$$I_k = I''_{k2}$$

$$I_k = \lambda \cdot \sqrt{3} \cdot I_{rG} \tag{11.34}$$

The factor λ depends on I''_{kG}/I_{rG}, the excitation, and the type of synchronous machine.

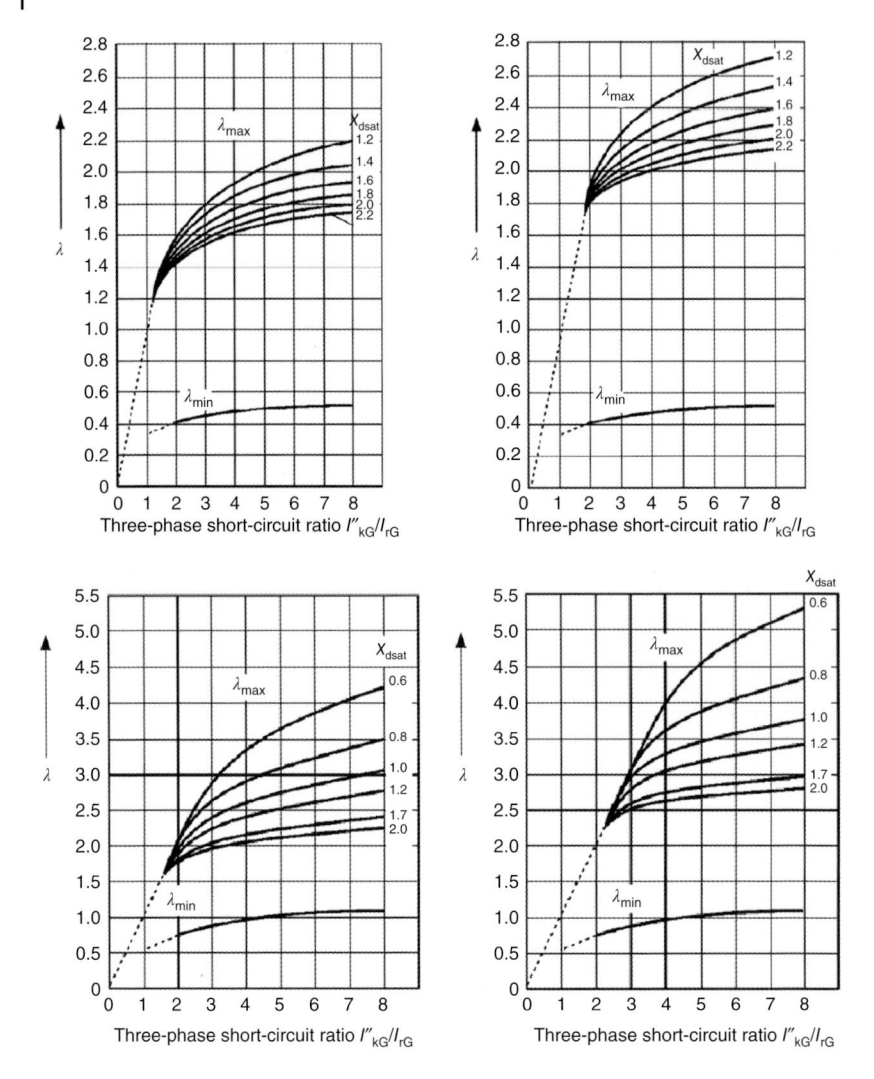

Figure 11.12 Factors λ_{min} and λ_{max} for calculating the steady-state short-circuit current I_k (see also Ref. [1]).

For the steady-state short-circuit current, we distinguish between:

$I_{kmax} = A_{max} \cdot I_{rG}$ (maximum excitation).
$I_{kmin} = A_{min} \cdot I_{rG}$ (constant unregulated excitation).

The upper and lower limits of A can be taken from Figure 11.12. It should also be pointed out that the A curves depend on the ratio of the maximum excitation voltage to the excitation voltage under normal load conditions (Series 1and 2).

The following statements can be made for Series 1and 2 [1]:

Series 1: The largest possible excitation voltage is 1.3 times the rated excitation voltage for the rated apparent power factor for turbo generators or 1.6 times the rated excitation voltage for salient-phase generators.

Series 2: The largest possible excitation voltage is 1.6 times the rated excitation voltage for the rated apparent power factor for turbo generators or 1.3 times the rated excitation voltage for salient-phase generators.

12

Motors in Electrical Networks

Asynchronous motors are used mostly in industry and for the internal consumption of power stations. For a short circuit, these deliver a part of the initial short-circuit current, peak short-circuit current, and symmetrical breaking current and for single-phase faults, part of the steady-state short-circuit current as well.

The peak short-circuit part of the asynchronous motors must be considered. For the calculation of squirrel-cage motors and wound rotor motors, there is no difference, since the starting resistors of wound rotor motors are short-circuited during operation. In the following cases, asynchronous motors can be neglected for the calculation of short-circuit currents [1]:

- Asynchronous motors in public low-voltage networks.
- Contributions of asynchronous motors or motor groups, which are less than 5% of the initial symmetrical short-circuit current without motors.
- Asynchronous motors which due to interlocking or type of process control cannot operate at the same time.
- Groups of low-voltage asynchronous motors that can be combined through their connection cables to an equivalent motor with $\frac{I_{LR}}{I_{rM}} = 5$ and $\frac{R_M}{X_M} = 0.42$.
- Asynchronous motors can be neglected for single-phase short circuits.

Various connection variants for asynchronous motors in industrial networks will be discussed in the following.

12.1 Short Circuits at the Terminals of Asynchronous Motors

The contribution of an asynchronous machine (Figure 12.1) can, in accordance with IEC 60909, be neglected provided that

$$I''_{kM} \leq 0.05 \cdot I''_{kQ} \tag{12.1}$$

$$I''_{kM} = \frac{c \cdot U_n}{\sqrt{3} \cdot Z_M} = \frac{I_{LR}}{I_{rM}} \frac{c \cdot U_n}{U^2_{rM}} \cdot I_{rM} \tag{12.2}$$

Short Circuits in Power Systems: A Practical Guide to IEC 60909-0, Second Edition. Ismail Kasikci.
© 2018 Wiley-VCH Verlag GmbH & Co. KGaA. Published 2018 by Wiley-VCH Verlag GmbH & Co. KGaA.

12 Motors in Electrical Networks

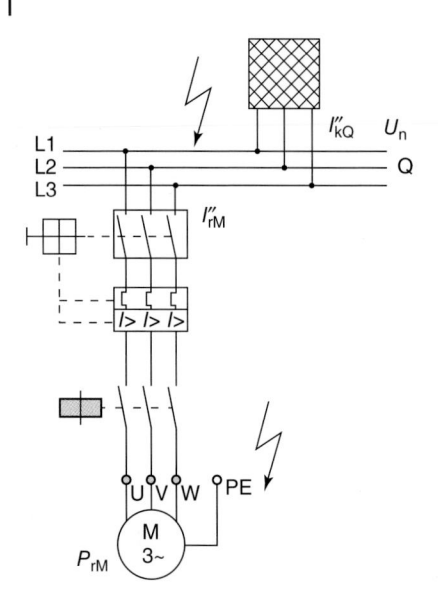

Figure 12.1 Short circuit at the terminals of an asynchronous motor.

The impedance Z_M of an asynchronous machine in the positive-sequence and negative-sequence systems is calculated from the relationships:

$$Z_M = \frac{1}{\frac{I_{LR}}{I_{rM}}} \cdot \frac{U_{rM}}{\sqrt{3} \cdot I_{rM}} = \frac{1}{\frac{I_{LR}}{I_{rM}}} \frac{U_{rM}^2}{S_{rM}} \tag{12.3}$$

$$Z_M = \frac{\eta_{rM} \cdot \cos\varphi_{rM}}{\frac{I_{LR}}{I_{rM}}} \frac{U_{rM}}{P_{rM}} \tag{12.4}$$

Motor groups (Equation (12.1)) can be neglected with the assumptions $I_{LR}/I_{rM} = 5$ and $c \cdot U_n/U_{rM} \approx 1$:

$$\sum I_{rM} \leq 0.01 \cdot I_k'' \tag{12.5}$$

The reactances and resistances can be calculated from the following equations [1]:

For medium voltages with effective power P_{rM} per phase pair ≥ 1 MW:

$$\frac{R_M}{X_m} = 0.10 \quad \text{mit} \quad X_M = 0.995 \cdot Z_M \tag{12.6}$$

For medium-voltage motors with effective power P_{rM} per phase pair < 1 MW:

$$\frac{R_M}{X_m} = 0.15 \quad \text{mit} \quad X_M = 0.989 \cdot Z_M \tag{12.7}$$

For low-voltage motors, including motor connection cable:

$$\frac{R_M}{X_m} = 0.42 \quad \text{mit} \quad X_M = 0.922 \cdot Z_M \tag{12.8}$$

Figure 12.2 Motor groups supplied from transformers with two windings.

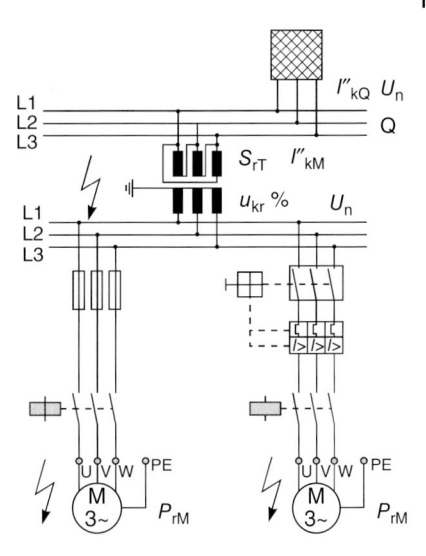

12.2 Motor Groups Supplied from Transformers with Two Windings

High-voltage and low-voltage motors that supply short-circuit currents to the short-circuit location Q (Figure 12.2) can be neglected provided that

$$\frac{\sum P_{rM}}{\sum S_{rT}} \leq \frac{0.8}{\dfrac{c \cdot 100 \sum S_{rT}}{S''_{kQ}} - 0.3} \tag{12.9}$$

When this condition is satisfied, asynchronous motors contribute less than 5% to the short-circuit current without motors.

For Equation (12.9): $I_{LR}/I_{rM} = 5$, $\cos \varphi \, \eta_r = 0.8$ and $u_{kr} = 6\,\%$.

12.3 Motor Groups Supplied from Transformers with Different Nominal Voltages

High-voltage and low-voltage motors supplied from transformers with different nominal voltages must be considered (Figure 12.3).

$$\sum I''_{kM} = I''_{kM1} + I''_{kM2} + I''_{kMn} \tag{12.10}$$

$$\sum S_{rT} = S_{rT1} + S_{rT2} + \cdots + S_{rTn} \tag{12.11}$$

$$\sum P_{rM} = P_{rM1} + P_{rM2} + \cdots + P_{rMn} \tag{12.12}$$

$$\frac{\sum R_{rM}}{\sum S_{rT}} \leq \frac{\cos\varphi_r \cdot \eta_r}{\left|\dfrac{I_{LR}}{I_{rM}}\left(\dfrac{c \cdot \sum S_{rT}}{0.05 \cdot S''_{kQ}} - \dfrac{u_{kr}}{100\%}\right)\right|} \qquad (12.13)$$

Equation (12.13) is true only when the conditions of Section 12.2 no longer apply.

The meaning of the symbols is as follows:

- I_{LR} locked rotor current of motor
- I''_k initial symmetrical short-circuit current without influence of motors
- S''_{kQ} initial symmetrical short-circuit power without influence of motors
- $\sum I_{rM}$ sum of rated currents of motors
- U_{rM} rated voltage of motor
- I_{rM} rated current of motor
- I_{LR}/I_{rM} ratio of locked rotor current to rated current for motor
- $\sum P_{rM}$ sum of rated effective powers of motors
- $\sum S_{rM}$ sum of rated apparent powers of transformers is between 4 and 8

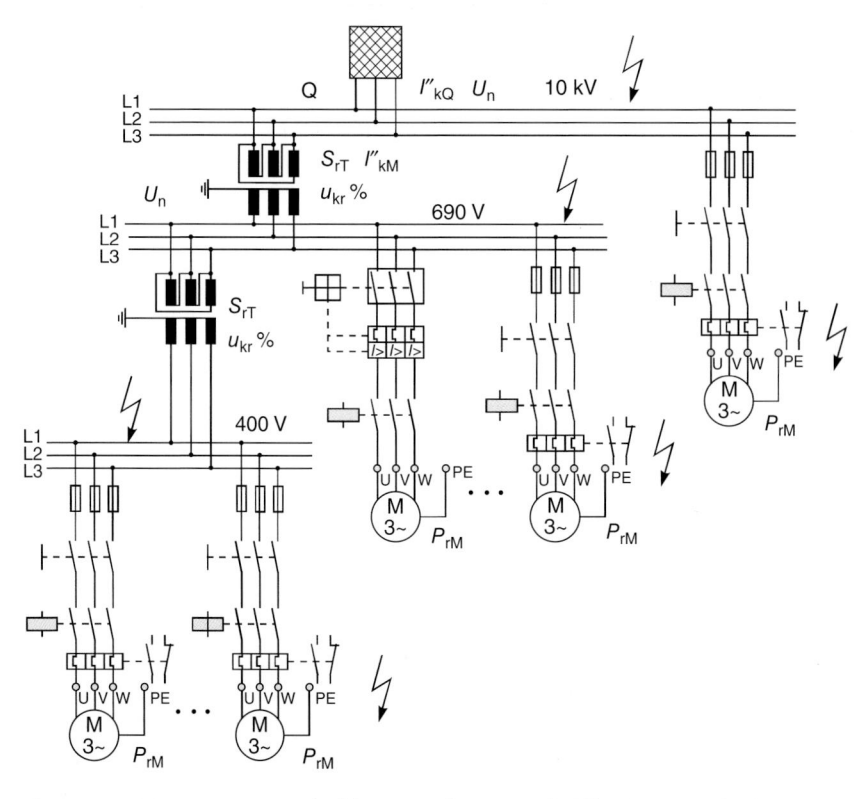

Figure 12.3 Motor groups supplied from transformers with different nominal voltages.

R_M/X_M 0.42 (when motor power is neglected)
Z_M short-circuit reactance.

Table 12.1 gives a summary of short circuits at the terminals of induction motors.

Table 12.1 Calculation of short circuits at the terminals of motors.

Short-circuit type	Three-phase short circuit	Line-to-line short circuit	Line-to-earth short circuit
Initial symmetrical short-circuit current	$I''_{k3M} = \dfrac{c \cdot U_n}{\sqrt{3} Z_M}$	$I''_{k2M} = \dfrac{\sqrt{3}}{2} I''_{k3M}$	—
Peak short-circuit current	$i_{p3M} = \kappa_M \sqrt{2} I''_{k3M}$	$i_{p2M} = \dfrac{\sqrt{3}}{2} i_{p3M}$	$i_{p1M} = \kappa_M \cdot \sqrt{2} \cdot I''_{k1M}$
Symmetrical short-circuit breaking current	$I_{b3M} = \mu q \cdot I''_{k3M}$	$I_{b2M} \approx \dfrac{\sqrt{3}}{2} I''_{k3M}$	$I_{b1M} \approx I''_{k1M}$
Steady-state short-circuit current	$I_{k3M} \approx 0$	$I_{k2M} \approx \dfrac{\sqrt{3}}{2} I''_{k3M}$	$I_{k1M} \approx I''_{k1M}$

13

Mechanical and Thermal Short-Circuit Strength

The dimensioning of electrical power installations with respect to stability against mechanical and thermal stresses is described exactly in EN 60865-1 [17]. This chapter gives a brief account of this information.

Electrical systems are subjected to mechanical and thermal stresses because of short circuits.

Busbars, switchgear, and contactors can be destroyed. Due to the evolution of heat, the operational equipment and the insulation of conductors and cables are affected. Furthermore, the potential hazards and injuries to the operating personnel must be considered.

In order to keep short circuits under control, these systems must be designed and dimensioned so that the operational equipment is able to withstand short-circuit conditions.

Here, the construction of the system components and the short-circuit current strength are especially important. The short-circuit strength is the sum total of all considerations for the prevention and control over the mechanical consequences of short-circuit currents and the joule heating that results.

The amount of heat generated over the duration of the short circuit must not exceed the permissible values for cables, conductors, and operational equipment.

13.1 Mechanical Short-Circuit Current Strength

Collecting bars, parallel conductors, switchgear, and fuses must be able to withstand the short circuit and the resulting peak short-circuit current occurring.

As a result of the magnetic field generated, parallel conductors that are separated by a distance attract each other when their currents flow in the same direction and repel each other when their currents flow in the opposite directions, whereby these forces are distributed uniformly over the length of the conductors l.

The manufacturers of operational equipment test their products for normal load conditions and for stability under the faulty operating conditions caused by a short circuit. At this point, it is useful to briefly discuss the individual circuit breakers and devices used in electrical systems:

- *Circuit breakers*: The rated short-circuit making and breaking capacities are normally given by the manufacturer for the better assessment of short-circuit strength. These specified numerical values must be larger than the calculated short-circuit value of the short-circuit current at the location of the fault. The contacts and mechanical parts must be able to withstand the short-circuit current. The circuit breakers are provided for breaking overload and short-circuit currents, for which the numerical values can be set accordingly. EN 60947 (IEC 947) describes the characteristics of low-voltage switches and IEC 282-1 and IEC 17A the characteristics of high-voltage switches.
- *Fuses*: Fusible links are the oldest forms of protective equipment and are of great importance in certain cases. The principle of these protective devices is based on a rupture joint, for example, a piece of wire, which melts with the application of a certain amount of heat given by $\int i^2 dt$. The type of time–current characteristic (function class) and the construction determine the locations for use. They are often used as backup protection or as the main protection in a system. The new version of IEC 269-1 describes the regulations for all fuse types.
- *Disconnectors, load interrupter switches, and load break switches*: In the low-voltage range, load interrupters are mostly used as disconnect switches for the load circuits, which break only operating currents, that is, currents under normal load, with an inductive power factor of 0.7. These switches are not able to break short-circuit currents and are, therefore, used together with fuses. In contrast to load interrupter switches, load break switches, which produce a visible isolating distance, are used in the medium-voltage range.

The operational equipment (busbars, insulators, wires, or cables) must be protected against the effects of short-circuit currents), that is, the systems must be short-circuit proof. The effect of forces acting on conductors with current flowing is of great interest here. Figure 13.1 depicts the effects of forces acting on busbars and parallel conductors.

In the magnetic field, a force acts on the conductors while the current is flowing. This force depends on the flux density B, the current intensity I, and the length l of the conductor. Two parallel conductors in which current is flowing attract each other when the currents are flowing in the same direction and repel each other when the currents are flowing in the opposite directions.

$$F = B \cdot l \cdot I$$

This force is greatest for a double-phase short circuit without contact to ground and for a three-phase short circuit. The maximum force occurring between the two conductors is proportional to the square of the current, that is, $F \sim i^2$. The greatest force acting on the conductors is calculated from the equation. For the current, the effective peak short-circuit current i_p is used.

The following will be obtained

$$F_M = \frac{\mu_0}{2\pi} i_p^2 \frac{l_s}{a_m}$$

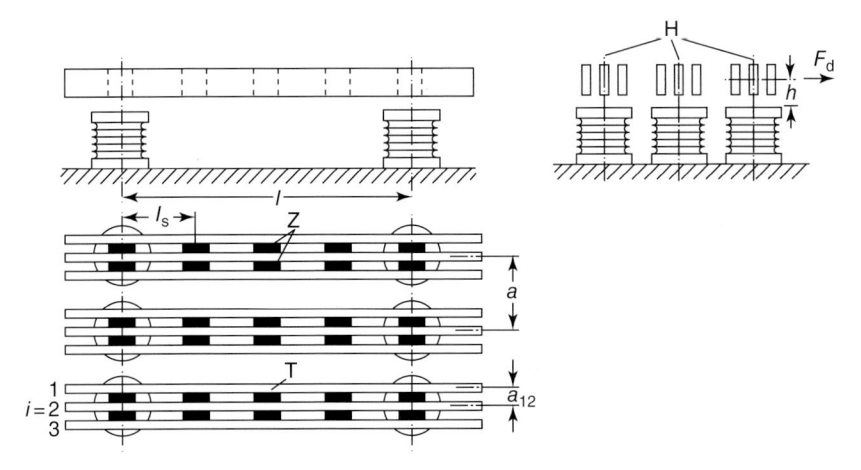

13.1 Mechanical Short-Circuit Current Strength

Figure 13.1 Effects of forces acting on busbars and parallel conductors (see also Ref. [6]).

or with the constant evaluated

$$F_M = 0.2\, i_{p2}^2 \frac{l_s}{a_m} \quad \text{or} \quad F_H = 0.17\, i_{p3}^2 \frac{l_0}{a_m}$$

The force between the neighboring conductor elements is

$$F_M = \frac{\mu_0}{2\pi} \left(\frac{i_p}{n}\right)^2 \frac{l_s}{a_s}$$

The meaning of the symbols is as follows:

F_M force for the main conductor in N
B magnetic induction in T
i peak short-circuit current in kA
a_m effective main conductor spacing in cm
μ_0 permeability constant ($4\pi \; 10^{-7}$ Vs/Am)
n number of conductors
l_s greatest center-to-center distance between the two neighboring conductor elements in cm
a_s effective spacing between the conductor elements in cm.

The forces between the conductor elements in which a short-circuit current flows depend on the geometrical arrangement and the profile of the conductors. This is why effective spacings have been introduced in the equation (EN 61660-1).

The effective spacing a_m is determined from:

$$a_m = \frac{a}{k_{12}}$$

The correction factor k_{12} is given in Figure 13.2.

As Figure 13.2 shows, the force is reduced because $k_{12} < 1$ for permanently installed flat ($b/d > 1$). On the other hand, the effect of the force increases slightly because $k_{12} > 1$ for neighboring bars.

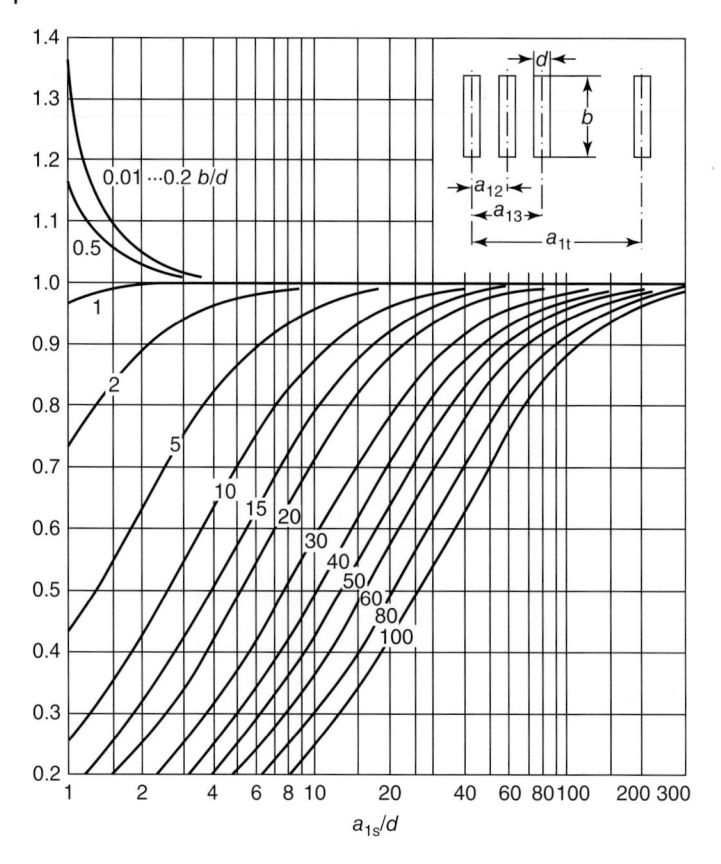

Figure 13.2 Correction factor k_{12} for the determination of the effective conductor spacing a_m (see also Ref. [1]).

The force leads to bending stress in rigid conductors, to tensile stress and deflection in conducting cables, and to transverse loading, compressive loading, or tensile loading of the support points. The conductors can be clamped, supported, or a combination of both and have several support points. The short-circuit loading of the busbars with respect to bending is calculated according to the laws of rigidity, so that

$$M = \frac{F_M \cdot l}{8}$$

The meaning of the symbols is as follows:
F_M force due to the short-circuit current in N
M bending moment in N cm
l support spacing in cm.

The bending stress can then be determined using the moment of resistance, that is,

$$\sigma_H = \frac{v_\sigma \cdot \beta \cdot M}{W}$$

13.1 Mechanical Short-Circuit Current Strength

The bending stress σ must always be less than the permissible bending stress σ_{per}.

$$\sigma_H \leq \sigma_{per}$$

The bending stress for several conductor elements can be calculated from

$$\sigma_M = \frac{v_\sigma \cdot \beta \cdot M}{W_T}$$

The permissible total bending stress of the main (σ_M) conductor and the conductor elements (σ_T) is then

$$\sigma_G = \sigma_M + \sigma_T$$

The dynamic force acting at each support point of the rigid conductors is calculated from

$$F_d = V_F V_r \alpha F$$

For d.c. systems:

$$V_F V_r = 2$$

For one-phase a.c. systems:

$$V_F V_r \leq 2$$

For three-phase systems:

$$V_F V_r \leq 2.7$$

The meaning of the symbols is as follows:
- σ bending stress in N/cm²
- γ frequency factor (Table 13.2)
- β factor for loading of the main conductor (Table 13.2)
- α depends on the type and number of support points (Table 13.2)
- M bending moment in N cm
- W moment of resistance in cm⁴.

The moments of resistance and moments of inertia of different tubular profiles can be obtained from Table 13.1.

The busbar arrangement must be checked for mechanical resonance. The natural mechanical oscillating frequency must not be close to the simple, double, or triple network frequency, since this would lead to damage as a result of resonance (Table 13.2).

For the natural mechanical oscillating frequency, then

$$f_0 = 112 \cdot \sqrt{\frac{E \cdot I}{G \cdot l^4}}$$

The meaning of the symbols is as follows:
- f_0 natural oscillating frequency in s^{-1}
- E modulus of elasticity of the busbar material for Cu: 1.1 \. 10⁶ kg/cm² and for Al: 0.65 \. 10⁶ kg/cm²
- I moment of inertia in cm⁴ (Table 13.1)
- G weight of bar in kg/cm.

Table 13.1 Moments of resistance and moments of inertia.

Representation	Moment of inertia (cm⁴)	Flexural modulus (m³)
(rectangular cross-section with width b and height h)	$I = \dfrac{b \cdot h^3}{12}$	$W = \dfrac{b \cdot h^2}{6}$
(solid circular cross-section with diameter d)	$I = \dfrac{\pi \cdot d^4}{64}$	$W = \dfrac{\pi \cdot d^3}{32}$
(tubular cross-section with inner diameter d and outer diameter D)	$I = \dfrac{\pi}{64}(D^4 - d^4)$	$W = \dfrac{\pi}{32} \dfrac{(D^4 - d^4)}{D}$

Table 13.2 Factors α and β for different support point arrangements (see also Ref. [6]).

Type of busbar and its clamping condition		Force on support Factor α	Main conductor stress Factor β	Relevant characteristic frequency Factor γ
Single-span beam	Both sides supported	A: 0.5 B: 0.5	1.0	1.57
	Fixed and supported	A: 0.625 B: 0.375	0.73	2.45
	Both sides fixed	A: 0.5 B: 0.5	0.50	3.56
Continuous beam with multiple supports and N equal or approximately equal support distances	$N = 2$	A: 0.375 B: 1.25	0.73	2.45
	$N \geq 3$	A: 0.4 B: 1.1	0.73	3.56

13.2 Thermal Short-Circuit Current Strength

Switches, conductors, and transformers must as a result of short-circuit currents be dimensioned with regard to thermal effects as well. It must be checked whether the thermal equilibrium short-time withstand current I_{th} is correctly determined. The magnitude and the behavior in time of the short-circuit current determine the effective short-time withstand current, the effective value of which produces the same quantity of heat as the changing short-circuit current during the duration of the short circuit T_K in its d.c. aperiodic and a.c. periodic current components.

The thermal short-circuit current can be calculated from the effective value of the initial symmetrical short-circuit current and the m and n factors (Figure 13.3). Here:

$$I_{th} = I_k'' \cdot \sqrt{m+n} \tag{13.1}$$

We can also determine the factor m using the relationship:

$$m = \frac{1}{2 f t_k \ln(\kappa - 1)} [e^{4 f T_k \ln(\kappa-1)} - 1] \tag{13.2}$$

The factor m takes account of the thermal effect of the d.c. aperiodic component for three-phase and polyphase currents and the factor n the thermal effect of the a.c. periodic component for a three-phase short circuit.

Electrical operational equipment is dimensioned according to either the permissible short-time withstand current I_{th} or the permissible duration of the short circuit T_K.

The duration of the short circuit and the short-time withstand current are given by the manufacturers of protective equipment.

$$I_{th} \leq I_{thz} \tag{13.3}$$

In accordance with IEC 76-1, I_{th} must not exceed 25 times the nominal current for longer than 3 s with short-circuit current-limiting reactors. With current transformers, this value is given on the nameplate. Here:

$$I_{th} \leq \sqrt{\frac{I_{kn}}{(T_K + 0.05\,\text{s})}} \tag{13.4}$$

For short-circuit durations:

$$T_K \geq 0.1\,\text{s} \tag{13.5}$$

the following relationship holds true:

$$I_a = I_k'' \tag{13.6}$$

If the peak short-circuit current i_p is not known and the initial symmetrical short-circuit current I_k'' is given, we can then calculate using the withstand ratio $\kappa = 1.8$.

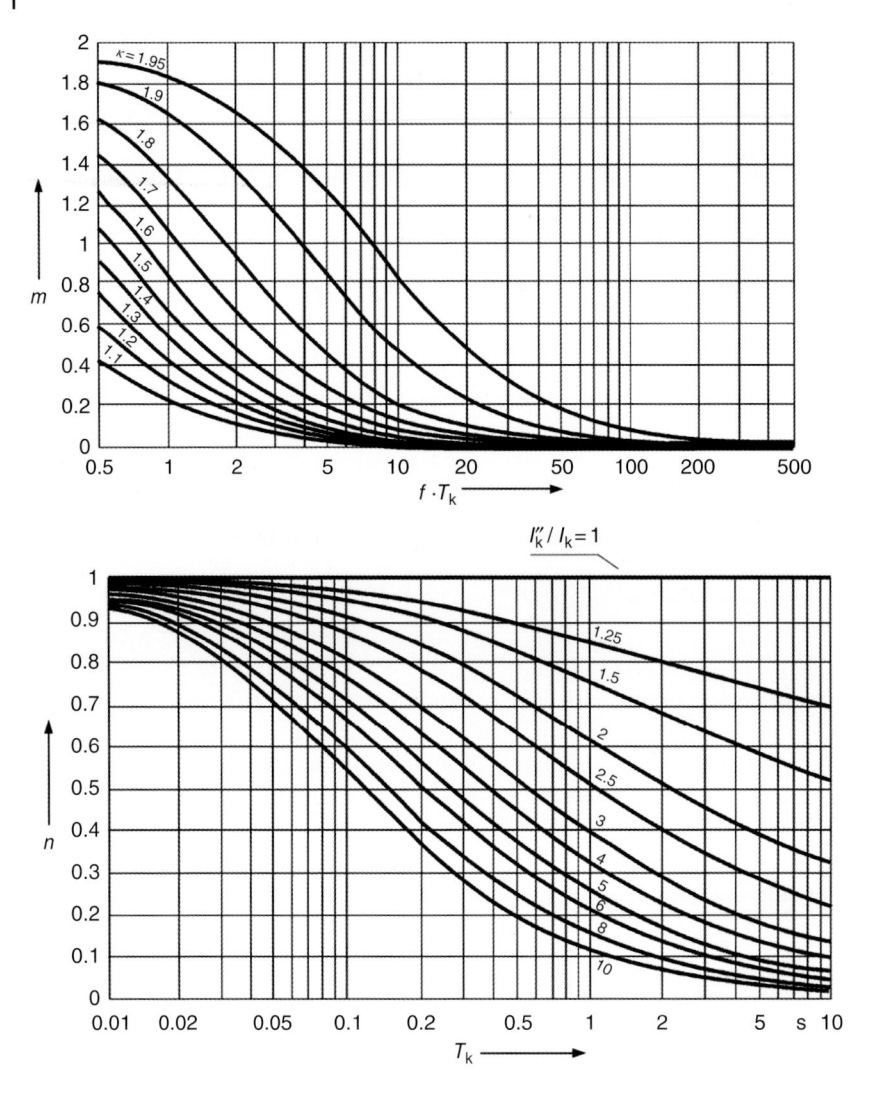

Figure 13.3 The factors m and n (see also Ref. [1]).

The thermal equivalent short-circuit current density S_{th} must be smaller than the rated short-time current density S_{thr}. The following equation must be satisfied:

$$S_{th} \leq S_{thr} \cdot \frac{1}{\eta} \cdot \sqrt{\frac{T_{kr}}{T_k}}$$

The factor η takes account of heat transfer for the insulation. The rated short-time current density S_{thr} can be obtained from Figure 13.4. The initial temperature v_b of a conductor is the maximum permissible continuous operating

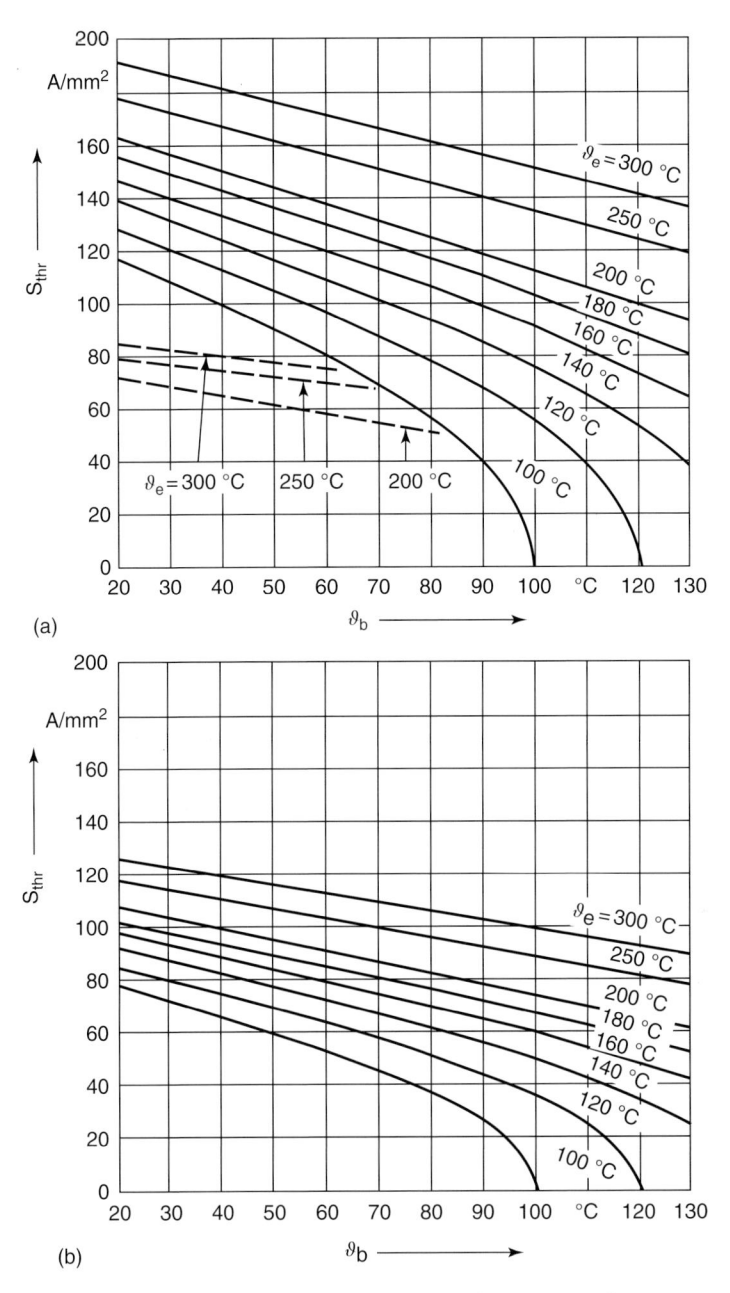

Figure 13.4 Rated short-time current density S_{thr} for $T_{kr} = 1$ s (a) for copper and steel (solid line curves), (b) aluminum (broken line curves) Aldrey and Al/St (see also Ref. [17]).

temperature. The ultimate temperature v_e is the maximum permissible temperature under short-circuit conditions for plastic-insulated cables.

13.3 Limitation of Short-Circuit Currents

It is necessary to control both the minimum and the maximum short-circuit currents in low-voltage systems without impairing the selectivity. Investigations have shown that the three-phase short-circuit currents are less than 9 kA, with an average value of 2.8 kA [18]. However, this value only applies for low-voltage networks. In low-voltage networks, the single-phase short circuits are generally less than 1 kA. In order to limit the short-circuit currents (Figure 13.5), special measures are possible in individual systems according to the various operational equipment.

Fuses are unsuitable for smaller short-circuit currents. This is why many power supply companies prescribe line protection circuit breakers in house installations. The short-circuit current limitation in high-voltage and medium-voltage networks can be achieved through economical (e.g., the choice of nominal network voltage or division of the network into individual groups) and technical (e.g., the use of i_p limiters and fuses) means.

Figures 13.6–13.11 illustrate the short-circuit-carrying capacity of different cables as a function of the break time.

13.4 Examples for Thermal Stress

13.4.1 Feeder of a Transformer

At the end of a cable, three-phase short circuit is calculated (Figure 13.12). Relay total breaking time is $T_k = 0.5$ s. Check it to see if the cable is thermally adequate.

Factor m for the d.c. component m and factor n for the thermal effect of a.c. component are given.

$m = 0.09$
$n = 0.73$

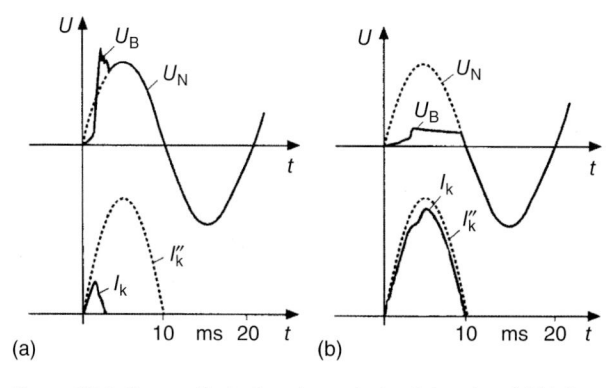

Figure 13.5 Current limitation through circuit breaker: (a) high power current limitation through circuit breaker and (b) line protection circuit breaker with neutral point quencher, without definite current limitation.

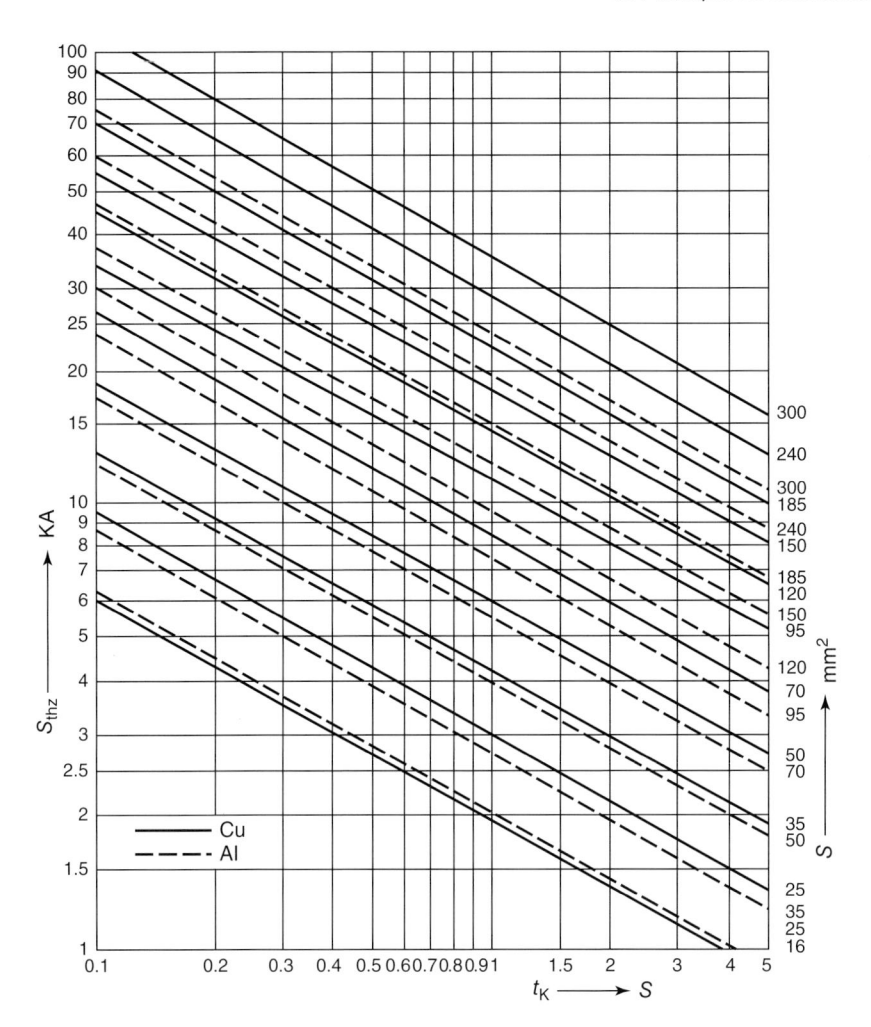

Figure 13.6 Thermally permissible short-circuit current from paper-insulated cables at 1–10 kV (see also Ref. [10]).

Then, we can calculate the thermal equivalent short-circuit current:

$$I_{th} = I_k'' \cdot \sqrt{m+n}$$
$$I_{th} = 21.5 \, \text{kA} \cdot \sqrt{0.09 + 0.73} = 19.46 \, \text{kA}$$

The permissible rated short-time current density at the beginning of a short circuit at a temperature $\vartheta_b = 70\,°C$ and end temperature $\vartheta_e = 250\,°C$ is shown in Figure 13.4, $S_{thr} = 143 \, \text{A/mm}^2$. Therefore, the current density is

$$S_{th} = S_{thr} \cdot \sqrt{\frac{T_{kr}}{T_k}}$$

$$S_{th} = 143 \cdot \frac{A}{\text{mm}^2} \cdot \sqrt{\frac{1\,\text{s}}{0.5\,\text{s}}} = 202.2 \, \text{A/mm}^2$$

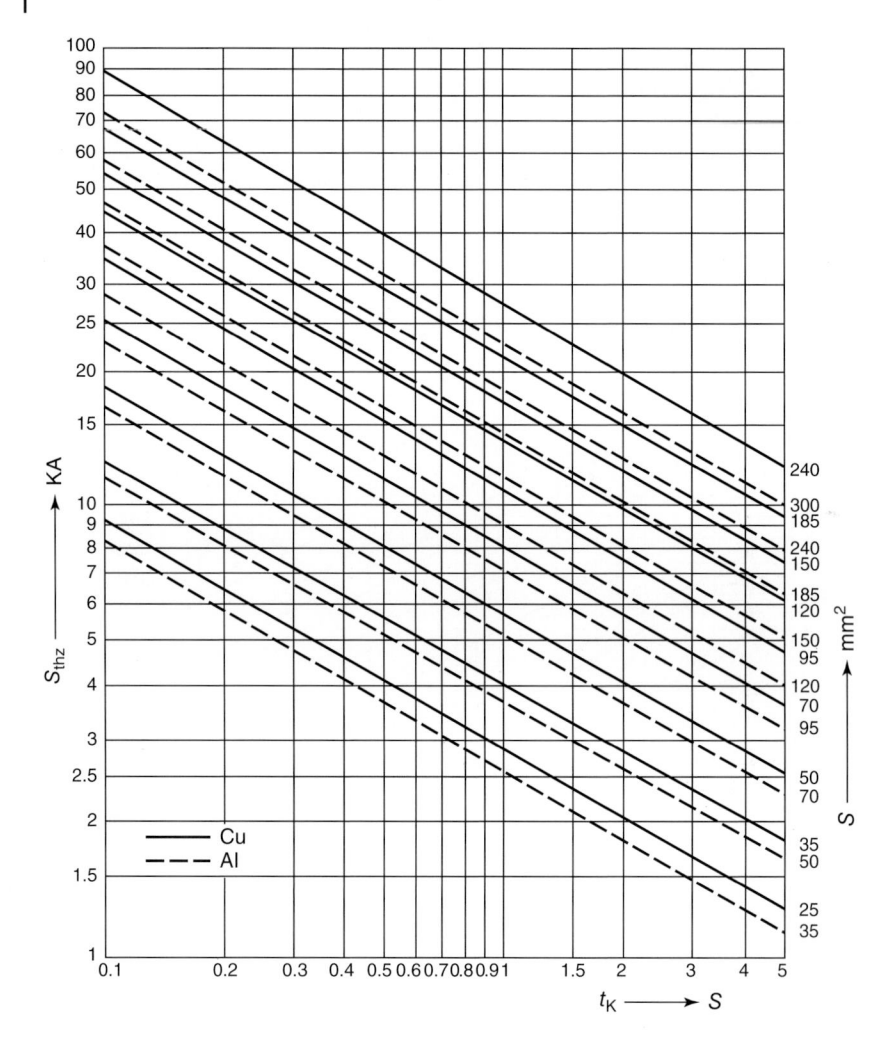

Figure 13.7 Thermally permissible short-circuit current from paper-insulated cables at 12/20 kV (see also Ref. [10]).

Minimum cross-section of cable can then be calculated as shown below:

$$A = \frac{I_{th}}{S_{th}} = \frac{19.46\,\text{kA}}{202.2\,\text{A/mm}^2} = 96.24\,\text{mm}^2$$

The feeder conductor cross-section is enough which is

$$120 > 96.24\,\text{mm}^2$$

13.4.2 Mechanical Short-Circuit Strength

The short-circuit current at the busbar of a transformer station is calculated with Simaris, which gives $I''_{kmax} \approx 16\,\text{kA}$ (Figure 13.13).

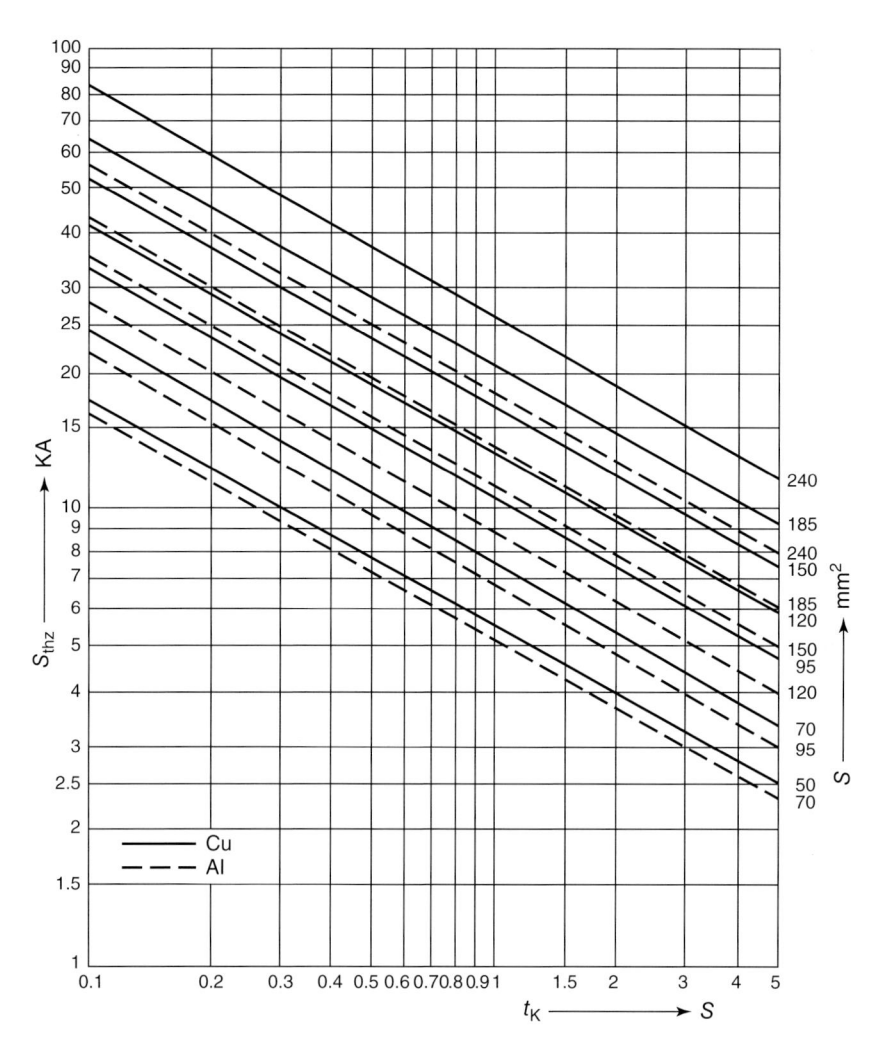

Figure 13.8 Thermally permissible short-circuit current from paper-insulated cables at 18/30 kV (see also Ref. [10]).

Figure 13.14 shows the busbar configuration with three main conductors with following data:

$$a = 10\,\text{cm}, \quad l = 100\,\text{cm}, \quad b = 10\,\text{cm}, \quad h = 50\,\text{cm}$$

First, we calculate the peak current with

$$i_p = \kappa \cdot \sqrt{2} \cdot I''_{k\max} = 1.8 \cdot \sqrt{2} \cdot 16\,\text{kA} = 40.73\,\text{kA}$$

The electrodynamic force between conductors

$$F = 0.2 \cdot i_p^2 \cdot \frac{l}{a} = 0.2 \cdot (40.73\,\text{kA})^2 \cdot \frac{100\,\text{cm}}{10\,\text{cm}} = 3317.86\,\text{N}$$

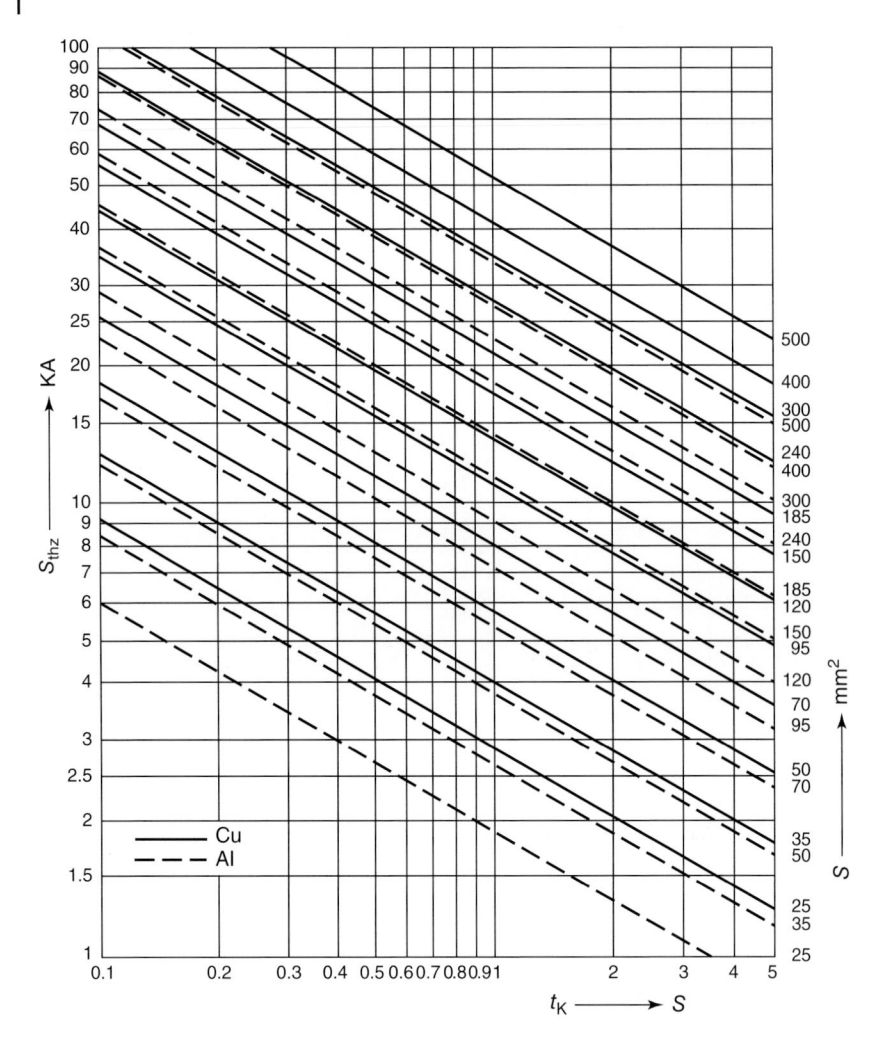

Figure 13.9 Thermally permissible short-circuit current from PVC-insulated cables at 1–10 kV (see also Ref. [10]).

Bending moment:
$$M = \frac{F \cdot l}{8} = \frac{3317.86 \, \text{N} \cdot 100 \, \text{cm}}{8} = 41\,473.25 \, \text{N cm}$$

Flexural modulus on the vertical track assembly (Table 13.1):
$$W = \frac{b \cdot h^2}{6} = \frac{10 \, \text{cm} \cdot (50 \, \text{cm})^2}{6} = 4166.6 \, \text{cm}^3$$

Bending stress by the vertical track assembly:
$$\sigma = \frac{M}{W} = \frac{41\,473.25 \, \text{N cm}}{4166.6 \, \text{cm}^3} = 9.953 \, \text{N/cm}^2$$

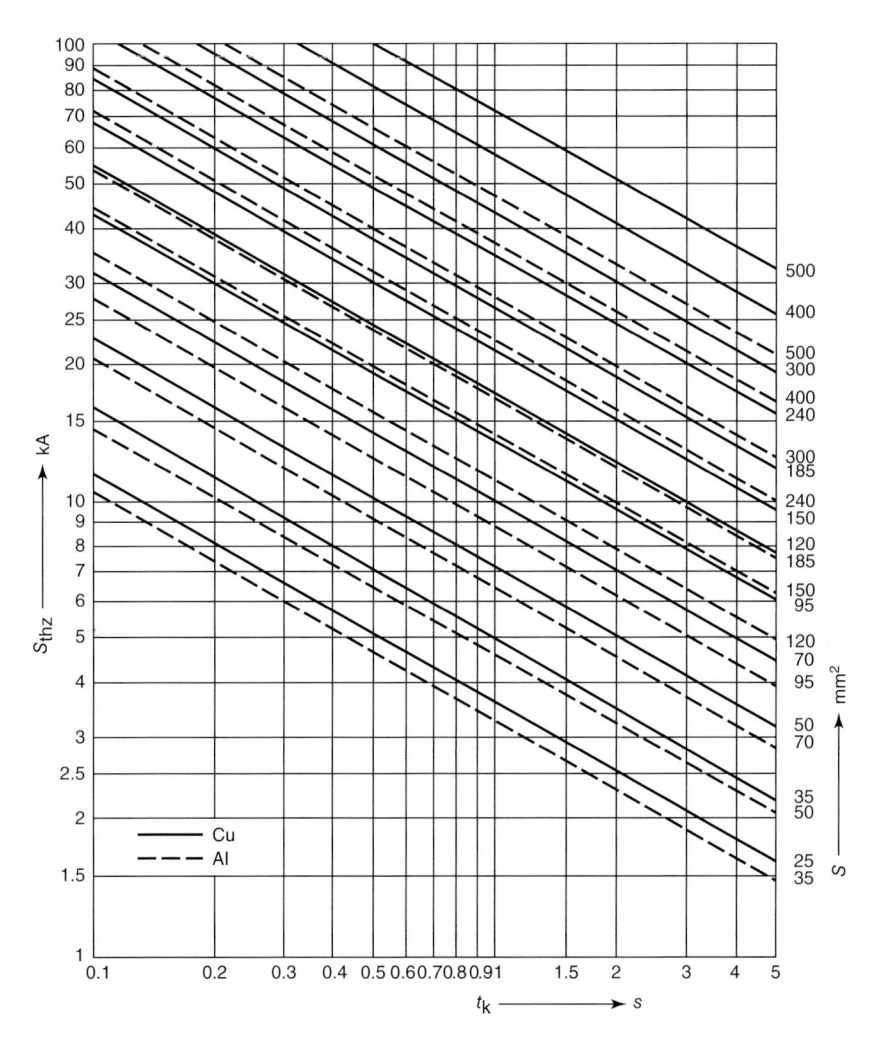

Figure 13.10 Thermally permissible short-circuit current from XLPE-insulated cables (see also Ref. [10]).

Moment of inertia (Table 13.1):
$$I = \frac{b \cdot h^3}{12} = \frac{10\,\text{cm} \cdot (50\,\text{cm})^3}{12} = 104\,166\,\text{cm}^4$$

System frequency of the main conductor:
$$f = 112 \cdot \sqrt{\frac{E \cdot I}{G \cdot l^4}} = 112 \cdot \sqrt{\frac{1.1 \cdot 10^6 \cdot 10.4166}{13.5 \cdot 10^{-3} \cdot 100^4}}\,\text{s}^{-1} = 250\,\text{s}^{-1}$$

There is no harm to be expected because $5 \times 50 = 250$ Hz.

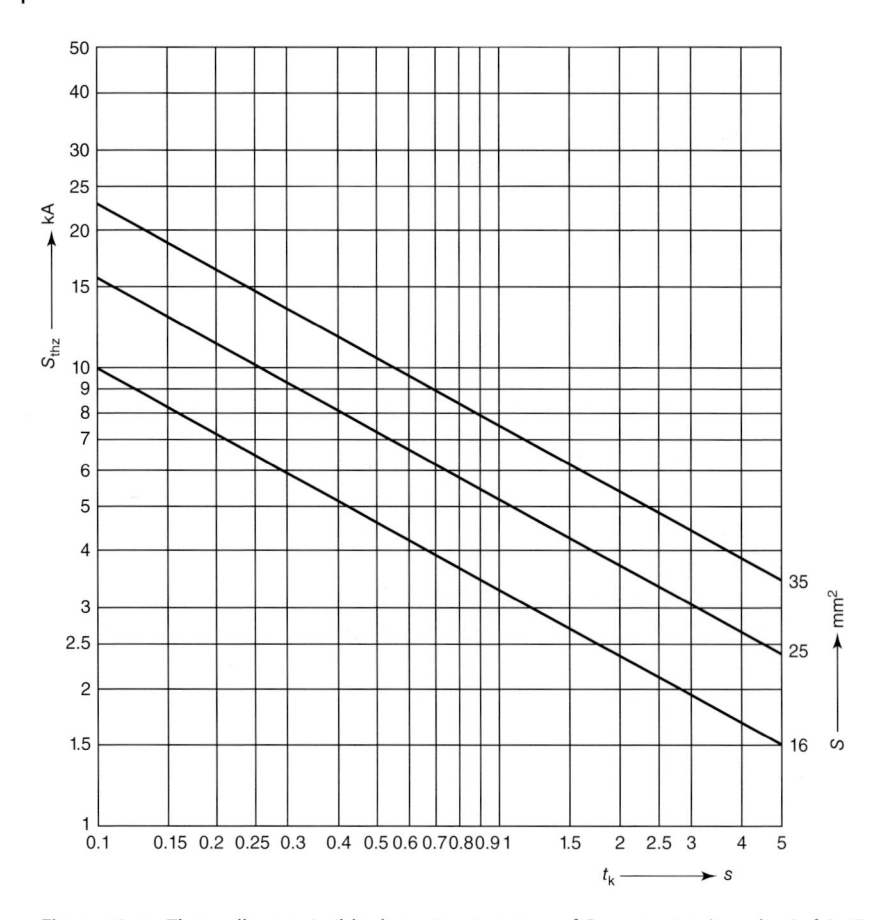

Figure 13.11 Thermally permissible short-circuit current of Cu screening (see also Ref. [10]).

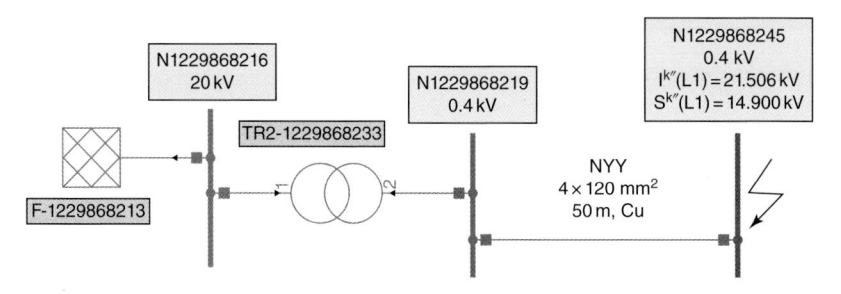

Figure 13.12 Feeder of a transformer.

13.4 Examples for Thermal Stress | 183

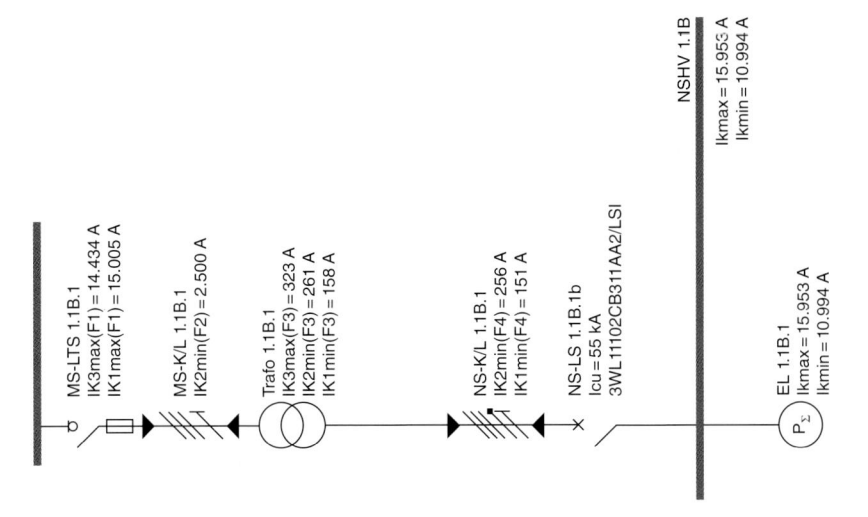

Figure 13.13 Mechanical short-circuit strength.

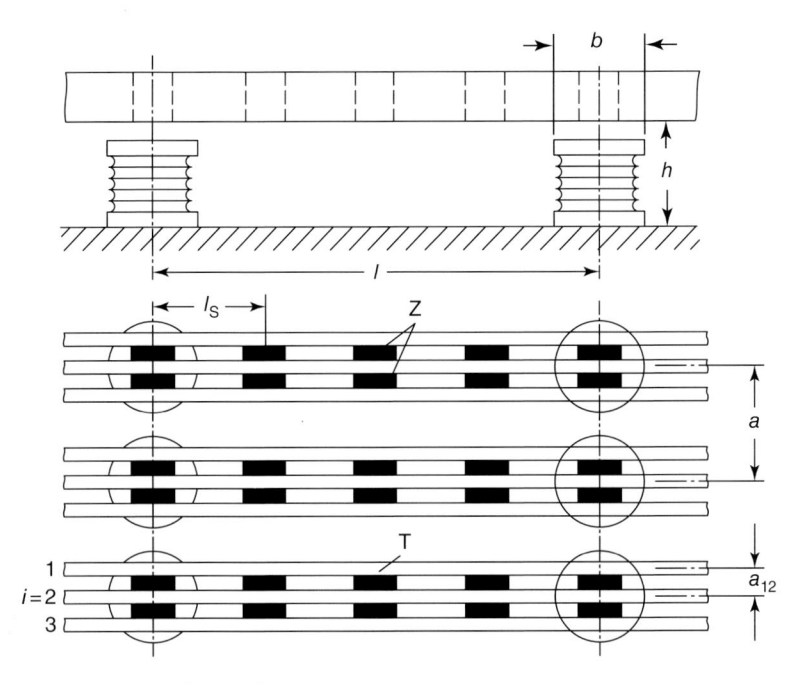

Figure 13.14 Busbar configuration.

14

Calculations for Short-Circuit Strength

To determine the short-circuit strength for switchgear and control gear, the magnitude of the prospective short-circuit current is decisive.

The choice of switchgear follows from:

- the short-circuit strength
- the rated breaking capacity.

14.1 Short-Circuit Strength for Medium-Voltage Switchgear

The following selection criteria are important for the short-circuit strength of medium-voltage switchgear (EN 61936-1:2011-11):

- The rated short-circuit breaking current I_{sc}: this is the effective value of the breaking current at the terminals of the switchgear. The following condition must hold true:

$$I_{sc} \geq I_k'' \tag{14.1}$$

Standardized values for middle-voltage systems are as follows:

8, 12.5, 16, 20, 25, 31.5, 40, 50, and 63 kA

- The rated short-circuit making current I_{ma}: this is the peak value of the making current for a short circuit at the terminals of the switchgear. The following condition must apply for circuit breakers, load (break) switches, and grounding switches of medium-voltage switchgear:

$$I_{ma} \geq i_p \tag{14.2}$$

Standardized values for medium-voltage systems are as follows:

20, 25, 31.5, 40, 50, 63, 80, 100, 125, and 160 kA

- The rated short-time current I_{th}: this is the effective value of the short-circuit current which the switchgear can carry in the closed state during the rated short-circuit duration t_{th} under the specified conditions for use and operation.

Short Circuits in Power Systems: A Practical Guide to IEC 60909-0, Second Edition. Ismail Kasikci.
© 2018 Wiley-VCH Verlag GmbH & Co. KGaA. Published 2018 by Wiley-VCH Verlag GmbH & Co. KGaA.

- The thermal short-circuit capacity I_{thz}:

$$I_{thz} \geq I_{thm} \tag{14.3}$$

$$I_{thz} = I_{th}\sqrt{\frac{t_{th}}{t_k}} \tag{14.4}$$

The symbols have the meaning:
I_{sc} rated short-circuit breaking current
I_k'' initial symmetrical short-circuit current
I_{thz} thermal short-circuit carrying capacity
I_{thm} thermal equivalent short-circuit current
I_{ma} rated short-circuit making current
i_p peak short-circuit current
I_{th} rated short-time current
t_{th} rated short-circuit duration (1 and 3 s)
t_k maximum short-circuit duration.

14.2 Short-Circuit Strength for Low-Voltage Switchgear

The following selection criteria must be considered for the short-circuit strength of low-voltage systems (IEC 60947):

- The rated ultimate short-circuit breaking current (breaking capacity) I_{cn}: this is the largest current which a switchgear can break without incurring damage. It is given as an effective value. Standardized values for low-voltage switchgear are 18, 25, 40, 70, and 100 kA.

$$I_{cn} \geq I_k'' \tag{14.5}$$

- The rated short-circuit making current (making capacity) I_{cm}: this is the largest current which a switchgear can make without incurring damage. It is given as a peak value. The power factor $\cos\phi$ of the short-circuit path depends primarily on the reactance of the input transformer. The greater its power, the smaller is the power factor. For this relationship, IEC 60947-2 gives a minimum value for circuit breakers:

$$I_{cm} = n \cdot I_{cn} = \kappa \cdot \sqrt{2} \cdot I_{cn} \tag{14.6}$$

According to the type, the circuit breakers must be able to interrupt the following short circuit and operating currents:

1) Circuit breakers with $\cos\phi \geq 0.1$: short-circuit currents
2) Circuit breakers with $\cos\phi \geq 0.7$: load currents
3) Circuit breakers with $\cos\phi < 0.1$: operating currents.

14.2 Short-Circuit Strength for Low-Voltage Switchgear

- The rated short-time withstand current I_{cw}: this is the permissible effective value of the a.c. periodic component of the uninfluenced short-circuit current which the switchgear can carry for a certain length of time without any noticeable effect (e.g., due to excessive evolution of heat), for example, from 0.05 to 1 s.

In accordance with IEC 60947, circuit breakers are tested with two currents for breaking the short-circuit current.

The I_{cu} test sequence is $0-t-C_0$
The I_{cs} test sequence is $0-t-C_0-t-C_0$.

The current I_{cs} represents the more severe condition for the circuit breakers, because switching takes place one additional time and is stipulated as 25%, 50%, 75%, and 100% of I_{cu} in order to distinguish the function of the switch following a short-circuit break.

The symbols have the meaning:
I_{cu} rated short-circuit breaking current
I_k'' initial symmetrical short-circuit current
I_{cm} rated short-circuit making current
i_p peak short-circuit current.

15

Equipment for Overcurrent Protection

Overcurrent protective devices are used in electrical power installations for the control and protection of electrical equipment. The selection of these devices is subject to the responsibilities and the processes in the plant. This chapter describes the most important overcurrent short-circuit devices (Figure 15.1).

The time–current characteristics of limit switch fuses (Figure 15.2) are plotted in logarithmic scale as a function of current. At higher short-circuit currents, they break faster so that the peak short-circuit current i_p can no longer occur. The cut-off current is reached during the break process and can be obtained from Figure 15.3.

The effective current limitation and the very high breaking capacity are the good features of these fuses. The cut-off energy or I^2t values of the fuses are decisive for the selectivity, which is normally guaranteed to be 1.6 times the rated current (better is two times). The high voltage–high power fuses protect the equipment and system components from the dynamic and thermal effects of short-circuit currents. The HH fuses are used in the distribution transformers, high-voltage motors, capacitors, and middle-voltage transformers. The combination of these fuses with load break switches, load switches, and vacuum protectors is possible. Figures 15.4 and 15.5 show the time–current characteristics and Figure 15.6 the cut-off characteristics.

The miniature circuit breakers according to IEC 60898 (Figures 15.7 and 15.8) are used in house installations and systems against overloading and short circuits. Internationally, the characteristics B, C, and D are standardized. The function values in the overload region are the same for all miniature circuit breakers, but in the range of short circuits, the breaking current is a multiple of the rated current for the breaker.

The assessment of capacity is according to the following criteria:

1) Rated breaking capacity 3, 6, 10, or 25 kA
2) Short-circuit stress and selectivity
3) Limiting class 1, 2, or 3
4) Backup protection.

Short Circuits in Power Systems: A Practical Guide to IEC 60909-0, Second Edition. Ismail Kasikci.
© 2018 Wiley-VCH Verlag GmbH & Co. KGaA. Published 2018 by Wiley-VCH Verlag GmbH & Co. KGaA.

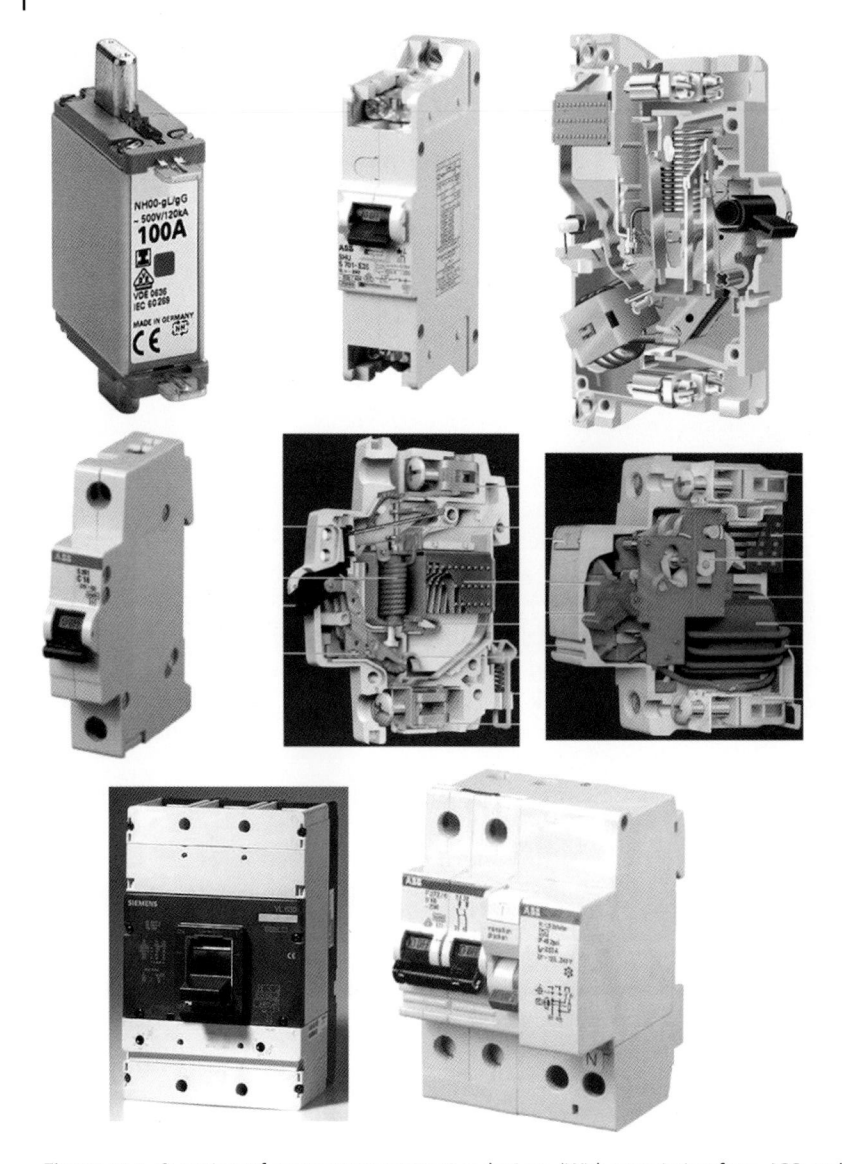

Figure 15.1 Overview of overcurrent protective devices. (With permission from ABB and Siemens.)

The characteristics of circuit breakers are derived from IEC 60947 and are valid for the cold state with a three-phase load for which in accordance with IEC 60898 the deviation from the release time from three times the current setting may be a maximum of ±3%. In the warm operational state, the release times of the thermal tripping devices are reduced by about 25%. The circuit breakers are compact circuit breakers up to 63 A and function according to the principle of current limitation. These devices are used for breaking and protecting motors, cables and conductors, and other operational equipment with undelayed overcurrent

Equipment for Overcurrent Protection | 191

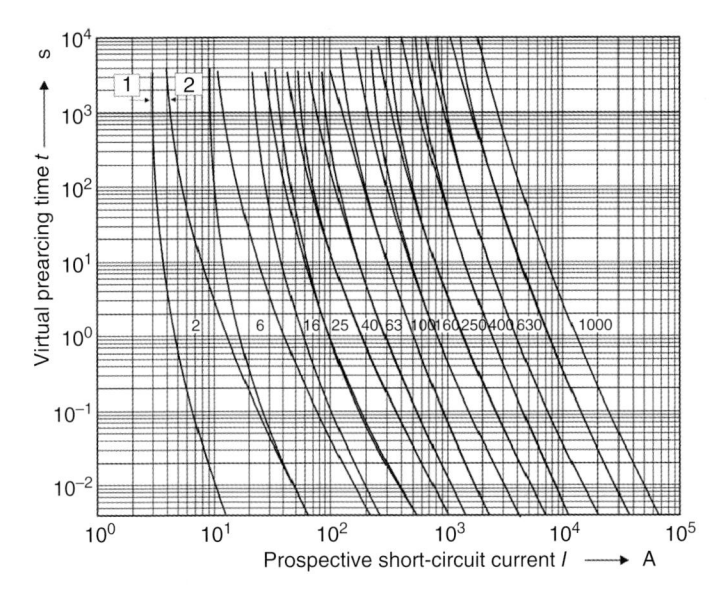

Figure 15.2 Time–current characteristics of limit switch fuses in accordance with IEC 60269-2 (2–1 000 A). (1) Peak current without d.c. part and (2) peak current with d.c. part.

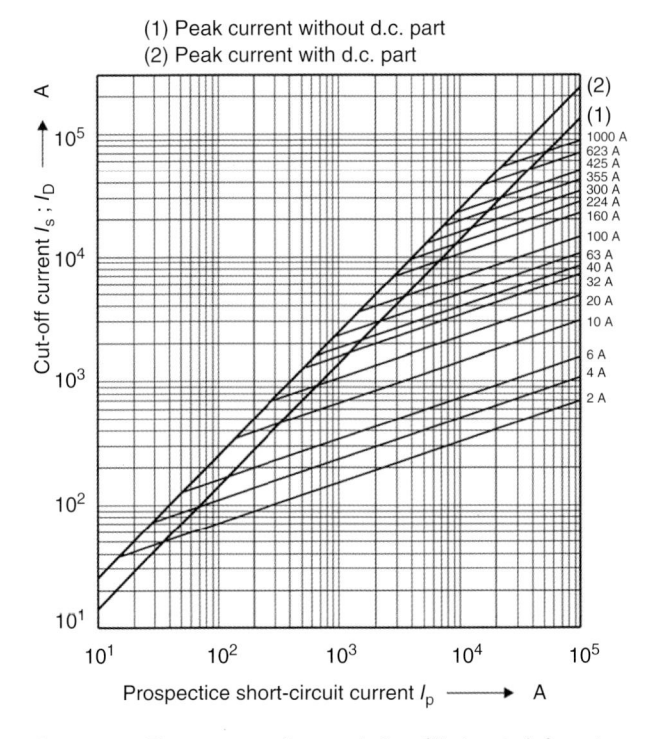

Figure 15.3 Time–current characteristics of limit switch fuses in accordance with IEC 60269-2. (1) Small test current and (2) large test current.

192 | *15 Equipment for Overcurrent Protection*

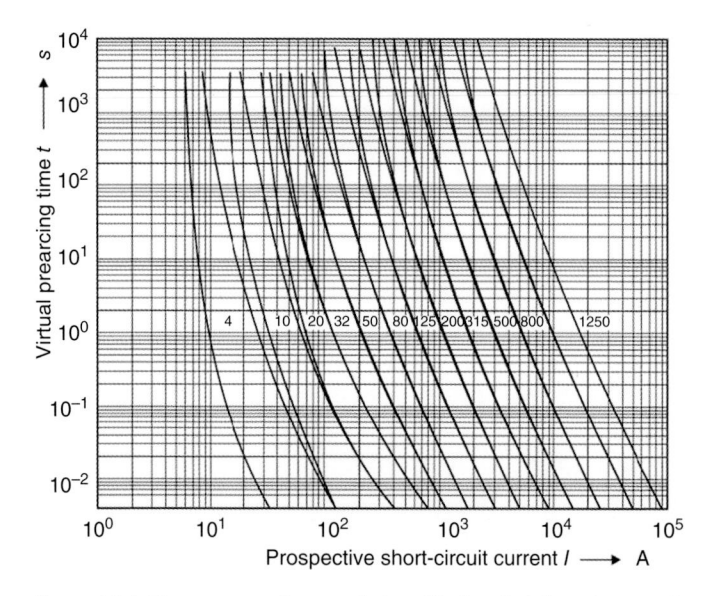

Figure 15.4 Time–current characteristics of limit switch fuses in accordance with IEC 60269-2 (4–1250 A).

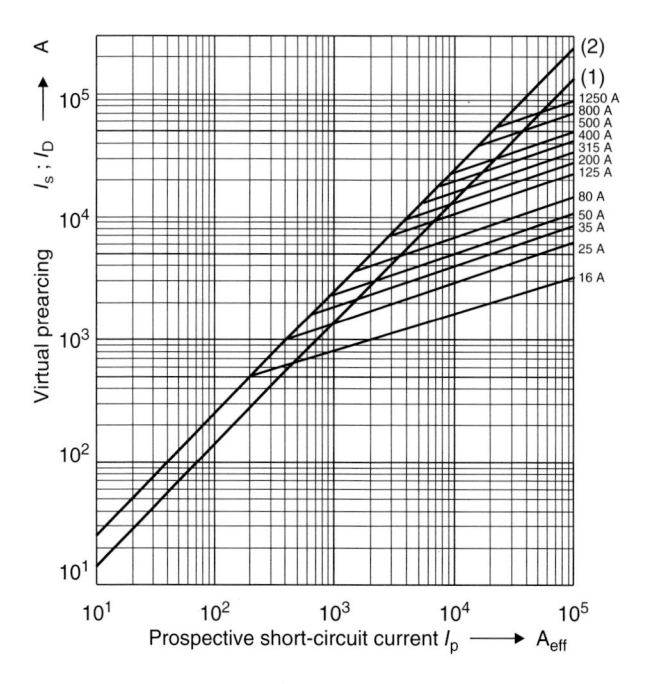

Figure 15.5 Time–current characteristics of limit switch fuses in accordance with IEC 60269-2. (1) Small test current and (2) large test current.

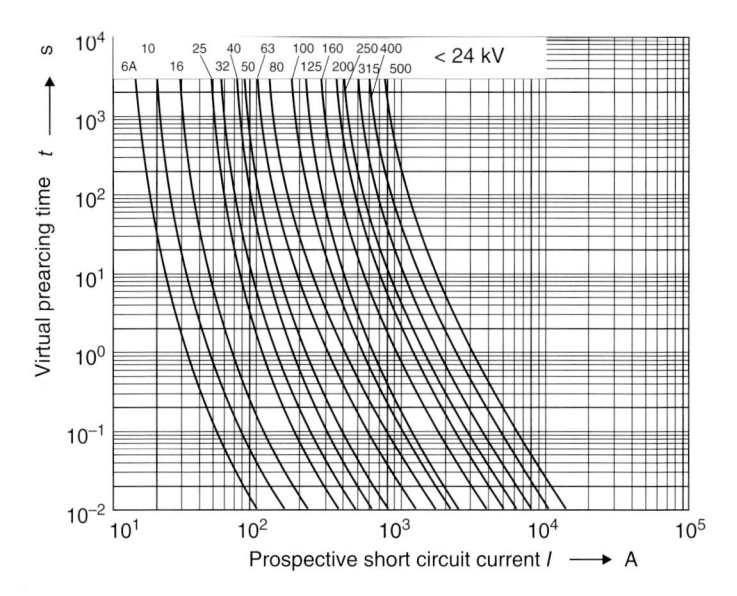

Figure 15.6 Time–current characteristics of high voltage–high power (HH) fuses in accordance with IEC 60282-1.

tripping devices and current-dependent overload tripping devices. The thermal tripping device a is set to the rated current of the motor to be protected and the magnetic tripping device n to 12 times the current (Figures 15.9 and 15.10).

For the protection of control transformers, the current is set 19 times. The dependence of the rated short-circuit breaking capacity I_{cu} and the rated service short-circuit breaking capacity I_{cn} on the rated current I_r and the rated normal current I_r is taken from the respective manufacturer. Whether backup fuses are required must be calculated from the prospective three-phase short-circuit current.

The circuit breakers that protect the motors, cables and conductors, transformers, and other system components satisfy the increased requirements for overloading and short-circuit current protection. They are climatically stable and suitable for operation in closed rooms in which no extreme operating conditions (e.g., dust, caustic vapors, or aggressive gases) are present. Otherwise, metal enclosing is required. Time–current characteristics of circuit breakers are shown in Figure 15.11.

Protective functions and setting possibilities:

1) Current-dependent delayed overload tripping
   The current setting I_e can be set to between 0.5 and 1 times the rated switchgear current I_n in four steps.
2) Short-time delay release of short circuit
   The pickup value I_d can be set in seven steps between two and eight times the value of I_r. The delay time can be selected between 0 and 500 ms and in some cases higher. This allows a time selectivity.
3) Undelayed release of short circuit

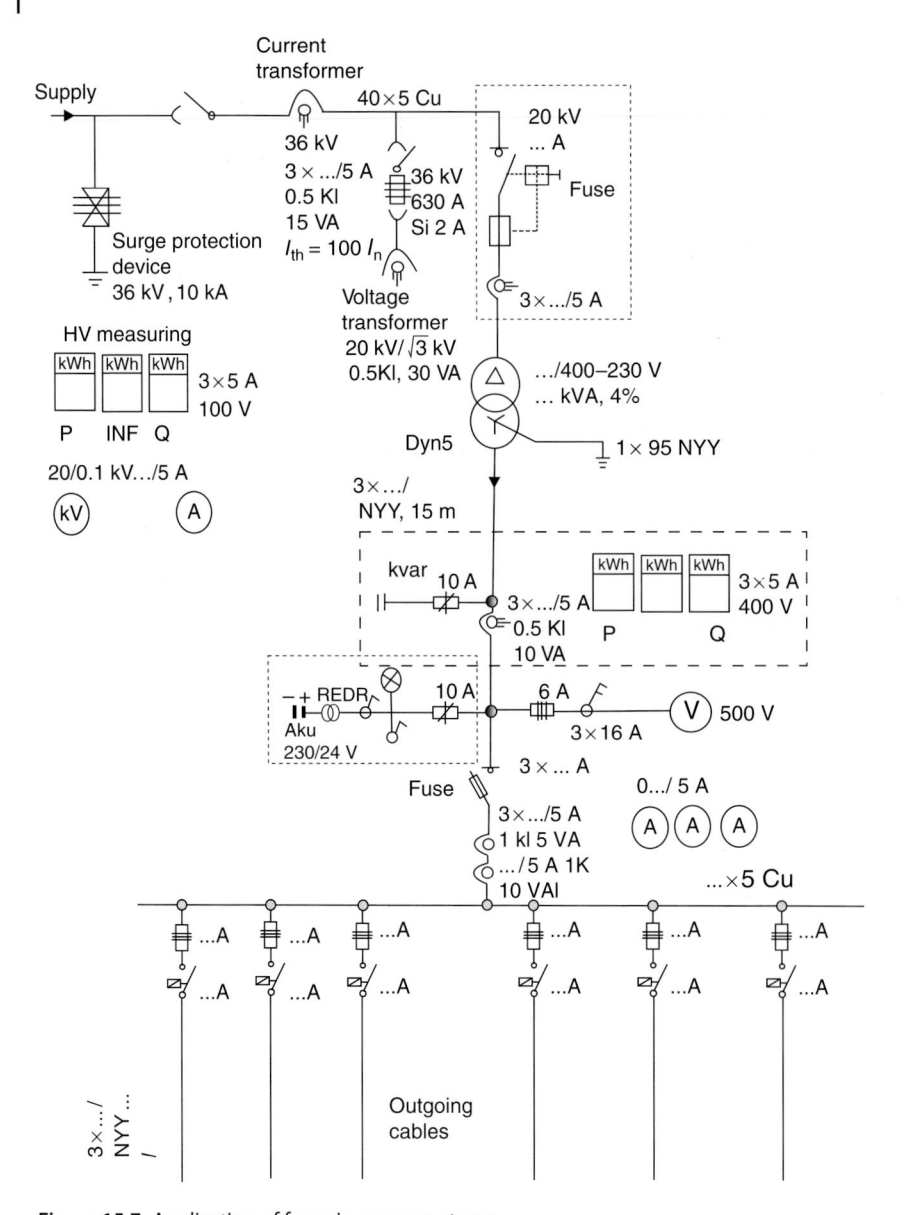

Figure 15.7 Application of fuses in power systems.

The pickup value I_i for an undelayed short-circuit release is set to 15 or to 20 kA.

4) Ground fault tripping

The ground fault tripping device senses fault currents that flow to ground and can cause fires in the system. The adjustable delay time allows the use of several switching steps sequentially.

Equipment for Overcurrent Protection

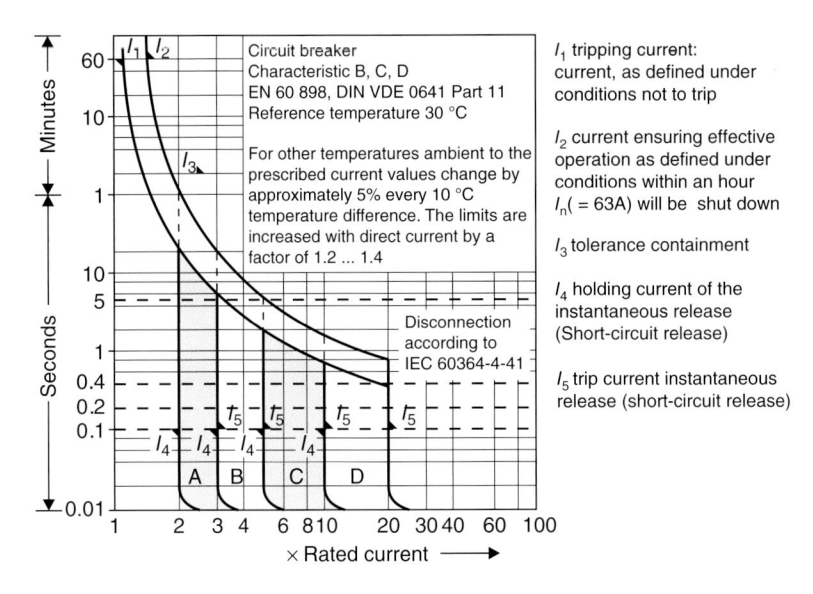

Figure 15.8 Time–current characteristics of miniature circuit breakers in accordance with IEC 60898-1.

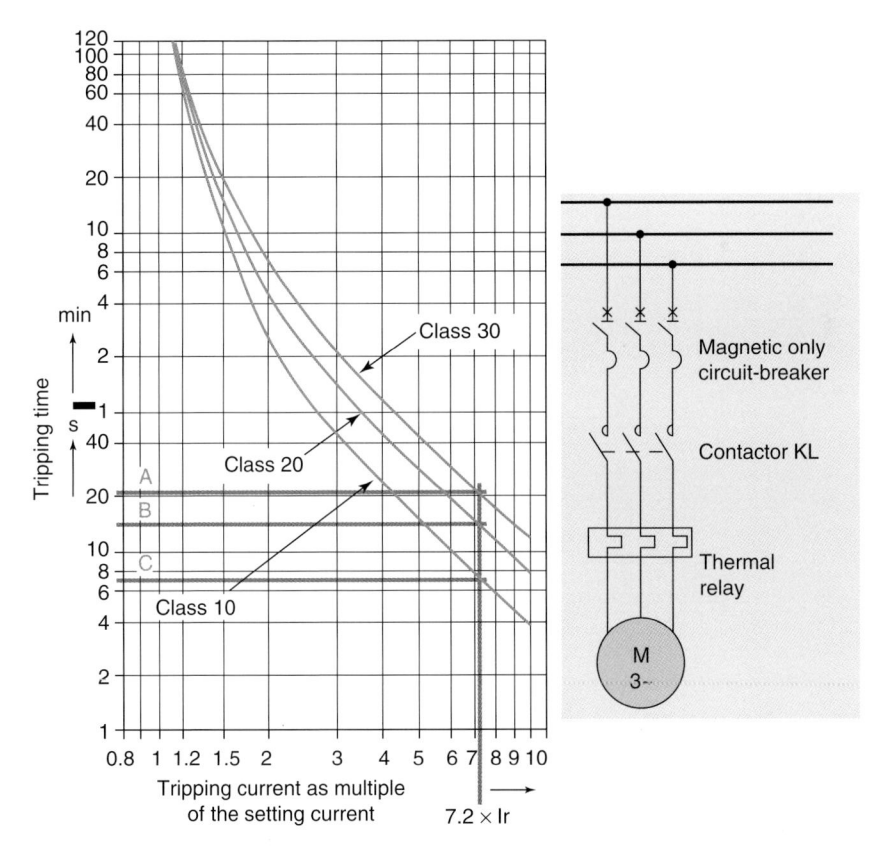

Circuit breaker Characteristic B, C, D
EN 60 898, DIN VDE 0641 Part 11
Reference temperature 30 °C

For other temperatures ambient to the prescribed current values change by approximately 5% every 10 °C temperature difference. The limits are increased with direct current by a factor of 1.2 ... 1.4

Disconnection according to IEC 60364-4-41

I_1 tripping current: current, as defined under conditions not to trip

I_2 current ensuring effective operation as defined under conditions within an hour $I_n(= 63A)$ will be shut down

I_3 tolerance containment

I_4 holding current of the instantaneous release (Short-circuit release)

I_5 trip current instantaneous release (short-circuit release)

Figure 15.9 Tripping curves with thermal relays for different starting classes. (With permission from Siemens.)

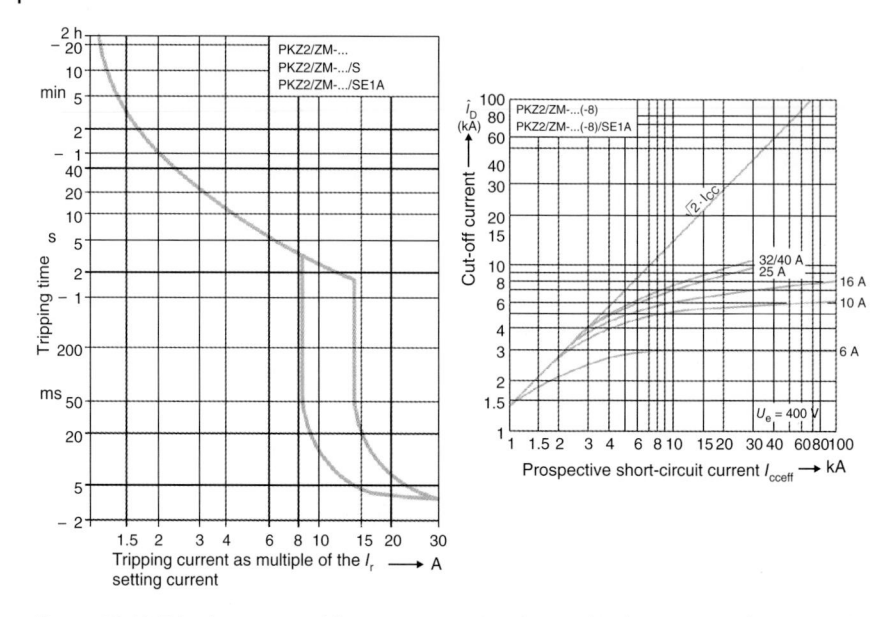

15 Equipment for Overcurrent Protection

Figure 15.10 Tripping curves with motor protection device. (With permission from Eaton.)

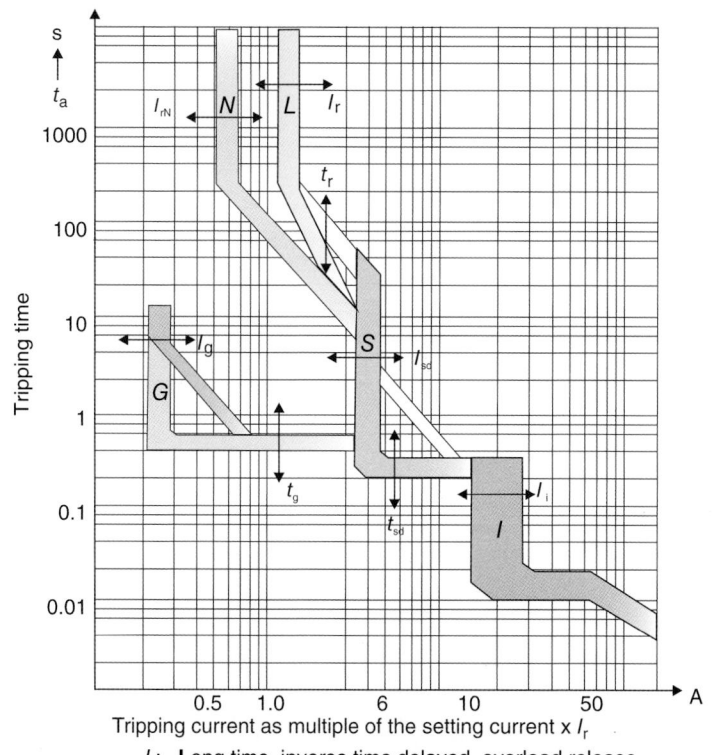

Figure 15.11 Time–current characteristics of circuit breakers. (With permission from Siemens.)

L: **L**ong time, inverse time delayed, overload release
S: **S**hort-time delay short-circuit release
I: **I**nstantaneous short-circuit release
G: **G**round fault protection

Circuit breakers are used as:

1) Input and branch circuit breakers in three-phase distribution systems.
2) Main circuit breakers for machining and processing systems in accordance with IEC 204-1.
3) Emergency Off circuit breakers in accordance with IEC 204-1 with undervoltage release and in conjunction with an Emergency Off control station.
4) Switching devices and protective equipment for motors, transformers, generators, and capacitors.
5) Meshed network circuit breakers in meshed low-voltage networks with several high-voltage inputs.
6) Ground fault protection.

Applications of fuses in power systems are shown in Figure 15.7.
Cut off currents of miniature circuit breakers:
For the overloading range:

A: $I_a = 1.45 \cdot I_n$
B: $I_a = 1.45 \cdot I_n$
C: $I_a = 1.45 \cdot I_n$
D: $I_a = 1.45 \cdot I_n$

For the short-circuit range:

A: $I_a = 3 \cdot I_n$
B: $I_a = 5 \cdot I_n$
C: $I_a = 10 \cdot I_n$
D: $I_a = 20 \cdot I_n$

The meaning of the symbols is as follows:

I_n	nominal current
I_r	rated current
I_e	current setting
I_d	threshold current
t_d	delay time
z	short-time delay release
n	undelayed electromagnetic tripping device
a	current-dependent delayed overload tripping device
t_{vs}	virtual prearcing time
I_c	maximum cut-off current
I_{ts}^2	$I^2 t$ value of fuse
I_{ta}^2	$I^2 t$ break value
I_{cn}	rated short-circuit breaking capacity
I_{eff}	effective value of uninfluenced short-circuit current.

16

Short-Circuit Currents in DC Systems

The most important power generator is the three-phase synchronous generator. Consequently, short-circuit currents in electrical networks are supplied from the generator feed-ins. In power stations and in industry DC systems such as DC motors, permanently installed lead batteries, smoothing capacitors, and converter arms are frequently found. The short-circuit current calculation for these systems is described in IEC 61660-1 and will be discussed briefly here. This standard contains a method for the calculation of short-circuit currents in these systems [18]. However, it is also possible to use another method.

The short-circuit currents, pure resistances, and inductances can be measured or obtained from model experiments. In DC systems, it is also necessary to distinguish between short-circuit currents of different magnitudes:

1) The smallest short-circuit current is used as the basis for the selection of overcurrent protection equipment. For the calculation of the smallest short-circuit currents, we must consider under which circuit and operating conditions the smallest short-circuit current flows:
   - The resistances of the conductors must be considered in relation to the maximum operating temperature.
   - The transition resistances must be considered.
   - The short-circuit current of the current converter is the rated value of the current limiting.
   - The batteries are discharged by up to the discharge voltage of 1.8 V/cell, unless otherwise specified by the respective manufacturer.
   - Diodes used to decouple power supply units must be considered.
2) The largest short-circuit current serves to dimension the electrical operational equipment. For the calculation of the largest short-circuit currents, we must consider under which circuit and operating conditions the largest short-circuit current flows:
   - The resistances of the conductors must be considered in relation to a temperature of 20 °C.
   - The transition resistances of the busbars can be neglected.

Short Circuits in Power Systems: A Practical Guide to IEC 60909-0, Second Edition. Ismail Kasikci.
© 2018 Wiley-VCH Verlag GmbH & Co. KGaA. Published 2018 by Wiley-VCH Verlag GmbH & Co. KGaA.

16 Short-Circuit Currents in DC Systems

- The regulation for limiting the current converter current is not functioning.
- The batteries are fully charged.

Figure 16.1 illustrates the typical short-circuit currents for different sources. Figure 16.2 shows the standardized approximation functions for all current paths. Figure 16.3 represents a DC system with four sources.

The calculation procedure for this circuit (Figure 16.3) will now be briefly explained in the following section (see IEC 61660-1 for more detailed information).

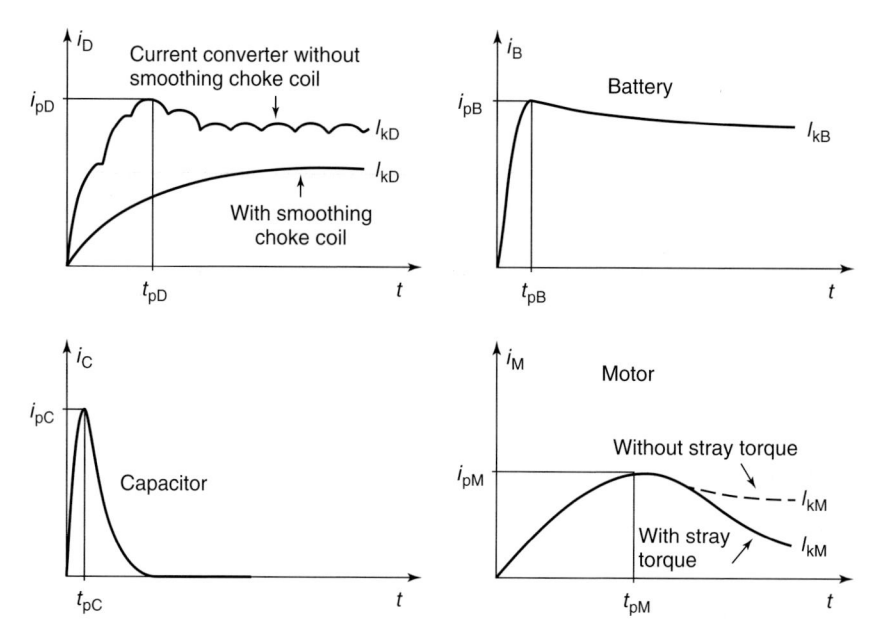

Figure 16.1 Typical paths for short-circuit currents. (See also Ref. [18].)

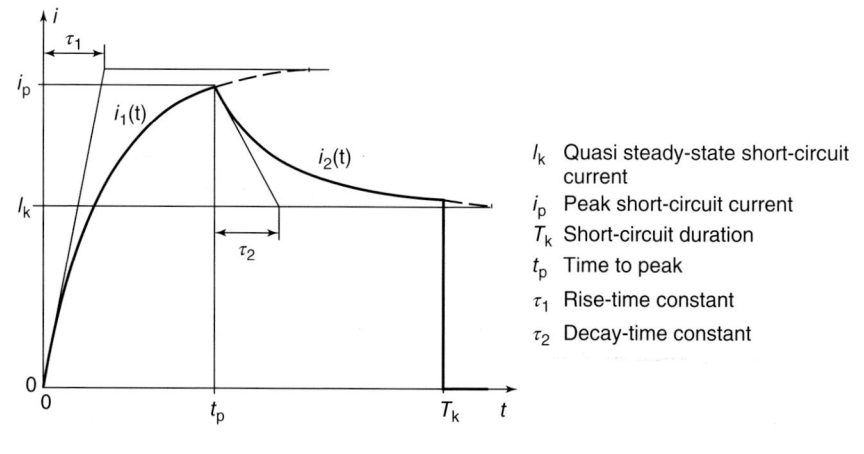

Figure 16.2 Standardized approximation functions. (See also Ref. [18].)

I_k Quasi steady-state short-circuit current
i_p Peak short-circuit current
T_k Short-circuit duration
t_p Time to peak
τ_1 Rise-time constant
τ_2 Decay-time constant

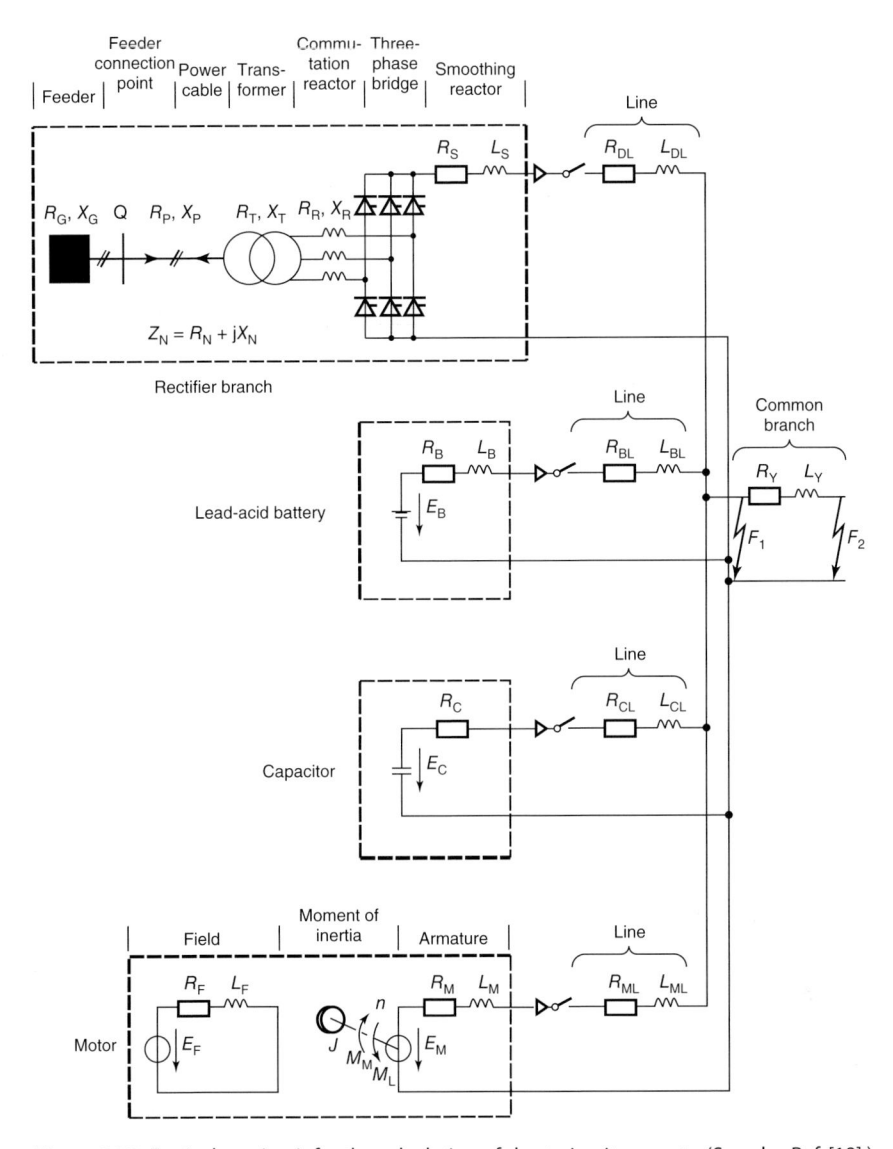

Figure 16.3 Equivalent circuit for the calculation of short-circuit currents. (See also Ref. [18].)

16.1 Resistances of Line Sections

The pure resistance per unit length is

$$R' = \frac{2 \cdot \rho}{S} \tag{16.1}$$

The touch resistance of screw connections for the smallest short-circuit current is

$$R = \frac{14 \cdot \rho \cdot d}{S} \tag{16.2}$$

where:
S   cross-section in mm²
R'   resistance per unit length in Ω/m
ρ   resistivity in Ω · mm²/m
d   width of the rectangular conductor.

The inductance per unit length of the individual cable is calculated from

$$L' = \frac{\mu_0}{\pi}\left(\ln\frac{a}{r} + \frac{1}{4}\right) \tag{16.3}$$

The inductance per unit length of busbar is calculated, for $a > b$, from

$$L' = \frac{\mu_0}{\pi}\left(\frac{3}{2} + \ln\frac{a}{b+h}\right) \tag{16.4}$$

where
b   width of the conductor in m,
h   height of the conductor in m
r   diameter in m
μ_0 $4\pi \times 10^{-7}$ H/m permeability
a   average distance between the conductors in m
L'   inductance per unit length in H/m.

16.2 Current Converters

The network impedances for the determination of the smallest short-circuit current can be obtained from Chapter 8. The resistance and the inductance of the current converter are as follows (Figure 16.4):

$$R_{DBr} = R_S + R_{DL} + R_Y \tag{16.5}$$
$$L_{DBr} = L_S + L_{DL} + L_Y \tag{16.6}$$

where
R_S, L_S resistance and inductance of the saturated choke coil
R_Y, L_Y resistance and inductance of the coupling branch
R_{DL}, L_{DL} resistance and inductance of the conductor in current converter.

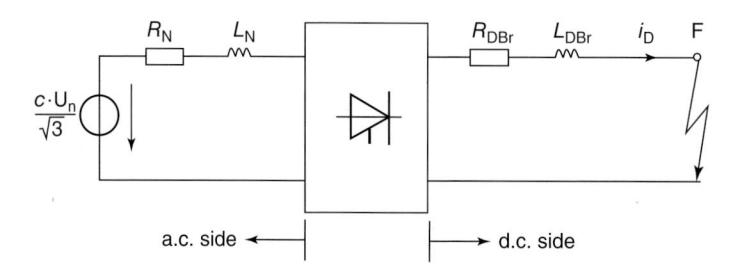

Figure 16.4 Equivalent circuit of a converter arm. (See also Ref. [18].)

The peak short-circuit current is

$$i_{pD} = \kappa \cdot I_{kD} \quad \text{with} \quad \kappa = f\left[\frac{R_N}{X_N}\left(1 + \frac{2}{3} \cdot \frac{R_{DBr}}{R_N}\right); \frac{L_{DBr}}{L_N}\right] \tag{16.7}$$

The steady-state short-circuit current is

$$I_{kD} = \lambda_D \cdot \frac{1}{\sqrt{3}} \cdot \frac{\sqrt{2} \cdot c \cdot U_n}{Z_N} \quad \text{with} \quad \lambda_D = f\left[\frac{R_N}{X_N}; \frac{R_{DBr}}{R_N}\right] \tag{16.8}$$

16.3 Batteries

The simplified equivalent circuit of a permanently installed lead battery for the short-circuit current is shown in Figure 16.5.

For the nominal voltage U_{nb} of a battery:

$$U_{nB} = 2.0\,\text{V/cell} \tag{16.9}$$

No-load voltage of the charged battery:

$$E_B = 1.05 \cdot U_{nB} \tag{16.10}$$

No-load voltage of the uncharged battery:

$$E_B = 0.9 \cdot U_{nB} \tag{16.11}$$

Resistance of the battery:

$$R_{BBr} = 0.9 \cdot R_B + R_{BL} + R_Y \tag{16.12}$$

Inductance of the battery:

$$L_{BBr} = L_B + L_{BL} + L_Y \tag{16.13}$$

Peak short-circuit current of the battery:

$$I_{pB} = \frac{E_B}{R_{BBr}} \tag{16.14}$$

Steady-state short-circuit current of the battery:

$$I_{kB} = 0.95 \cdot \frac{E_B}{R_{BBr} + 0.1 \cdot R_B} \tag{16.15}$$

where
E_B no-load voltage of the battery
L_B inductance of the battery

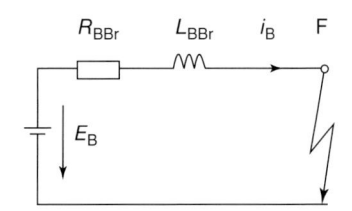

Figure 16.5 Equivalent circuit of a battery. (See also Ref. [18].)

L_{BBr} total inductance of the battery
R_{BL} resistance of a battery conductor
L_{BL} inductance of a battery conductor
R_{BY} resistance of the battery coupling branch
L_Y inductance of the battery coupling branch
R_B resistance of the battery
R_{BBr} total resistance of the battery
U_{nB} nominal voltage of the battery.

16.4 Capacitors

The equivalent circuit and the short-circuit parameters of a capacitor are shown in Figure 16.6.
Here:

$$R_{CBr} = R_C + R_{CL} + R_Y \tag{16.16}$$

$$L_{CBr} = L_C + L_{CL} + L_Y \tag{16.17}$$

Peak short-circuit current of the capacitor:

$$i_{pC} = \kappa_C \cdot \frac{E_C}{R_{CBr}} \quad \text{with} \quad R_C = f\left[\frac{2 \cdot L_{CB}}{R_{CB}}; \frac{1}{\sqrt{L_{CE} \cdot C}}\right] \tag{16.18}$$

For $L_{CBr} = 0$, $R_C = 1$ and $I_{kC} = 0$, where
E_C voltage of capacitor before the occurrence of short circuit
L_C inductance of the capacitor
L_{CBr} total inductance of the capacitor
R_{CL} resistance of a capacitor conductor
L_{CL} inductance of a capacitor conductor
R_{CY} resistance of the coupling branch for capacitor
L_{CY} inductance of the coupling branch for capacitor
R_C resistance of the capacitor
R_{CBr} total resistance of the capacitor.

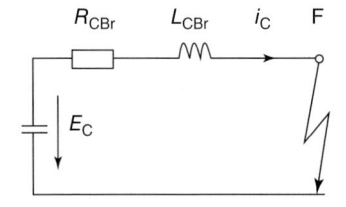

Figure 16.6 Equivalent circuit of a capacitor. (See also Ref. [18].)

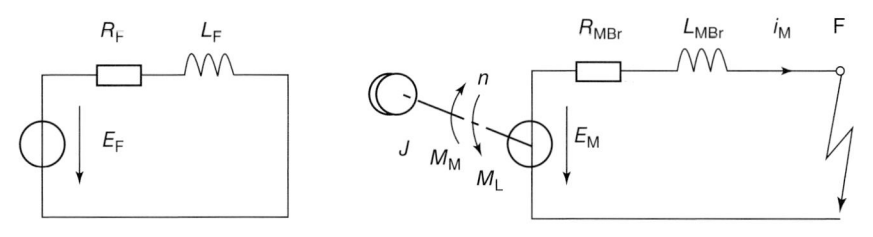

Figure 16.7 Equivalent circuit of an externally excited DC motor. (See also Ref. [18].)

16.5 Direct Current Motors

The equivalent circuit of an externally excited DC motor is shown in Figure 16.7. For the resistances and inductances, the following hold true:

$$R_{MBr} = R_M + R_{ML} + R_Y \tag{16.19}$$

$$L_{MBr} = L_M + L_{ML} + L_Y \tag{16.20}$$

$$\tau_M = \frac{L_{MBr}}{R_{MBr}} \tag{16.21}$$

DC motors can be neglected when

$$\sum I_{rM} < 0.01 \cdot I_{kD} \tag{16.22}$$

Peak short-circuit current of the DC motor:

$$i_{pM} = \kappa_M \cdot \frac{U_{rM} - I_{rM} \cdot R_M}{R_{MBr}} \tag{16.23}$$

Steady-state short-circuit current of the DC motor:

$$I_{kM} = \frac{L_F}{I_{OF}} \cdot \frac{U_{rM} - I_{rM} \cdot R_M}{R_{MBr}} \tag{16.24}$$

where
L_M inductance of the DC motor
L_{MBr} total inductance of the DC motor
R_{ML} resistance of the DC motor conductor
L_{ML} inductance of the DC motor conductor
R_{MY} resistance of the coupling branch for DC motor
L_Y inductance of the coupling branch for DC motor
R_M resistance of the DC motor
U_{rM} rated voltage of the DC motor
R_{MBr} total resistance of the DC motor

17

Power Flow Analysis

The calculation of networks is performed in stationary operations. The power flow analysis considers the active and reactive power flows in a network of predetermined infeed and loads. The power flow enables network planning, optimal network design with respect to voltage drop, and selection of cable cross-sections. Further, network operation allows the economic and technical point of optimal network management in the system.

A simple power flow calculation specifically provides the following:

- node voltages and currents of the resources in terms of magnitude and phase
- active and reactive power flows on the lines
- transmission losses in cables and transformers.

Electric power systems are complex, multiphase networks in which the capacitive and magnetic coupling exist, and in equivalent circuits, they are considered by capacitances and mutual inductances. Mathematical modeling results then coupled linear equations. To simplify the calculations, the multiphase system was initially decoupled by a similarity transformation, thereby the task in the balanced operation of calculating a single-phase network was reduced. Based on the two Kirchhoff's laws, the mathematical models can be established in the form of linear equations. They contain all known information on topology, branch impedances, admittances, supply voltages, and loads. The solution delivers the unknown branch currents and voltages. As a result, the current node process is obtained, for example, the unknown node voltages, from their differences, the branch voltages, and over the Ohm's law, the branch currents. The mesh method yields the mesh currents from which the branch current superimposes itself, and through Ohm's law, the branch voltages can be calculated. In weakly meshed networks, the mesh method has certain significance. It provides a lower number of equations and allows manual calculation of small networks. The majority of the calculations are performed using Kirchhoff's current law (KCL) for networks on the node process, which sets the predetermined or desired quantities in relationship. Ohm's current law leads to the so-called nodal admittance matrix, from which the node hybrid matrix and the nodal impedance matrix can be derived. According to the method, the analysis of a network node leads to a linear equation

Short Circuits in Power Systems: A Practical Guide to IEC 60909-0, Second Edition. Ismail Kasikci.
© 2018 Wiley-VCH Verlag GmbH & Co. KGaA. Published 2018 by Wiley-VCH Verlag GmbH & Co. KGaA.

system with the node voltages as unknown and the loads as input variables. The elements of the associated node matrix are the admittances of the network.

The analytical calculation of the three-phase systems based on simplifications leads to real solutions. This procedure is described in low-voltage power systems and even in medium-power systems. However, high-voltage power systems and a large part of medium-voltage networks are strongly intermeshed with very different line impedances between nodes. Therefore, complex calculations are performed. An important characteristic of a meshed network with lines and nodes (without the reference node) is the degree of meshing [14].

For the load flow calculation of the individual network nodes, specifications on the type of loads need to be made. These nodes are divided into load or consumer nodes and feed or generator nodes. At least one node, the slack node, is available where the voltage magnitude and angle are set. At this node, active and reactive power is set up so that a balance exists in all networks between the feed-in and detached services. The slack node must be able to take over the allocated power account. The system of equations is not linear because the power depends on the square of the voltage. For load flow calculation, an iteration process can be used for solving the nonlinear system of equations.

17.1 Systems of Linear Equations

For the solution of linear and nonlinear algebraic equations, a variety of different methods are available, which differ by storage space, speed, accuracy, and programming. We differentiate between the direct and iterative processes. Iterative methods are used in solving the nonlinear equations [20, 21].

A distinction is made between the current iteration with the Gauss–Seidel method and the Newton–Raphson method.

In the following description, an introductory overview of the methods employed in the power calculation is introduced.

First, consider the following equation system:

$$\begin{aligned} a_{11}x_1 + a_{12}x_2 + a_{13}x_3 &= y_1 \\ a_{21}x_1 + a_{22}x_2 + a_{23}x_3 &= y_2 \\ a_{31}x_1 + a_{32}x_2 + a_{33}x_3 &= y_3 \\ &\vdots \\ a_{m1}x_1 + a_{m2}x_2 + a_{mn}x_3 &= y_m \end{aligned} \tag{17.1}$$

The variables include the designations:

x_1, x_2, x_3: unknowns
$a_{11}, a_{12}, a_{13},..., a_{33}$: coefficients of the variables
y_1, y_2, y_3: known parameters.

We summarize the coefficients a_{ij} of the linear system to a coefficient matrix **A**, the unknown variables x_1, x_2, x_3 to a column vector **x**, and absolute variables y_1,

y_2, y_3 to a column vector **y**. Then, the system can be written in a matrix form.

$$\mathbf{A} = [a_{ij}] = \begin{bmatrix} a_{11} & a_{12} & a_{13} \\ a_{21} & a_{22} & a_{23} \\ a_{31} & a_{32} & a_{33} \end{bmatrix} \quad \text{according to } \mathbf{x}$$

$$\mathbf{x} = \begin{bmatrix} x_1 \\ x_2 \\ x_3 \end{bmatrix} \quad \text{and} \quad \mathbf{y} \quad \mathbf{y} = \begin{bmatrix} y_1 \\ y_2 \\ y_3 \end{bmatrix} \tag{17.2}$$

The **m·n** matrix describes a matrix with **m** rows (row vectors) and **n** columns (column vectors).

Equation (17.1) can be written as:

$$A\mathbf{x} = \mathbf{y} \tag{17.3}$$

Special cases to note:

For row matrix: **m** = 1 **A** [1, **n**] = **A**[**n**]
For column matrix: **n** = 1 **A** [**m**, 1] = **A**[**m**].

If **m** = **n**, **A** is an **n** × **n** *square matrix*.

$$\mathbf{A} = \mathbf{A}[\mathbf{m}, \mathbf{m}] = \begin{bmatrix} a_{11} & a_{12} & a_{1m} \\ a_{21} & a_{22} & a_{2m} \\ a_{m1} & a_{m2} & a_{mm} \end{bmatrix} \tag{17.4}$$

The matrix **A** is of the order *m*. The elements for which the matrix rows and columns are identical, $i = j$, are called the main diagonal elements. All other elements for which the matrix rows and columns are not identical, $i \neq j$, are called non-diagonal elements or adjacent diagonal elements.

If we multiply square matrices **A** and **B**, a unit matrix **E** is obtained:

$$\mathbf{A} \cdot \mathbf{B} = \mathbf{E} \tag{17.5}$$

Or, the inverse of the matrices **A**:

$$\mathbf{B} = \mathbf{A}^{-1} \tag{17.6}$$

17.2 Determinants

A matrix can assign a certain number, namely, its determinant. For example, we need it in the implementation of multidimensional extreme value tasks where we have to look exactly like the one-dimensional case, if the calculated second derivative is positive. The second derivative will be given in the form of a matrix, and some kind of positivity matrices will be needed. This is not easy, and one can use determinants to solve linear equations.

For the introduction of the concept of determinants, we consider a system of linear equations with two equations and two unknowns as:

(1) $a_{11}x_1 + a_{12}x_2 = b_1$
(2) $a_{21}x_1 + a_{22}x_2 = b_2$ \hfill (17.7)

It is to be examined under which preconditions this equation system has a unique solution, exactly one solution.

Thus, Equation (17.1) is multiplied by a_{22} and Equation (17.2) by a_{12}, and all equations are added.

$$\begin{aligned}(1) \quad & a_{11}x_1 + a_{12}x_2 = b_1 \cdot a_{22} \\ (2) \quad & a_{21}x_1 + a_{22}x_2 = b_2 \cdot a_{12}\end{aligned} \tag{17.8}$$

$$\begin{aligned}& a_{11}a_{22}x_1 + a_{12}a_{22}x_2 = b_1 \cdot a_{22} \\ & a_{21}a_{12}x_1 + a_{12}a_{22}x_2 = b_2 \cdot a_{12}\end{aligned} \tag{17.9}$$

Now, subtract the second equation from the first. Then:

$$\begin{aligned}(a_{11}a_{22} - a_{12}a_{21})x_1 &= b_1 \cdot a_{22} - b_2 a_{12} \\ (a_{11}a_{22} - a_{12}a_{21})x_2 &= b_2 \cdot a_{11} - b_1 a_{21}\end{aligned} \tag{17.10}$$

If the expression in brackets is not zero, it follows the solution:

$$\begin{aligned}x_1 &= \frac{b_1 \cdot a_{22} - b_2 a_{12}}{a_{11}a_{22} - a_{12}a_{21}} \\ x_2 &= \frac{b_2 \cdot a_{11} - b_1 a_{21}}{a_{11}a_{22} - a_{12}a_{21}}\end{aligned} \tag{17.11}$$

If Equation (7) is written in a coefficient matrix, then:

$$\mathbf{A} = \begin{bmatrix} a_{11} & a_{12} \\ a_{21} & a_{22} \end{bmatrix} \tag{17.12}$$

Thus, there is a denominator by "crossover multiplying" the matrix entries, and it is called the determinant of A, and the *second order* is denoted and defined by the following:

$$\det A = \det \begin{bmatrix} a_{11} & a_{12} \\ a_{21} & a_{22} \end{bmatrix} = a_{11}a_{22} - a_{12}a_{21} \tag{17.13}$$

$$D = a_{11}a_{22} - a_{12}a_{21} \tag{17.14}$$

$$D_1 = \begin{vmatrix} b_1 & a_{12} \\ b_2 & a_{22} \end{vmatrix} = b_1 a_{22} - a_{22} b_2$$

$$D_2 = \begin{vmatrix} a_{11} & b_1 \\ a_{21} & b_2 \end{vmatrix} = a_{11} b_2 - b_1 a_{21} \quad D \neq 0 \tag{17.15}$$

$$x_1 = \frac{D_1}{D}, \quad x_2 = \frac{D_2}{D} \tag{17.16}$$

$$x_1 = \frac{\det \begin{bmatrix} b_1 & a_{12} \\ b_2 & a_{22} \end{bmatrix}}{\det A} \tag{17.17}$$

and

$$x_2 = \frac{\det \begin{bmatrix} a_{11} & b_1 \\ a_{21} & b_2 \end{bmatrix}}{\det A} \tag{17.18}$$

Where $a_{11}a_{22}$ is the main diagonal and $a_{12}a_{21}$ is the off-diagonal. Therefore, there is a solution formula that allows a clear notation for systems of equations with two equations and two unknowns. Each solution is the quotient of two determinants, where the denominator is always the determinant of the coefficient matrix **A** is in counter the *j*th solution, just x_j is the *j*th column of **A** by the right side $\begin{pmatrix} b_1 \\ b_2 \end{pmatrix}$ is replaced.

However, systems of equations have more than two unknowns, which is why we determine for **n × n** matrices. Consequently, one should be able to solve linear equations of *third-order determinants*, such as

$$a_{11}x_1 + a_{12}x_2 + a_{13}x_3 = b_1$$
$$a_{21}x_1 + a_{22}x_2 + a_{23}x_3 = b_2 \tag{17.19}$$
$$a_{31}x_1 + a_{32}x_2 + a_{33}x_3 = b_3$$

$$\mathbf{A} = \begin{bmatrix} a_{11} & a_{12} & a_{13} \\ a_{21} & a_{22} & a_{23} \\ a_{31} & a_{32} & a_{33} \end{bmatrix} \tag{17.20}$$

Then, the following solution results:

$$D = a_{11}a_{22}a_{33} + a_{12}a_{23}a_{31} + a_{13}a_{21}a_{32}$$
$$- a_{13}a_{22}a_{31} - a_{11}a_{23}a_{32} - a_{12}a_{21}a_{33} \tag{17.21}$$

$$x_1 = \frac{\det \begin{bmatrix} b_1 & a_{12} & \cdots & a_{1n} \\ b_2 & a_{22} & \cdots & a_{12} \\ \vdots & \vdots & & \vdots \\ b_n & a_{n2} & \cdots & a_{nn} \end{bmatrix}}{\det \mathbf{A}} \tag{17.22}$$

and

$$x_2 = \frac{\det \begin{bmatrix} a_{11} & \cdots & a_{12} & \cdots & b_1 \\ a_{21} & \cdots & a_{22} & \cdots & b_2 \\ \vdots & \vdots & & \vdots & \\ a_{n1} & \cdots & a_{n2} & \cdots & b_n \end{bmatrix}}{\det \mathbf{A}} \tag{17.23}$$

A system of linear equations can be solved with *n* equations and *n* unknowns by dividing two determinants for each unknown. This formula is called Cramer's rule.

$$x_1 = \frac{D_1}{D} \quad x_2 = \frac{D_2}{D} \quad x_3 = \frac{D_3}{D} \quad D \neq 0 \tag{17.24}$$

The determinants are given by

$$D_1 = \begin{bmatrix} b_1 & a_{12} & a_{13} \\ b_2 & a_{22} & a_{23} \\ b_3 & a_{32} & a_{33} \end{bmatrix} \quad D_2 = \begin{bmatrix} a_{12} & b_1 & a_{13} \\ a_{22} & b_2 & a_{23} \\ a_{32} & b_3 & a_{33} \end{bmatrix} \quad D_3 = \begin{bmatrix} a_{11} & a_{12} & b_1 \\ a_{21} & a_{22} & b_2 \\ a_{31} & a_{32} & b_3 \end{bmatrix} \tag{17.25}$$

In this section, the problem of linear and nonlinear equations in power systems was discussed, and the calculation methods were introduced. Cramer's rule is an especially helpful application in the power systems to solve networks. The next section deals with network matrices.

17.3 Network Matrices

In this section, the methods of admittance matrix, impedance matrix, and hybrid matrix will be discussed [14, 22].

17.3.1 Admittance Matrix

Figure 17.1 shows a network section with the nodes i and k and the reference point 0. Each node is associated with each node voltage. The load current in a node is denoted by $\underline{I}_{ii}$.

We look at the nodal point i, which is connected with other nodes n of the network. The node rule specifies that:

$$\sum_{k=0}^{n} \underline{I}_{ik} = \underline{I}_{i0} + \underline{I}_{i1} + \underline{I}_{i2} + \cdots + \underline{I}_{ii} + \cdots + \underline{I}_{in} = 0 \qquad (17.26)$$

The currents in the branches:

$$\underline{I}_{ik} = -\underline{Y}_{ik} \cdot (\underline{U}_i - \underline{U}_k)$$
$$\underline{Y}_{i0} \cdot (\underline{U}_i - \underline{U}_0) + \underline{Y}_{i1} \cdot (\underline{U}_i - \underline{U}_1) + \cdots + \underline{I}_{ii} + \cdots + \underline{Y}_{in} \cdot (\underline{U}_i - \underline{U}_n) = 0 \qquad (17.27)$$

If multiplied and summarized, the equation is as follows:

$$\underline{U}_i(\underline{Y}_{i0} + \underline{Y}_{i1} + \cdots + \underline{Y}_{in}) - \underline{Y}_{i0}\underline{U}_0 - \underline{Y}_{i1}\underline{U}_1 - \cdots - \underline{Y}_{in} \cdot \underline{U}_n = -\underline{I}_{ii} \qquad (17.28)$$

The equation is multiplied by (−1) and ordered by indices of voltages, which are then given by nodes' equation of the node i:

$$\underline{Y}_{i0}\underline{U}_0 + \underline{Y}_{i1}\underline{U}_1 + \underline{Y}_{i2}\underline{U}_2 + \cdots + \underline{Y}_{ii}\underline{U}_i + \cdots + \underline{Y}_{in}\underline{U}_n = \underline{I}_{ii} \qquad (17.29)$$

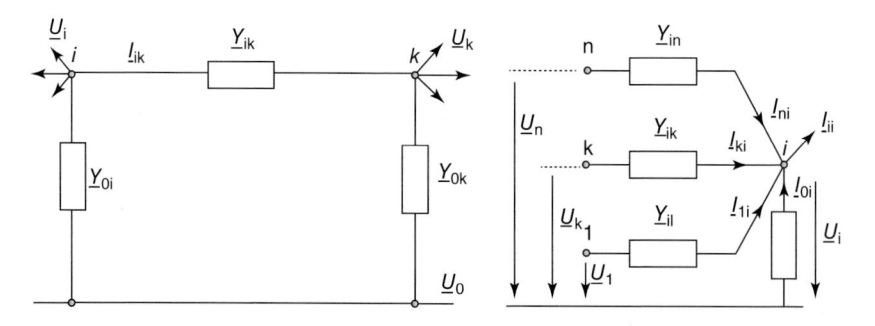

Figure 17.1 Network section with the nodes i and k and the reference point 0.

If this process is repeated for all other nodes, the following linear algebraic system of equations is used:

$$\underline{Y}_{00}\underline{U}_0 + \underline{Y}_{01}\underline{U}_1 + \underline{Y}_{02}\underline{U}_2 + \cdots + \underline{Y}_{0i}\underline{U}_i + \cdots + \underline{Y}_{0n}\underline{U}_n = \underline{I}_{00}$$
$$\underline{Y}_{10}\underline{U}_0 + \underline{Y}_{11}\underline{U}_1 + \underline{Y}_{12}\underline{U}_2 + \cdots + \underline{Y}_{1i}\underline{U}_i + \cdots + \underline{Y}_{1n}\underline{U}_n = \underline{I}_{11}$$
$$\vdots$$
$$\underline{Y}_{n0}\underline{U}_0 + \underline{Y}_{n1}\underline{U}_1 + \underline{Y}_{n2}\underline{U}_2 + \cdots + \underline{Y}_{ni}\underline{U}_i + \cdots + \underline{Y}_{nn}\underline{U}_n = \underline{I}_{nn} \quad (17.30)$$

A matrix notation of the equation:

$$\begin{bmatrix} \underline{Y}_{00} & \underline{Y}_{01} & \underline{Y}_{02} & \cdots & \underline{Y}_{0i} & \cdots & \underline{Y}_{0n} \\ \underline{Y}_{10} & \underline{Y}_{11} & \underline{Y}_{12} & \cdots & \underline{Y}_{1i} & \cdots & \underline{Y}_{1n} \\ \vdots & & & & & & \\ \underline{Y}_{n0} & \underline{Y}_{n1} & \underline{Y}_{n2} & \cdots & \underline{Y}_{ni} & \cdots & \underline{Y}_{nn} \end{bmatrix} \begin{bmatrix} \underline{U}_0 \\ \underline{U}_1 \\ \vdots \\ \underline{U}_n \end{bmatrix} = \begin{bmatrix} \underline{I}_{00} \\ \underline{I}_{11} \\ \vdots \\ \underline{I}_{nn} \end{bmatrix} \quad (17.31)$$

In vector notation:

$$\underline{\mathbf{Y}} \cdot \underline{\mathbf{U}} = \underline{\mathbf{I}} \quad (17.32)$$

$\underline{\mathbf{Y}}$ represents the square admittance matrix (coefficient matrix), $\underline{\mathbf{U}}$ represents the vector of node voltages, and $\underline{\mathbf{I}}$ represents the vector of nodal currents.

17.3.2 Impedance Matrix

The impedance matrix provides the inverse matrix $\underline{\mathbf{Z}} = \underline{\mathbf{Y}}^{-1}$ of the admittance matrix and corresponds to a complete exchange of variables. Then, the matrix equation of a network is

$$\underline{\mathbf{Z}} \cdot \underline{\mathbf{I}} = \underline{\mathbf{U}} \quad (17.33)$$

The impedance matrix allows the direct calculation of node voltages given by multiplication node currents with the matrix elements. In a voltage vector, the differences of the individual node voltages $\underline{U}_i$ to node voltage $\underline{U}_0$ of the reference node can be seen. The main application of the hybrid and the impedance matrices is located in the short-circuit and power flow calculation. This method allows a faster convergence of iterative solution but needs very large storage requirements since all elements are different from zero.

The resolution of Equation (17.33) to the node voltages leads to the equation system with the node impedance matrix $\underline{\mathbf{Z}}$, which is in matrix form:

$$\begin{bmatrix} \underline{U}_0 \\ \underline{U}_1 \\ \vdots \\ \underline{U}_n \end{bmatrix} = \begin{bmatrix} \underline{Z}_{00} & \underline{Z}_{01} & \underline{Z}_{02} & \cdots & \underline{Z}_{0i} & \cdots & \underline{Z}_{0n} \\ \underline{Z}_{10} & \underline{Z}_{11} & \underline{Z}_{12} & \cdots & \underline{Z}_{1i} & \cdots & \underline{Z}_{1n} \\ \vdots & & & & & & \\ \underline{Z}_{n0} & \underline{Z}_{n1} & \underline{Z}_{n2} & \cdots & \underline{Z}_{ni} & \cdots & \underline{Z}_{nn} \end{bmatrix} \begin{bmatrix} \underline{I}_{00} \\ \underline{I}_{11} \\ \vdots \\ \underline{I}_{nn} \end{bmatrix} \quad (17.34)$$

17.3.3 Hybrid Matrix

In the calculation of electrical networks at different nodes, different sizes are often specified. In a part of the network nodes, the node voltages are given in other node currents. In these, the admittance matrix is reshaped so that all known sizes in the column vector occur and that all required quantities to the right-hand side of the equal sign form a column. This is reached by the partial inversion of the matrix. The principles will be explained briefly.

The matrix form of

$$\begin{bmatrix} i_1 \\ i_2 \end{bmatrix} = \begin{bmatrix} A_{11} & A_{12} \\ A_{22} & A_{22} \end{bmatrix} \cdot \begin{bmatrix} u_1 \\ u_2 \end{bmatrix} \quad (17.35)$$

A hybrid matrix can be written as follows:

$$\begin{bmatrix} u_1 \\ i_2 \end{bmatrix} = \begin{bmatrix} H_{11} & H_{12} \\ H_{22} & H_{22} \end{bmatrix} \cdot \begin{bmatrix} -i_1 \\ u_2 \end{bmatrix} \quad (17.36)$$

By introduction of the negative sign for i_1, the hybrid matrix becomes symmetrical. Submatrices are determined by

$$\begin{aligned} H_{11} &= -A_{11}^{-1} \\ H_{12} &= -A_{11}^{-1} A_{12} \\ H_{22} &= -A_{21} A_{11}^{-1} \\ H_{22} &= A_{22} - A_{21} A_{11}^{-1} A_{12} \end{aligned} \quad (17.37)$$

The calculation of a hybrid matrix is tedious and time consuming. Therefore, this method will not be explained further.

17.3.4 Calculation of Node Voltages and Line Currents at Predetermined Load Currents

Figure 17.2 shows a single-line diagram for a simple case of the calculation of node voltage and node currents by knowing the load current and node power.

We obtain the matrix equation by the node method for the mathematical model of a network with given load currents, node voltages, and node currents.

$$\begin{bmatrix} \underline{Y}_{11} & \underline{Y}_{12} & \underline{Y}_{1i} & \cdots & \underline{Y}_{1n} \\ \underline{Y}_{21} & \underline{Y}_{22} & \underline{Y}_{2i} & \cdots & \underline{Y}_{2n} \\ \vdots & & & & \\ \underline{Y}_{n1} & \underline{Y}_{n2} & \underline{Y}_{ni} & \cdots & \underline{Y}_{ni} \end{bmatrix} \cdot \begin{bmatrix} \underline{U}_1 \\ \underline{U}_2 \\ \vdots \\ \underline{U}_n \end{bmatrix} = \begin{bmatrix} \underline{I}_0 \\ \underline{I}_1 \\ \vdots \\ \underline{I}_n \end{bmatrix} \quad (17.38)$$

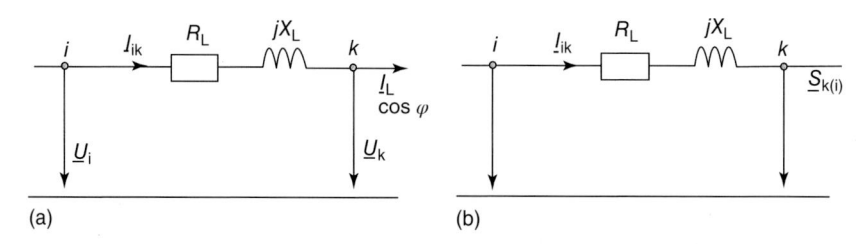

Figure 17.2 Calculation of the line currents between two nodes i, k from the node voltages $\underline{U}_i$, $\underline{U}_k$ and the branch impedance $\underline{Z}_{ik}$ with (a) a constant current and (b) constant power.

Respectively,

$$\underline{Y} \cdot \underline{U} = \underline{I} \qquad (17.32)$$

We get voltage by inversion of $\underline{Y}$:

$$\underline{U} = \underline{Y}^{-1} \cdot \underline{I} \qquad (17.39)$$

The outgoing line current from node i:

$$\underline{I}_{ik} = \underline{Y}_{ik} \cdot (\underline{U}_i - \underline{U}_k) + \underline{U}_i \cdot \underline{Y}_{i0(k)} \qquad (17.40)$$

With this current, the power flow can be calculated from i to k:

$$\underline{S}_{ik} = \underline{U}_i \cdot \underline{I}_{ik}^* \qquad (17.41)$$

17.3.5 Calculation of Node Voltages at Predetermined Node Power

The problem of a constant power is that it always leads to an iterative solution of power flow equations. For first approximation, the current is calculated with a voltage from the power relation:

$$\underline{I}_{ii} = \frac{\underline{S}_i^*}{\underline{U}_i^*} = \frac{P_i - jQ_i}{\underline{U}_i^*} = \underline{Y}_{i1}\underline{U}_1 + \underline{Y}_{i2}\underline{U}_2 + \cdots + \underline{Y}_{in}\underline{U}_n \qquad (17.42)$$

$$\underline{S}_i^* = P_i - jQ_i = \underline{Y}_{i1}\underline{U}_1\underline{U}_i^* + \underline{Y}_{i2}\underline{U}_2\underline{U}_i^*, \cdots = \underline{U}_i^* \sum_i \underline{Y}_{ik}\underline{U}_k \qquad (17.43)$$

By the iteration, a start vector of nodal currents is calculated from an estimated node voltage vector $\underline{U}^{(0)}$ and the known node power $\underline{S}$.

An improved voltage vector $\underline{U}^{(1)}$ can be calculated accurately using an inverted admittance matrix with the initial vector $\underline{I}^{(0)}$.

$$\underline{U}^{(1)} = \underline{Y}^{-1} \cdot \underline{I}^{(0)} \qquad (17.44)$$

17.3.6 Calculation of Power Flow

The design of networks, both in normal operation and under fault conditions, is of great interest. The dimensioning of the electrical equipment in consideration of the transferred services, the cross-sections, the voltage differences, and short-circuit currents is carried out with simpler equations by hand or with the aid of computer programs. The magnitude of the networks and the large number of systems of equations of the networks make it impossible to repeatedly set up new programs. The power distribution is often unknown and must be determined beforehand.

The load flow calculation is an important part of the electric power supply. The objective is to determine the voltages and power at all nodes of magnitude and phase as well as the utilization of all network elements in the system.

The power direction is chosen in the consumer system. Here, the inflowing in a resource active power and inductive reactive power counted positively. The load flow calculation is assumed that the system is constructed in a stationary and symmetrical way. In this section, the basic designs of the load flow calculations will be explained.

17.3.6.1 Type of Nodes

In the load flow calculation, three different network nodes can be distinguished:

Slack node: At this node, only the voltage (amplitude and phase system) is specified. The slack node covers the power difference, after the current and voltage distribution results.

Load nodes: The complex power is specified, which is constant during the calculation (*PQ*-node). Moreover, the behavior of the consumers has to be observed at a change of voltage (constant current and constant impedance).

Node generator: The amplitude of the voltage and the active power are fixed (*PU*-node). In many cases, the boundaries of the reactive power must be taken into consideration, arising from the operation diagram of the generator.

17.3.6.2 Type of Loads and Complex Power

At a certain load condition, the following questions must be answered:

1) Which loads are connected to the node?
2) What is the line and transformer load throughout the system?
3) What is the voltage stability in the whole system?

To answer these questions, the reproduction of loads will be considered (Figure 17.3).

For each node, the complex power in sinusoidal steady-state circuits can be defined as:

$$\underline{i}_k = \underline{A}_k \cdot \underline{u}_k \tag{17.45}$$

The complex voltage vector $\mathbf{u}_k$ can be split into a real and imaginary part:

$$\underline{u}_k = e_k + jf_k = u_k \cdot \cos\varphi_k + ju_k \cdot \sin\varphi_k \tag{17.46}$$

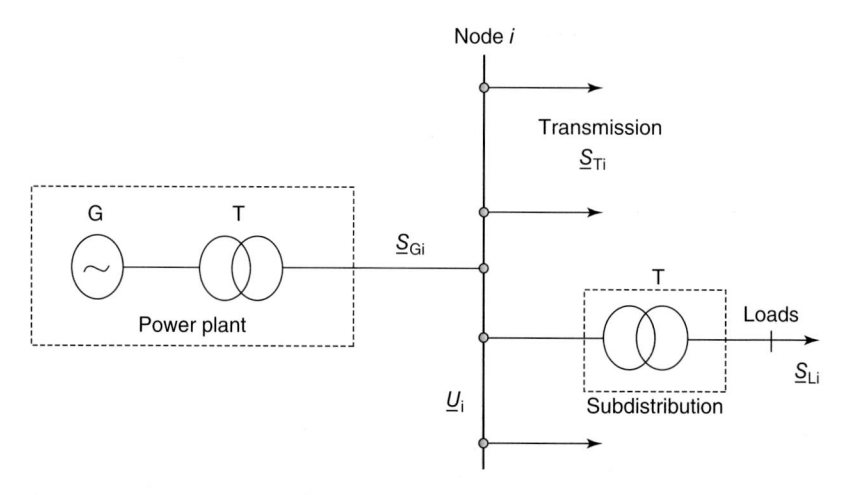

Figure 17.3 Representation of a power system.

Correspondingly, we get the elements of a node admittance matrix:

$$\underline{a}_{ij} = g_{ij} + jb_{ij} \tag{17.47}$$

The apparent power can be given for each node:

$$\underline{S}_i = \underline{U}_i \cdot \underline{I}_i^* = P_i + jQ_i \tag{17.48}$$

For all other elements:

$$\underline{S}_{Gi} = \underline{S}_{Li} + \underline{S}_{Ti} \tag{17.49}$$
$$\underline{S}_{Gi} = P_{Gi} + jQ_{Gi} \tag{17.50}$$
$$\underline{S}_{Li} = P_{Li} + jQ_{Li} \tag{17.51}$$
$$\underline{S}_{Ti} = P_{Ti} + jQ_{Ti} \tag{17.52}$$
$$\underline{U}_i = U_i \cdot e^{j\varphi_i} \tag{17.53}$$

The node types are as follows:

a) Load node (e.g., motors, loads, and pipes) (PQ-node)

Given : P and Q;   wanted : U and δ

b) Infeed node (e.g., mains' supplies and generators) (PU-node)

Given : P and U;   wanted : Q and δ

c) Balance sheet or reference node (slack node)

Given : U and δ;   wanted : P and Q

The load modeling plays a very important role. Several consumers are grouped into a load. They can be emulated by the following variables:

- constant impedances
- constant currents
- constant performance
- voltage-dependent services.

Measurements have shown the voltage dependence of the loads with the equations:

$$P_i = P_{i0} \cdot \left(\frac{U_i}{U_{i0}}\right)^p \tag{17.54}$$

$$Q_i = Q_{i0} \cdot \left(\frac{U_i}{U_{i0}}\right)^q \tag{17.55}$$

Substituting p and q with the values 0, 1, and 2, the measured exponents are

$p = q = 0$: load with constant power
$p = q = 1$: load as a constant current
$p = q = 2$: load as a constant impedance.

Figure 17.4 shows the circuit diagrams for loads of power flow calculations.

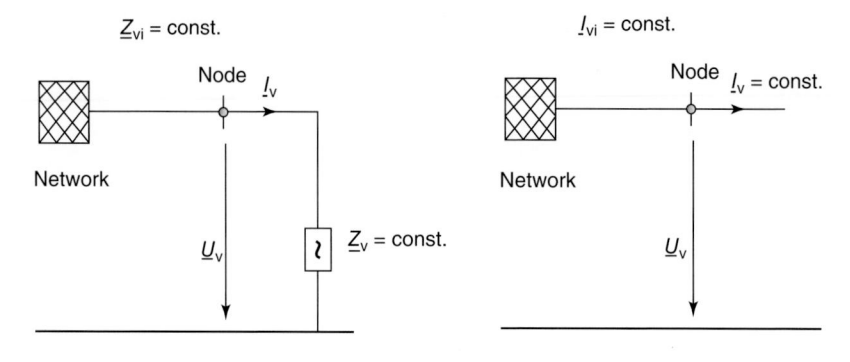

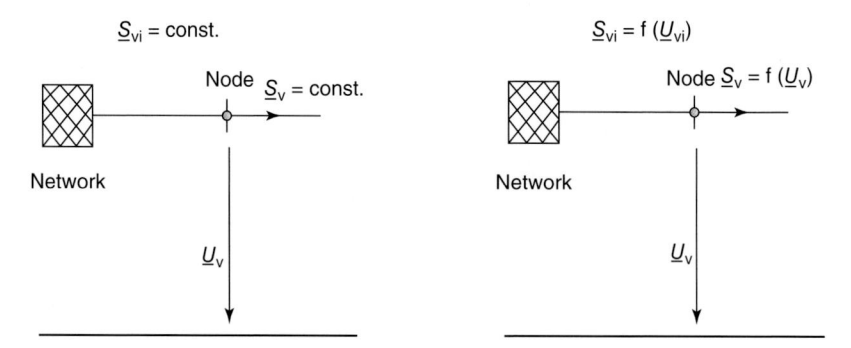

Figure 17.4 Circuit diagrams for power flow calculations.

17.3.7 Linear Load Flow Equations

Figure 17.5 shows the one-line and equivalent circuit diagram for the network.

The relationship between the node currents and the node voltages can be set up by the linear load flow equations, which is given by

$$\begin{bmatrix} \underline{I}_1 \\ \underline{I}_2 \\ \cdot \\ \cdot \\ \underline{I}_n \end{bmatrix} = \begin{bmatrix} \underline{Y}_{11} & \cdot & \cdot & \underline{Y}_{1n} \\ \underline{Y}_{21} & \cdot & \cdot & \underline{Y}_{2n} \\ \cdot & \cdot & \cdot & \cdot \\ \cdot & \cdot & \cdot & \cdot \\ \underline{Y}_{n1} & \cdot & \cdot & \underline{Y}_{nn} \end{bmatrix} \begin{bmatrix} \underline{U}_1 \\ \underline{U}_2 \\ \cdot \\ \cdot \\ \underline{U}_n \end{bmatrix} \quad (17.56)$$

$$\underline{i} = \underline{Y} \cdot \underline{u} \quad (17.57)$$

The elements of the node $\underline{Y}$ are obtained with the following rule:

- The diagonal item $\underline{Y}_{ii}$ arises from the sum of all the nodes i and connected admittances $\underline{y}_{ik}$:

$$\underline{Y}_{ii} = \sum_{k=0, i \neq k}^{n} \underline{y}_{ik} \quad (17.58)$$

- The off-diagonal element $\underline{Y}_{ik}$ has the negative value of the nodes i and connected k admittance $\underline{y}_{ik}$:

$$\underline{Y}_{ik} = -\underline{y}_{ik} \quad (17.59)$$

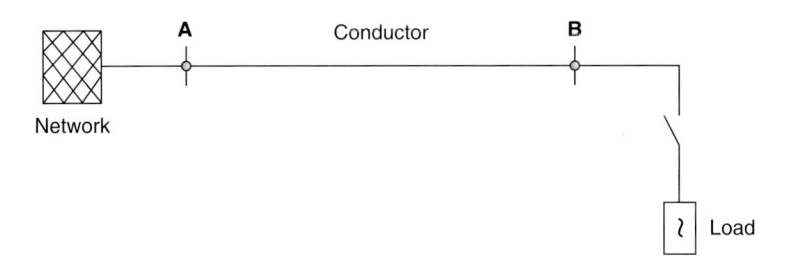

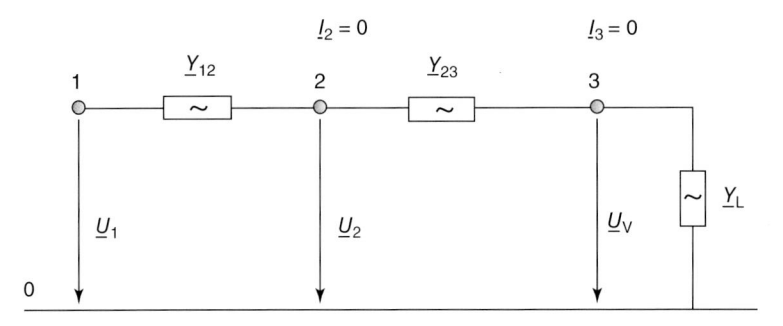

Figure 17.5 One-line diagram and equivalent circuit.

It is important to ensure that with $\underline{Y}_{ik}$, the elements of the node $-\underline{Y}$ and y_{ik} admittances of the network elements are designated.

The equivalent power equation is

$$\begin{bmatrix} \underline{S}_1 \\ \underline{S}_2 \\ \cdot \\ \cdot \\ \underline{S}_n \end{bmatrix} = 3 \cdot \begin{bmatrix} \underline{U}_1 & \cdot & \cdot & 0 \\ \cdot & & & \cdot \\ \cdot & \underline{U}_2 & \cdot & \cdot \\ \cdot & & & \cdot \\ 0 & \cdot & \cdot & \underline{U}_n \end{bmatrix} \begin{bmatrix} \underline{I}_1^* \\ \underline{I}_2^* \\ \cdot \\ \cdot \\ \underline{I}_n^* \end{bmatrix} \quad (17.60)$$

For the real power:

$$\underline{P}_i + j\underline{Q}_i = 3 \cdot \underline{U}_i \cdot \underline{I}_i^* \quad (17.61)$$

In matrix form:

$$p + jq = 3 \cdot \text{diag}(u)\underline{Y}^* \cdot \underline{u}^* \quad (17.62)$$

The method of the load modeling is based on the calculation of the corrections for the node voltages. To solve the nonlinear equations, different iterations are used.

The Newton–Raphson method has proven it in practice. Therefore, the mathematical formulation is discussed in more detail in the next section.

17.3.8 Load Flow Calculation by Newton–Raphson

The power equation is used by the Newton–Raphson method. This is obtained from the current equation of the KCL node methods, in which first conjugated this complex and then multiplied on the left by the factor 3 and a diagonal matrix $\underline{U}$ multiplied by the node voltages as elements.

17 Power Flow Analysis

$$3 \cdot \underline{U} \cdot \underline{Y} \cdot \underline{u}^* = 3 \cdot \underline{U} \cdot \underline{i}^* = s + jq \tag{17.63}$$

The power is defined for each node i:

$$\underline{S}_i = P_i + jQ_i = \underline{U}_i \cdot \underline{I}_i^* \tag{17.64}$$

$$\underline{i}_k = \underline{A}_k \cdot \underline{u}_k \tag{17.65}$$

And then:

$$s_k = \text{diag}(u_k) \cdot i_k^* \quad \text{and} \quad s_k = \text{diag}(u_k) \cdot A_k^* u_k^* \tag{17.66}$$

Substituting for the node voltages:

$$\underline{U}_i = E_i + jF \tag{17.67}$$

And the elements of the node admittance matrix:

$$\underline{a}_{ik} = g_{ik} - jb_{ik} \tag{17.68}$$

for each node

$$P_i + jQ_i = (E_i + jF_i) \cdot \sum_{k=1}^{n}(g_{ik} + jb_{ik}) \cdot (E_k - jF_k) \tag{17.69}$$

The division into the real and imaginary parts results in the following:

$$P_i = \sum_{k=1}^{n}[E_i(E_k g_{ik} + F_k jb_{ik}) + F_i(F_k g_{ik} - E_k jb_{ik})] \tag{17.70}$$

$$Q_i = \sum_{k=1}^{n}[F_i(E_k g_{ik} + F_k jb_{ik}) - E_i(F_k g_{ik} - E_k jb_{ik})] \tag{17.71}$$

Thus, the complex load flow equations have been split into two real systems. For each node, there are two equations. At each node, the active and reactive power are given. Calculation is performed for the real and imaginary parts of the voltages at all nodes except for the slack nodes. There, the voltage is known and held in its size and phase position. The equation system with $2(n-1)$ must be solved to get the load flow in a network with n nodes.

The Newton–Raphson method requires the setting of a system equation that changes sets in the active and reactive power in relation to changes in node voltages. The following is applied:

$$\begin{bmatrix} \Delta P_1 \\ \vdots \\ \Delta P_{n-1} \\ \Delta Q_1 \\ \vdots \\ \Delta Q_{n-1} \end{bmatrix} = - \begin{bmatrix} \frac{\partial P_1}{\partial E_1} & \cdots & \frac{\partial P_1}{\partial E_{n-1}} & \frac{\partial P_1}{\partial F_1} & \cdots & \frac{\partial P_1}{\partial F_{n-1}} \\ \frac{\partial P_{n-1}}{\partial E_1} & \cdots & \frac{\partial P_{n-1}}{\partial E_{n-1}} & \frac{\partial P_{n-1}}{\partial F_1} & \cdots & \frac{\partial P_{n-1}}{\partial F_{n-1}} \\ \frac{\partial Q_1}{\partial E_1} & \cdots & \frac{\partial Q_1}{\partial E_{n-1}} & \frac{\partial Q_1}{\partial F_1} & \cdots & \frac{\partial Q_1}{\partial F_{n-1}} \\ \frac{\partial Q_{n-1}}{\partial E_1} & \cdots & \frac{\partial Q_{n-1}}{\partial E_{n-1}} & \frac{\partial Q_{n-1}}{\partial F_1} & \cdots & \frac{\partial Q_{n-1}}{\partial F_{n-1}} \end{bmatrix} \cdot \begin{bmatrix} \Delta E_1 \\ \vdots \\ \Delta E_{n-1} \\ \Delta F_1 \\ \vdots \\ \Delta F_{n-1} \end{bmatrix}$$

$$(17.72)$$

17.3 Network Matrices

The coefficient matrix is called the Jacobian matrix, which is defined as a matrix in which the elements are the partial representative derivatives of the independent variable. The Jacobian matrix offers a breakdown into four partial matrices.

$$\begin{bmatrix} \Delta p \\ \Delta q \end{bmatrix} = - \begin{bmatrix} H & N \\ J & L \end{bmatrix} \cdot \begin{bmatrix} \Delta e \\ \Delta f \end{bmatrix} \tag{17.73}$$

whereby:

$$\Delta P_i = P_i - P_{ig} \quad \text{and} \quad \Delta Q_i = Q_i - Q_{ig} \tag{17.74}$$

From the equation $P_i = \sum_{k=1}^{n}[E_i(E_k g_{ik} + F_k jb_{ik}) + F_i(F_k g_{ik} - E_k jb_{ik})]$, the elements of Jacobian matrix can be derived:

$$P_i = E_i(E_i g_{ii} + F_i b_{ii}) + F_i(F_i g_{ii} - E_i b_{ii})$$

$$+ \sum_{\substack{k=1 \\ k \neq i}}^{n} [E_i(E_k g_{ik} + F_k b_{ik}) + F_i(F_k g_{ik} - E_k b_{ik})] \tag{17.75}$$

For the calculation of the elements of the matrix **H**, the active power needs to be differentiated according to the real part of voltages. One can distinguish between main diagonal elements and off-diagonal elements. The main diagonal elements are as follows:

$$\frac{\Delta P_i}{\partial E_i} = 2E_i g_{ii} + F_i b_{ii} - F_i b_{ii} + \sum_{\substack{k=1 \\ k \neq i}}^{n}(E_k g_{ik} + F_k b_{ik}) \tag{17.76}$$

For the off-diagonal elements $k \neq 1$:

$$\frac{\Delta P_i}{\partial E_k} = E_i g_{ik} - F_i b_{ik} \tag{17.77}$$

Accordingly, the elements of the submatrix **N** can be obtained:

$$\frac{\Delta P_i}{\partial F_i} = E_i b_{ii} + 2F_i g_{ii} - E_i b_{ii} + \sum_{\substack{k=1 \\ k \neq i}}^{n}(F_k g_{ik} - F_k b_{ik}) \tag{17.78}$$

For the off-diagonal elements $k \neq 1$:

$$\frac{\Delta P_i}{\partial F_k} = E_i b_{ik} + F_i g_{ik} \tag{17.79}$$

If the equation $Q_i = \sum_{k=1}^{n}[F_i(E_k g_{ik} + F_k jb_{ik}) - E_i(F_k g_{ik} - E_k jb_{ik})]$ is rewritten, the following is obtained:

$$Q_i = F_i(E_i g_{ii} + F_i b_{ii}) - E_i(F_i g_{ii} - E_i b_{ii})$$

$$+ \sum_{\substack{k=1 \\ k \neq 1}}^{n} [F_i(E_k g_{ik} + F_k b_{ik}) - E_i(F_k g_{ik} - E_k b_{ik})] \tag{17.80}$$

Thus, the elements of the submatrices **J** and **L** can be determined. For the main diagonal elements of **J**, the following is obtained:

$$\frac{\Delta Q_i}{\partial E_i} = F_i g_{ii} - F_i g_{ii} - E_i b_{ii} + 2E_i b_{ii} - \sum_{\substack{k=1 \\ k \neq i}}^{n}(F_k g_{ik} - E_k b_{ik}) \tag{17.81}$$

For the off-diagonal elements $k \neq 1$:

$$\frac{\Delta Q_i}{\partial E_k} = E_i b_{ik} + F_i g_{ik} \qquad (17.82)$$

Accordingly, the elements of the main matrix **L** can be obtained:

$$\frac{\Delta Q_i}{\partial F_i} = E_i g_{ii} + 2F_i b_{ii} - E_i g_{ii} + \sum_{\substack{k=1 \\ k \neq i}}^{n} (E_k g_{ik} + F_k b_{ik}) \qquad (17.83)$$

For the off-diagonal elements $k \neq 1$:

$$\frac{\Delta Q_i}{\partial F_k} = -E_i g_{ik} + F_i b_{ik} \qquad (17.84)$$

The following is obtained by simplifying the equations if the node currents are used:

$$\underline{I}_i = C_i + jD_i = (g_{ii} - jb_{ii})(E_i + jF_i) + \sum_{\substack{k=1 \\ k \neq 1}}^{n} (g_{ik} - jb_{ik})(E_k + jF_k) \qquad (17.85)$$

By separation of the real and imaginary parts, the following is obtained:

$$C_i = E_i g_{ii} + F_i b_{ii} + \sum_{\substack{k=1 \\ k \neq 1}}^{n} (E_k g_{ik} + F_k b_{ik}) \qquad (17.86)$$

$$D_i = F_i g_{ii} - E_i b_{ii} + \sum_{\substack{k=1 \\ k \neq 1}}^{n} (F_k g_{ik} - E_k b_{ik}) \qquad (17.87)$$

All equations for the calculation of Jacobi elements are summarized as follows:

Submatrix	Main diagonal elements	Off-diagonal elements	
H	$\dfrac{\Delta P_i}{\partial E_i} = E_i g_{ii} - F_i b_{ii} + C_i$	$\dfrac{\Delta P_i}{\partial E_k} = E_i g_{ik} - F_i b_{ik}$	
N	$\dfrac{\Delta P_i}{\partial F_i} = E_i b_{ii} + F_i g_{ii} + D_i$	$\dfrac{\Delta P_i}{\partial F_k} = E_i b_{ik} + F_i g_{ik}$	(17.88)
J	$\dfrac{\Delta Q_i}{\partial E_i} = E_i b_{ii} + F_i g_{ii} - D_i$	$\dfrac{\Delta Q_i}{\partial E_k} = E_i b_{ik} + F_i g_{ik}$	
L	$\dfrac{\Delta Q_i}{\partial F_i} = -E_i b_{ii} + F_i g_{ii} + C_i$	$\dfrac{\Delta Q_i}{\partial F_k} = -E_i g_{ik} + F_i b_{ik}$	

The simplified mathematical description of the Newton–Raphson method is shown in Figure 17.6. The complex system of equations can be split into two real components (real and imaginary parts of the voltages at all nodes). The so-called slack node or node compensation is not considered here, since the voltage is known and held in its size and phase position. The nominal voltages at the nodes, or the rated voltage of the generator nodes, are taken as initial values for the voltages and determined.

17.3 Network Matrices

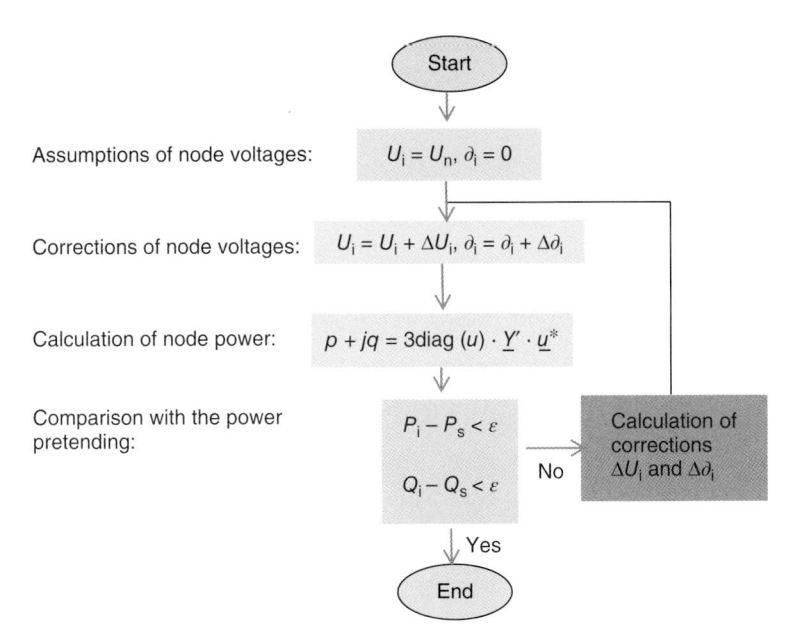

Figure 17.6 Flow diagram.

Flow diagram steps:
- Start
- Assumptions of node voltages: $U_i = U_n$, $\partial_i = 0$
- Corrections of node voltages: $U_i = U_i + \Delta U_i$, $\partial_i = \partial_i + \Delta \partial_i$
- Calculation of node power: $p + jq = 3\text{diag}(\underline{u}) \cdot \underline{Y}' \cdot \underline{u}^*$
- Comparison with the power pretending: $P_i - P_s < \varepsilon$, $Q_i - Q_s < \varepsilon$
- If No → Calculation of corrections ΔU_i and $\Delta \partial_i$ (loop back)
- If Yes → End

The active and reactive power at the nodes are calculated and determined. If the deviation is less than a predetermined threshold, then the iteration is terminated, and the load flow can be calculated in the entire network. Otherwise, the node currents are determined from the currently calculated performance that determines the Jacobian method, and a corrected voltage vector is formed so that new values can be formed at the nodes. This is continued until the end of the iteration.

A second possibility is that the load flow equations can be set up as a function of the voltage magnitude and angle magnitude.

17.3.9 Current Iteration

A distinction is made in the power flow calculation between the current iteration and the Newton–Raphson method. The former is prepared from the given node active power $\underline{S}_1, \underline{S}_2, \underline{S}_3, \ldots, \underline{S}_n$ using an estimated starting vector for the node voltages $(\underline{U}_1, \underline{U}_2, \underline{U}_3, \ldots, \underline{U}_n$ and currents $\underline{I}_1, \underline{I}_2, \underline{I}_3, \ldots, \underline{I}_n$ calculated), from which finally by solving a linear system of equations for given load currents $\underline{Y} \cdot \underline{U} = \underline{I}$ iteratively improved node voltages are calculated. In the second calculation step, replacing the initial vector of node voltages by the calculated, improved node voltages and receives there again new, improved load currents and node voltages, and so on. By the Newton–Raphson method, the node voltages are iteratively determined directly from the power of their first derivatives.

17.3.9.1 Jacobian Method

The Jacobian method for current iteration is called a total step process, where the whole equation is determined in each step for a fixed node voltage vector. The equation for the voltage $\underline{U}_i$ can be solved for each node i, and the following can be written as follows:

$$\underline{Y}_{i1} \cdot \underline{U}_1 + \underline{Y}_{i2} \cdot \underline{U}_2 + \underline{Y}_{ii} \cdot \underline{U}_i + \cdots + \underline{Y}_{in} \cdot \underline{U}_n = \underline{I}_{ii} \tag{17.89}$$

Accordingly:

$$\underline{Y}_{ii} \cdot \underline{U}_i = \underline{I}_{ii} - \sum_{k=1, k \neq 1}^{n} \underline{Y}_{ik} \cdot \underline{U}_k \tag{17.90}$$

The following algorithm results from the recursion equation:

$$\underline{U}_i^{v+1} = \frac{1}{\underline{Y}_{ii}} \left[\underline{I}_{ii} - \sum_{k=1, k \neq 1}^{n} \underline{Y}_{ik} \cdot \underline{U}_k^{v} \right] \tag{17.91}$$

Or

$$\underline{U}_i^{v+1} = \frac{1}{\underline{Y}_{ii}} \left[\frac{P_i - jQ_i}{\underline{U}_i^{(v)*}} - \sum_{k=1, k \neq 1}^{n} \underline{Y}_{ik} \cdot \underline{U}_k^{v} \right] \tag{17.92}$$

This equation can be calculated successively for the node voltages of all equation rows.

17.3.10 Gauss–Seidel Method

The Gauss–Seidel method for current iteration is called a single-step process. The already iterated equations obtained from the improved node voltages are considered in the iteration of the subsequent rows of equations, which result in faster convergence leads. One divides the square parenthesized sum into two subtotals and uses in the first, the already improved node voltages.

$$\underline{U}_i^{v+1} = \frac{1}{\underline{Y}_{ii}} \left[\underline{I}_{ii} - \sum_{k=1}^{i-1} \underline{Y}_{ik} \cdot \underline{U}_k^{(v+1)} - \sum_{k=i+1}^{i-1} \underline{Y}_{ik} \cdot \underline{U}_k^{(v)} \right] \tag{17.93}$$

or

$$\underline{U}_i^{v+1} = \frac{1}{\underline{Y}_{ii}} \left[\frac{P_i - jQ_i}{\underline{U}_i^{(v)*}} - \sum_{k=1}^{i-1} \underline{Y}_{ik} \cdot \underline{U}_k^{(v+1)} - \sum_{k=i+1}^{n} \underline{Y}_{ik} \cdot \underline{U}_k^{(v)} \right] \tag{17.94}$$

The convergence of the Gauss–Seidel method can be improved by further introducing an acceleration factor by operating at an iteration value is multiplied by a factor α from 1.5 to 1.7. The updated voltage value is then calculated as:

$$\underline{U}_i^{(v+1)'} \underline{U}_i^{(v)} + \alpha [\underline{U}_i^{(v+1)} - \underline{U}_i^{(v)}] \tag{17.95}$$

17.3.11 Newton–Raphson Method

The Newton–Raphson method is used extensively in the electrical power systems to solve power flows. It has a great assurance of convergence and speed. The idea is that we approximate a graph of f by suitable tangents. Using an approximate value x^0 obtained from the graph of f, x^1 is the point of intersection of the x-axis and then the tangent to the curve of f at x^0 (Figure 17.7).

Figure 17.7 Illustration of the Newton–Raphson algorithm.

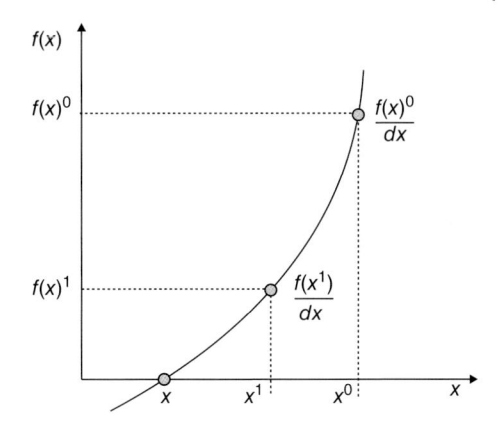

The Newton–Raphson method is an iteration algorithm for the solution of nonlinear equations of n equations with n unknowns. Consider the equation of a function of one variable x expressed as follows:

$$f(x) = 0 \tag{17.96}$$

The nonlinear equation is developed into a Taylor series and is then dissolved by x. The following applies

$$f(x) = f(x^0) + \frac{1}{1!}\frac{df(x^0)}{dx}(x - x^0) + \frac{1}{2!}\frac{df(x^1)}{dx}(x - x^0) + \cdots = 0 \tag{17.97}$$

$$f(x) = f(x^0) + \frac{1}{1!}f'(x^{(0)})(x - x^0) + \cdots = f(x^{(0)}) + f'(x^{(0)})\Delta x^{(0)} + \cdots \tag{17.98}$$

$$\Delta x^{(0)} = (x - x^0) \approx -\frac{f(x^{(0)})}{f'(x^{(0)})} = -[f'(x^{(0)})]^{-1} \cdot f(x^{(0)}) \tag{17.99}$$

The solution of the linearized problem and iteration rule give

$$x \approx x^{(0)} + \Delta x^{(0)} = x^{(0)} - [f'(x^{(0)})]^{-1} \cdot f(x^{(0)})$$
$$\Delta x^{(0)} = -[f'(x^{(0)})]^{-1} \cdot f(x^{(0)}) \tag{17.100}$$

For any particular iteration, the following is obtained:

$$x^{(j+1)} = x^{(j)} + \Delta x^{(j)} = x^{(j)} - [f'(x^{(j)})]^{-1} \cdot f(x^{(j)}) \tag{17.101}$$

During an iteration, each value of $\Delta x^{(x)}$ will be determined and added to the last approximation. The next question is how to apply this with n equations and n unknowns. Again, the function $f(x)$ is expanded in a Taylor series as follows:

$$f(x^{(0)} + \Delta x^{(0)}) = f(x^{(0)}) + f'(x^{(0)})\Delta x^{(0)} + \frac{1}{2}f''(x^{(0)})(\Delta x^{(0)})^2 + \cdots = 0 \tag{17.102}$$

Taking the second term to the right-hand side, the following can be written as follows:

$$f(x^{(0)} + \Delta x^{(0)}) = f(x^{(0)}) + f'(x^{(0)})\Delta x^{(0)} = 0 \tag{17.103}$$

To solve $f'(x^{(0)})$, the Jacobian matrix is used as

$$J = \begin{bmatrix} \dfrac{\partial f_1(x^{(0)})}{\partial x_1} & \dfrac{\partial f_1(x^{(0)})}{\partial x_2} & \cdots & \dfrac{\partial f_1(x^{(0)})}{\partial x_n} \\ \dfrac{\partial f_2(x^{(0)})}{\partial x_1} & \dfrac{\partial f_2(x^{(0)})}{\partial x_2} & \cdots & \dfrac{\partial f_2(x^{(0)})}{\partial x_n} \\ \vdots & \vdots & \vdots & \vdots \\ \dfrac{\partial f_n(x^{(0)})}{\partial x_1} & \dfrac{\partial f_n(x^{(0)})}{\partial x_2} & \cdots & \dfrac{\partial f_n(x^{(0)})}{\partial x_n} \end{bmatrix} \quad (17.104)$$

In Equation (17.103), taking $f(x^{(0)})$ to the right-hand side, the following is obtained:

$$f'(x^{(0)})\Delta x^{(0)} = -f(x^{(0)}) \quad (17.105)$$

Or, in terms of the Jacobian matrix **J**, the following is obtained:

$$J\Delta \underline{x}^{(0)} = -\underline{f}(\underline{x}^{(0)}) \quad (17.106)$$

Finally, solving Equation (17.103) for $\Delta x^{(0)}$, the following is given:

$$\Delta x^{(0)} = -[f'(x^{(0)})]^{-1} f(x^{(0)}) = -J^{-1} f(x^{(0)}) \quad (17.107)$$

Equation (17.55) provides the basis for the update formula to be used in the first iteration of the multidimensional case. This update formula is

$$x^{(1)} = x^{(0)} + \Delta x^{(0)} = x^{(0)} - J^{-1} f(x^{(0)}) \quad (17.108)$$

And, from Equation (17.107), the update formula for any particular iteration can be inferred as:

$$x^{(i+1)} = x^{(i)} + \Delta x^{(i)} = x^{(i)} - J^{-1} f(x^{(i)}) \quad (17.109)$$

Matrix inversion must be avoided because of high-dimensional problems and large-scale power networks, and matrix inversion is *very* time consuming. To do this, Equation (17.109) is written as:

$$x^{(i+1)} = x^{(i)} + \Delta x^{(i)} \quad (17.110)$$

Where $\Delta x^{(i)}$ is found from

$$-J\Delta x^{(i)} = f(x^{(i)}) \quad (17.111)$$

Equation (17.111) is simply the linear matrix equation because $\Delta x^{(i)}$ is an $\mathbf{n} \times 1$ vector of unknowns and $f(x^{(i)})$ is an $\mathbf{n} \times 1$ vector of knowns. **J** is just a constant $\mathbf{n} \times \mathbf{n}$ matrix.

$$A \cdot z = b \quad (17.112)$$

17.3.12 Power Flow Analysis in Low-Voltage Power Systems

The calculation of a three-phase system with the aim of determining the distribution of active and reactive power on the individual lines and nodes and determining the node voltages is called load flow calculation. This calculation requires special considerations at different voltage levels. In low-voltage power systems, many single-phase and three-phase consumers may occur, mostly with an ohmic-inductive load. In industrial networks, however, the motor loads are very common. A mixed load is available in public networks. The planning and installation of an electrical system requires the precise calculation of the power demands for which the system is to be designed. This is determined by the equation:

$$P_m = \sum_{i=1}^{n} P_i \tag{17.113}$$

where

$$P_i = n \cdot g_i \cdot P_L \tag{17.114}$$

If a Gaussian (normal) distribution is assumed for the coincidence, then the coincidence function can be calculated as:

$$g_n = g_\infty + \frac{1 - g_\infty}{\sqrt{2}} \tag{17.115}$$

Often this value can be given approximately for flats by the context of [16]

$$g = 0.07 + \frac{0.93}{n} \tag{17.116}$$

The coincidence factor or demand factor g_i indicates how many consumers are in operation at the same time. It is an important factor for determining the feed-ins. n is the number of connections or residential units. When more motor drives are connected in the system, it is also necessary to consider the utilization factor a_i and the efficiency η_i in the calculation.

The main interest of load flow calculation in distribution is usually to determine the maximum branch currents, power losses, and maximum voltage drop. In low-voltage power systems, the R/X ratio is considerably greater than 1. The voltage drop, therefore, depends mainly on the active power flow. Reactive power flow in low-voltage systems is of less interest. A realistic representation of the loads results from modeling the problem. Temporal changes are usually simulated by load curves, which represent the time course of a load in the course of a day. If the load curves are entered only once, it is possible to perform a load flow calculation for any point in time or even to start sequential load flow calculations in which any size can be recorded as a function of time of day. This type of load flow calculation allows for a very accurate estimation of line loads or voltage levels [11].

In low-voltage power systems, the calculation of power flows is not necessary in many cases. In particular, the dimensioning of conductor sizes, voltage drop, and single-phase and three-phase short-circuit calculations are most important for the engineers. This entire topic is not the subject of this book, but it is presented extensively in other literature [7].

17.3.13 Equivalent Circuits for Power Flow Calculations

In this section, equivalent circuits for power flow calculations in positive-sequence components will be shown (Figure 17.8).

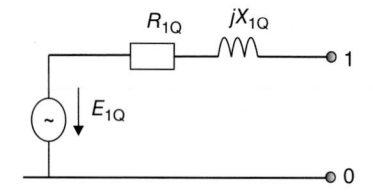

Figure 17.8 Equivalent circuits for power flow calculations.

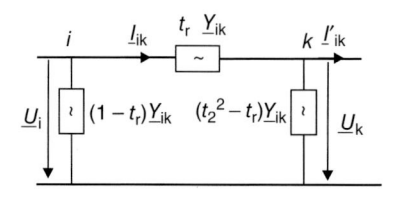

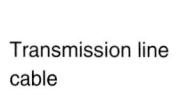

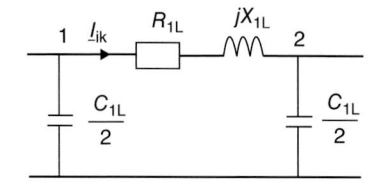

17.3.14 Examples

17.3.14.1 Calculation of Reactive Power

Calculate the voltage and reactive power at the end of the line.

Given: line impedance, $z_L = 0.148\,\Omega$; load impedance, $z_V = 2.2\,\Omega$; and $U_L = 400$ V.

$$U_V = \frac{z_V}{z_V + z_L} \cdot 100\% = 84.94\%$$

$$i = \frac{U_V}{\sqrt{3} \times z_V} = \frac{339.76\,\text{V}}{\sqrt{3} \times 2.2\,\Omega} = 89.16\,\text{A}$$

$$Q = \frac{U_V^2}{z_V} = \frac{(339.76\,\text{V})^2}{2.2\,\Omega} = 52.471\,\text{kvar}$$

17.3.14.2 Application of Newton Method

Given is the following equation:

$$f(x) = x + \sin x - 2 = 0$$

The derivation of this equation is

$$f(x) = 1 + \cos x$$

Recursion is

$$x^{k+1} = x^k - \frac{f_x^k}{f_x^k} = x^k - \frac{x^k + \sin x^k - 2}{1 + \cos x^k}$$

The solution of this equation with $x^0 = 0$ is to

Iteration	x^k
0	0
1	1
2	1.103
3	1.110606

17.3.14.3 Linear Equations
Solve the linear equation.

$$2x - 3y = 12$$
$$5x + 2y = 11$$

Solution:

$$2x - 3y = 12 | \cdot 2$$
$$\underline{5x + 2y = 11 | \cdot 3}$$
$$4x - 6y = 24$$
$$\underline{15x + 6y = 33|+}$$
$$19x = 57| : 19$$
$$x = 4$$

The result used in the first equation to determine y:

$$2x - 3y = 12$$
$$\underline{2 \cdot 3 - 3y = 12|}$$
$$6 - 3y = 12| - 6$$
$$\underline{-3y = 6 \ |:(-3)}$$
$$y = -2$$

17.3.14.4 Application of Cramer's Rule
Given is the following circuit (Figure 17.9). Calculate the currents.
KCL node rule specifies that

$$I_1 - I_2 - I_3 = 0$$

And Kirchhoff's voltage law (KVL) voltage rule specifies that

$$R_1 I_1 + R_2 I_2 = 0$$
$$R_2 I_2 - R_3 I_3 = 0$$
$$I_1 - I_2 - I_3 = 0$$
$$R_1 I_1 + R_2 I_2 = 0$$
$$R_2 I_2 - R_3 I_3 = 0$$

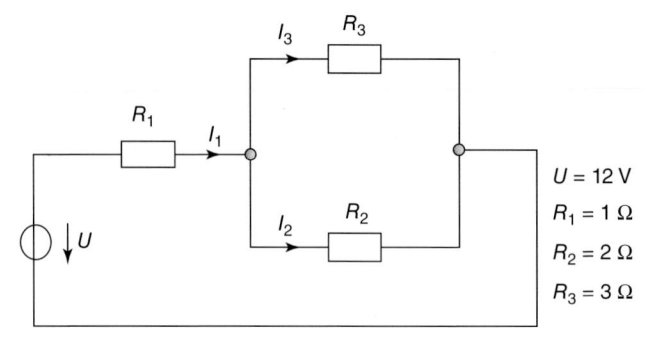

Figure 17.9 Application of Cramer's rule.

$U = 12\,\text{V}$
$R_1 = 1\,\Omega$
$R_2 = 2\,\Omega$
$R_3 = 3\,\Omega$

Coefficients give

$$A = \begin{pmatrix} 1 & -1 & -1 \\ R_1 & R_2 & 0 \\ 0 & R_2 & -R_3 \end{pmatrix}$$

Or

$$\begin{pmatrix} 1 & -1 & -1 \\ R_1 & R_2 & 0 \\ 0 & R_2 & -R_3 \end{pmatrix} \begin{pmatrix} I_1 \\ I_2 \\ I_3 \end{pmatrix} = \begin{pmatrix} 0 \\ U \\ 0 \end{pmatrix}$$

$$D = \det A = \begin{vmatrix} 1 & -1 & -1 \\ R_1 & R_2 & 0 \\ 0 & R_2 & -R_3 \end{vmatrix} = -(R_1 R_2 + R_1 R_3 + R_2 R_3) \neq 0$$

$$D_1 = \begin{vmatrix} 0 & -1 & -1 \\ U & R_2 & 0 \\ 0 & R_2 & -R_3 \end{vmatrix} = -R_2 U - R_3 U = -(R_2 + R_3)U$$

$$D_2 = \begin{vmatrix} 1 & 0 & -1 \\ R_1 & U & 0 \\ 0 & 0 & -R_3 \end{vmatrix} = -R_3 U$$

$$D_3 = \begin{vmatrix} 1 & -1 & 0 \\ R_1 & R_2 & U \\ 0 & R_2 & 0 \end{vmatrix} = -R_2 U$$

So that we can obtain the following currents:

$$I_1 = \frac{D_1}{D} = \frac{(R_2 + R_3)U}{R_1 R_2 + R_1 R_3 + R_2 R_3} = \frac{6\,\Omega \cdot 12\,\text{V}}{11\,\Omega} = 6.54\,\text{A}$$

$$I_2 = \frac{D_2}{D} = \frac{R_3 U}{R_1 R_2 + R_1 R_3 + R_2 R_3} = \frac{3\,\Omega \cdot 12\,\text{V}}{11\,\Omega} = 3.27\,\text{A}$$

$$I_3 = \frac{D_3}{D} = \frac{R_2 U}{R_1 R_2 + R_1 R_3 + R_2 R_3} = \frac{2\,\Omega \cdot 12\,\text{V}}{14\,\Omega} = 2.18\,\text{A}$$

17.3.14.5 Power Flow Calculation with NEPLAN

Figure 17.10 shows a part of a project that was calculated with NEPLAN. All possible information are given in the figure.

LIN 7-6

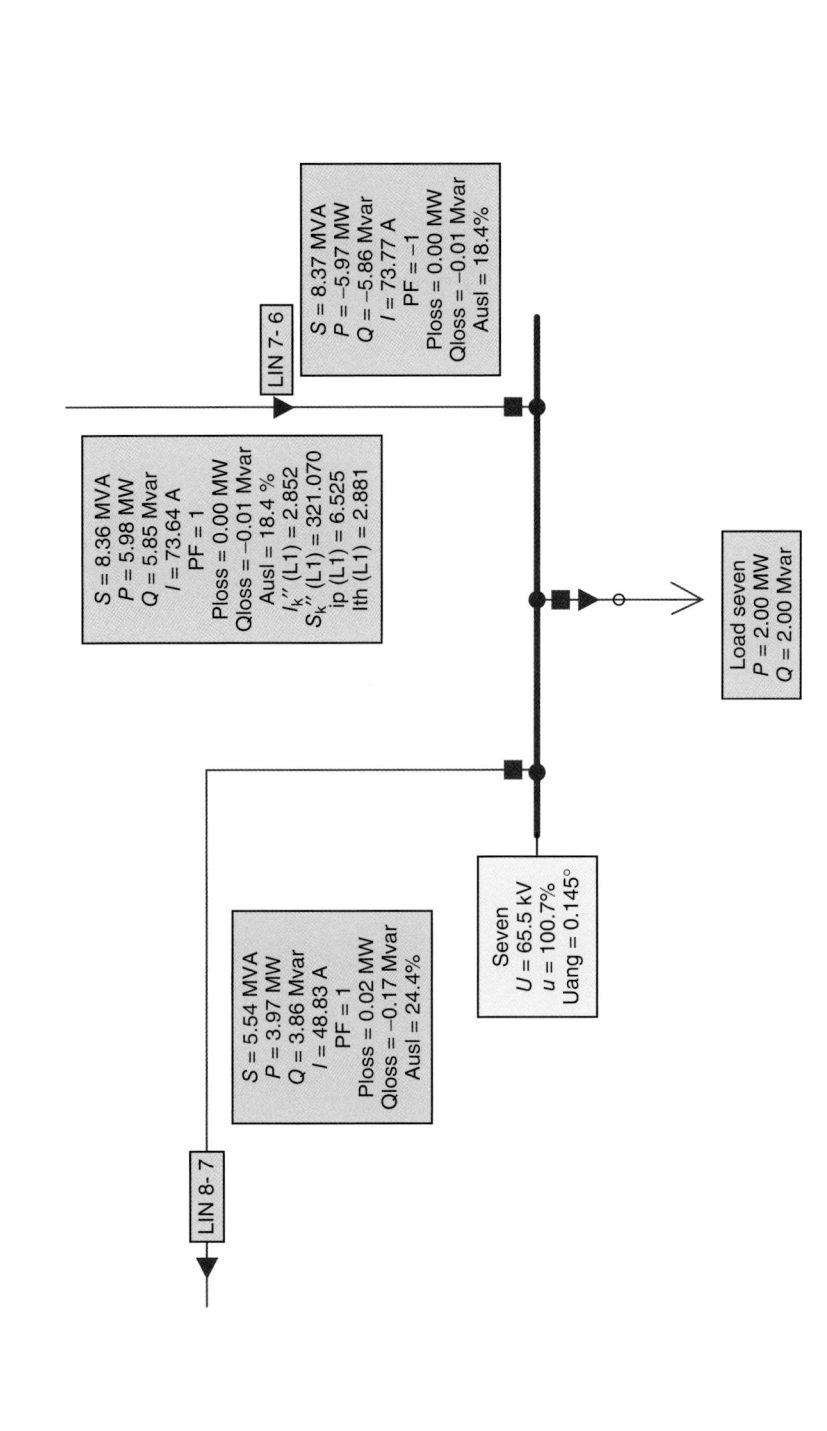

S = 8.37 MVA
P = −5.97 MW
Q = −5.86 Mvar
I = 73.77 A
PF = −1
Ploss = 0.00 MW
Qloss = −0.01 Mvar
Ausl = 18.4%

S = 8.36 MVA
P = 5.98 MW
Q = 5.85 Mvar
I = 73.64 A
PF = 1
Ploss = 0.00 MW
Qloss = −0.01 Mvar
Ausl = 18.4 %
I_k'' (L1) = 2.852
S_k'' (L1) = 321.070
ip (L1) = 6.525
Ith (L1) = 2.881

Load seven
P = 2.00 MW
Q = 2.00 Mvar

Seven
U = 65.5 kV
u = 100.7%
Uang = 0.145°

S = 5.54 MVA
P = 3.97 MW
Q = 3.86 Mvar
I = 48.83 A
PF = 1
Ploss = 0.02 MW
Qloss = −0.17 Mvar
Ausl = 24.4%

LIN 8-7

Figure 17.10 Power flow calculation with NEPLAN.

18

Examples: Calculation of Short-Circuit Currents

This chapter presents a large number of examples taken from practice, worked by hand through the corresponding equations and also an example calculated with NEPLAN.

18.1 Example 1: Radial Network

Given a 400-V network, as shown in Figure 18.1.

1) Calculate the resistances and reactances.
2) Calculate the single-pole and three-pole short-circuit currents for the circuit with residual current device (RCD)-operated circuit breaker.

Smallest short-circuit current:
Resistances and reactances of the transformer:

$$R_T = 3.5\,m\Omega;\quad X_T = 13\,m\Omega$$

Resistances and reactances of the cable:

$$R_{K_{L1}} = 1.24 \cdot \frac{l}{\kappa \cdot S}$$

$$R_{K_{L1}} = 1.24 \cdot \frac{85\,m}{56\frac{m}{\Omega mm^2} \cdot 120\,mm^2} = 15.68\,m\Omega$$

$$R_{K_{PEN}} = 1.24 \cdot \frac{85\,m}{56\frac{m}{\Omega mm^2} \cdot 70\,mm^2} = 26.68\,m\Omega$$

$$X_{K_{L1}} = x'_{k_{L1}} \cdot l = 0.08\frac{m\Omega}{m} \cdot 85\,m = 6.8\,m\Omega$$

$$X_{K_{PEN}} = x'_{k_{PE}} \cdot l = 0.08\frac{m\Omega}{m} \cdot 85\,m = 6.8\,m\Omega$$

Short Circuits in Power Systems: A Practical Guide to IEC 60909-0, Second Edition. Ismail Kasikci.
© 2018 Wiley-VCH Verlag GmbH & Co. KGaA. Published 2018 by Wiley-VCH Verlag GmbH & Co. KGaA.

18 Examples: Calculation of Short-Circuit Currents

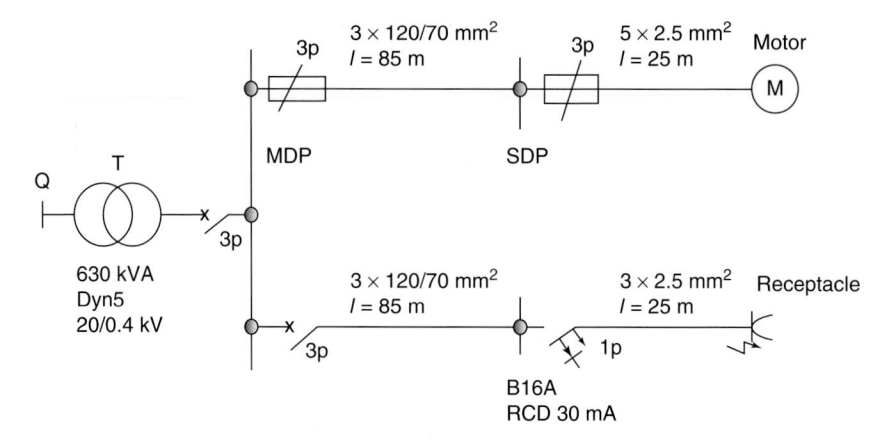

Figure 18.1 Example 1: network representation.

Resistances and reactances of the conductor:

$$R_{K_{L1}} = 1.24 \cdot \frac{l}{\kappa \cdot S_{L1}} = 1.24 \cdot \frac{25\,\text{m}}{56\frac{\text{m}}{\Omega\text{mm}^2} \cdot 2.5\,\text{mm}^2} = 221.4\,\text{m}\Omega$$

$$R_{K_{PE}} = 1.24 \cdot \frac{l}{\kappa \cdot S_{PE}} = 1.24 \cdot \frac{25\,\text{m}}{56\frac{\text{m}}{\Omega\text{mm}^2} \cdot 2.5\,\text{mm}^2} = 221.4\,\text{m}\Omega$$

$$X_{k_{L1}} = x'_{k_{L1}} \cdot l = 0.08\frac{\text{m}\Omega}{\text{m}} \cdot 25\,\text{m} = 2\,\text{m}\Omega$$

$$X_{k_{PE}} = x'_{k_{PE}} \cdot l = 0.08\frac{\text{m}\Omega}{\text{m}} \cdot 25\,\text{m} = 2\,\text{m}\Omega$$

Sum of the resistances and reactances:

$$\sum R = 488.86\,\text{m}\Omega; \quad \sum X = 30.6\,\text{m}\Omega$$

Calculation of short-circuit impedance:

$$Z_k = \sqrt{R^2 + X^2} = 489.8\,\text{m}\Omega$$

$$I''_{k1} = \frac{c \cdot U_n}{\sqrt{3} \cdot Z_k} = \frac{0.95 \cdot 400\,\text{V}}{\sqrt{3} \cdot 489.8\,\text{m}\Omega} = 448\,\text{A}$$

Three-pole short-circuit current:
Resistances and reactances of the transformer:

$$R_T = 3.5\,\text{m}\Omega, \quad X_T = 13\,\text{m}\Omega$$

Resistances and reactances of the cable:

$$R_k = \frac{l}{\kappa \cdot S}\,\text{m}\Omega = \frac{85\,\text{m}}{56\frac{\text{m}}{\Omega\text{mm}^2} \cdot 120\,\text{mm}^2} = 12.65\,\text{m}\Omega$$

$$X_k = x'_i l = 0.08\frac{\text{m}\Omega}{\text{m}} \cdot 85\,\text{m} = 6.8\,\text{m}\Omega$$

Resistances and reactances of the conductor:

$$R_L = 178.6 \, m\Omega$$
$$X_L = 2 \, m\Omega$$
$$\sum R = 204.25 \, m\Omega$$
$$\sum X = 21.8 \, m\Omega$$
$$Z_k = \sqrt{(R_k^2 + X_k^2)} = \sqrt{(204.25^2 + 21.8^2)} \, m\Omega = 205.4 \, m\Omega$$
$$I''_{k3} = \frac{c \cdot U_n}{\sqrt{3} \cdot Z_k} = \frac{1.1 \cdot 400 \, V}{\sqrt{3} \cdot 205.4 \, m\Omega} = 1.236 \, kA$$

18.2 Example 2: Proof of Protective Measures

Given a 230/400-V network, as shown in Figure 18.2, with a source impedance of 0.3 Ω, the protective measures must be proven.

1) Calculate the resistances and reactances of the network at the fault locations.
2) Calculate the single-pole short-circuit current.
3) Are the cut-off conditions fulfilled?

Calculation of impedances for supply conductors:
Conductor for NYM-J 4 × 25 mm², $l = 15$ m.

$$R = 1.24 \cdot \frac{2 \cdot l}{\kappa \cdot S}$$

$$R = 1.24 \frac{2 \cdot 15 \, m}{56 \frac{m}{\Omega mm^2} \cdot 25 \, mm^2} = 0.0265 \, \Omega$$

$$X = x' \cdot 2 \cdot l \approx 0.08 \frac{m\Omega}{m} \cdot 2 \cdot 15 \, m = 0.0024 \, \Omega$$

$$Z_1 = \sqrt{R^2 + X^2} = \sqrt{0.0265^2 + 0.00024^2} \, \Omega = 0.0266 \, \Omega$$

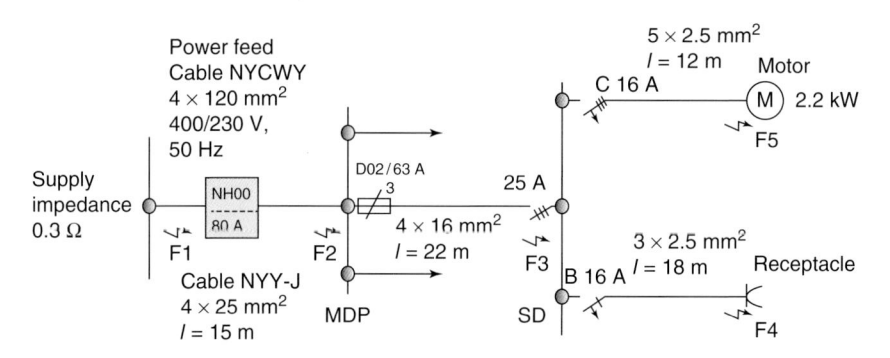

Figure 18.2 Example 2: calculation with source impedance.

Conductor for NYM-J $4 \times 16 \text{ mm}^2$, $l = 22 \text{ m}$.

$$R = 1.24 \cdot \frac{2 \cdot l}{\kappa \cdot S}$$

$$R = 1.24 \cdot \frac{2 \cdot 22 \text{ m}}{56 \frac{\text{m}}{\Omega \text{mm}^2} \cdot 16 \text{ mm}^2} = 0.0608 \, \Omega$$

$$X = x' \cdot 2 \cdot l \approx 0.08 \frac{\text{m}\Omega}{\text{m}} \cdot 2 \cdot 22 \text{ m} = 0.00352 \, \Omega$$

$$Z_2 = \sqrt{R^2 + X^2} = \sqrt{0.0608^2 + 0.00352^2} \, \Omega = 0.0609 \, \Omega$$

Conductors for outlet NYM-J $3 \times 2.5 \text{ mm}^2$, $l = 18 \text{ m}$.

$$R = 1.24 \cdot \frac{2 \cdot l}{\kappa \cdot S} = 1.24 \cdot \frac{2 \cdot 18 \text{ m}}{56 \frac{\text{m}}{\Omega \text{mm}^2} \cdot 2.5 \text{ mm}^2} = 0.318 \, \Omega$$

$$X = x' \cdot 2 \cdot l \approx 0.08 \frac{\text{m}\Omega}{\text{m}} \cdot 2 \cdot 18 \text{ m} = 0.00288 \, \Omega$$

$$Z_3 = \sqrt{R^2 + X^2} = \sqrt{0.318^2 + 0.00288^2} = 0.319 \, \Omega$$

Conductors for motor, $l = 18 \text{ m}$

$$R = 1.24 \cdot \frac{2 \cdot l}{\kappa \cdot S} = 1.24 \cdot \frac{2 \cdot 12 \text{ m}}{56 \frac{\text{m}}{\Omega \text{mm}^2} \cdot 2.5 \text{ mm}^2} = 0.212 \, \Omega$$

$$X = x' \cdot 2 \cdot l \approx 0.08 \frac{\text{m}\Omega}{\text{m}} \cdot 2 \cdot 12 \text{ m} = 0.00192 \, \Omega$$

$$Z_4 = \sqrt{R^2 + X^2} = \sqrt{0.212^2 + 0.00192^2} \, \Omega = 0.212 \, \Omega$$

Impedance at fault location F1
With $Z_V = 0.3 \, \Omega$
Single-pole short-circuit current:

$$I_{k1}'' = \frac{c \cdot U_n}{\sqrt{3} \cdot Z_V} = \frac{0.95 \cdot 400 \text{ V}}{\sqrt{3} \cdot 0.3 \, \Omega} = 731.3 \text{ A}$$

Impedance at fault location F2:

$$Z_A = Z_V + Z_1 = 0.3 \, \Omega + 0.0266 \, \Omega = 0.3266 \, \Omega$$

$$I_{k1}'' = \frac{0.95 \cdot 400 \text{ V}}{\sqrt{3} \cdot 0.3266 \, \Omega} = 671.7 \text{ A}$$

Impedance at fault location F3:

$$Z_B = Z_A + Z_2 = 0.3266 \, \Omega + 0.0609 \, \Omega = 0.3875 \, \Omega$$

$$I_{k1}'' = \frac{0.95 \cdot 400 \text{ V}}{\sqrt{3} \cdot 0.3875 \, \Omega} = 566.17 \text{ A}$$

Impedance at outlet F4:

$$Z_C = Z_B + Z_3 = 0.3875 \, \Omega + 0.318 \, \Omega = 0.7055 \, \Omega$$

$$I_{k1}'' = \frac{0.95 \cdot 400 \text{ V}}{\sqrt{3} \cdot 0.7055 \, \Omega} = 310.97 \text{ A}$$

Table 18.1 Summary of results.

Fault location	I''_{k1min} (A)	I_a (A)
F1	731.3	450
F2	671.7	450
F3	566.17	310
F4 outlet	310.97	80
F5 motor	365.96	160

Impedance at motor F5:

$$Z = Z_B + Z_4 = 0.3875\,\Omega + 0.212\,\Omega = 0.5995\,\Omega$$

$$I''_{k1} = \frac{0.95 \cdot 400\,\text{V}}{\sqrt{3} \cdot 0.5995\,\Omega} = 365.96\,\text{A}$$

Protection by cut off is the most important condition for satisfying the protective measures up to 1000 V. In accordance with IEC 60364, Part 41, the faults must be protected by cut off within 0.4 s for portable equipment of Protection Class I and in 5 s for permanently installed operational equipment. The cut-off currents of fuses and power breakers can be found in tables and diagrams (see IEC 60364, Part 43 and Part 600).

In this example, the cut-off currents I_a have the values:

$$\text{B16 A} \quad I_a = 5 \cdot I_n = 5 \cdot 16\,\text{A} = 80\,\text{A}$$

$$\text{C16 A} \quad I_a = 10 \cdot I_n = 10 \cdot 16\,\text{A} = 160\,\text{A}$$

$$\frac{\text{D02}}{63\,\text{A}} \quad I_{a(0.4\,s)} = 600\,\text{A}$$

$$I_{a(5\,s)} = 310\,\text{A}$$

$$\frac{\text{NH00}}{80\,\text{A}} \quad I_{a(0.4\,s)} = 800\,\text{A}$$

$$I_{a(5\,s)} = 450\,\text{A}$$

The condition $I'_{k1min} > I_a$ must always hold true.
Results from Table 18.1 show that the cut-off condition is satisfied.

18.3 Example 3: Connection Box to Service Panel

A power plant network as in Figure 18.3 supplies an on site connection box. The single-pole short-circuit current is 575 A.

1) Calculate the resistances and reactances.
2) Calculate the single-pole short-circuit current at the outlet and at the light.
3) Is protection by cut off guaranteed?

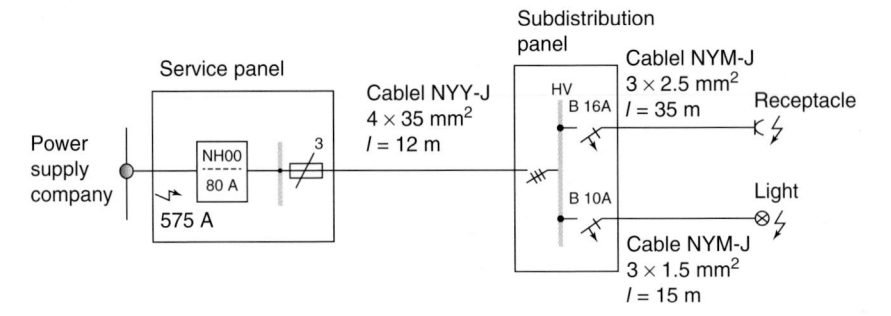

18 Examples: Calculation of Short-Circuit Currents

Figure 18.3 Example 3: power plant network with service panel.

The impedance of the connection box is

$$Z = \frac{c \cdot U_n}{\sqrt{3} \cdot I''_{k1}} = \frac{0.95 \cdot 400\,V}{\sqrt{3} \cdot 575\,A} = 381.6\,m\Omega$$

$$Z_{l1} = 2 \cdot z \cdot l_1 = 2 \cdot 0.654\,\Omega/km \cdot 0.012\,km = 15.7\,m\Omega$$

The total impedance of the subdistributor is

$$Z_k = Z + Z_{l1} = 397.3\,m\Omega$$

The single-pole short-circuit current is

$$I''_{k1} = \frac{c \cdot U_n}{\sqrt{3} \cdot Z_k} = \frac{0.95 \cdot 400\,V}{\sqrt{3} \cdot 397.3\,\Omega} = 542\,A$$

For the short circuit at the outlet:

$$z' = 2 \cdot z \cdot l_2 = 2 \cdot 9.02\,\Omega/km \cdot 0.035\,km + 0.3973\,\Omega = 1.0287\,\Omega$$

$$I''_{k1} = 213.27\,A$$

The cut-off current of a 16-A circuit breaker is 80 A. Since the single-pole short-circuit current is greater than the cut-off current, the cut-off condition is satisfied.

Short circuit on light:

$$z' = 2 \cdot z \cdot l_3 = 2 \cdot 15\,\Omega/km \cdot 0.015\,km + 0.3973\,\Omega = 0.8473\,\Omega$$

$$I''_{k1} = 258.9\,A$$

The cut-off current of a 10-A circuit breaker is 50 A. The cut-off condition is therefore again satisfied.

18.4 Example 4: Transformers in Parallel

Two transformers are connected in parallel, as shown in Figure 18.4.

1) Draw the equivalent circuit.
2) Calculate the resistances and reactances.
3) Calculate the three-pole short-circuit current.

18.4 Example 4: Transformers in Parallel

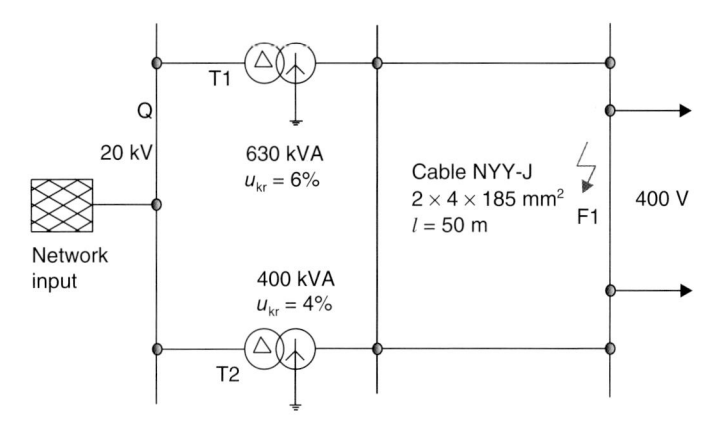

Figure 18.4 Example 4: transformers in parallel.

Total transformer power:

$$\sum S_{rT} = 630\,\text{kVA} + 400\,\text{kVA} = 1030\,\text{kVA}$$

$$u_{Rr} = \frac{u_{R1} + u_{R2}}{2} = 1.125\%$$

$$Z_T = \frac{u_{kr}}{100\%} \frac{U_{nT}^2}{S_{rT}} = \frac{5\%}{100\%} \frac{(400\,\text{V})^2}{1030\,\text{kVA}} = 7.77\,\text{m}\Omega$$

$$R_T = \frac{u_{Rr}}{100\%} \frac{U_{nT}^2}{S_{rT}} = \frac{1.125}{100\%} \frac{(400\,\text{V})^2}{1030\,\text{kVA}} = 1.75\,\text{m}\Omega$$

$$X_T = \sqrt{Z_T^2 - R_T^2} = 7.57\,\text{m}\Omega$$

Figure 18.5 shows equivalent circuit diagram.

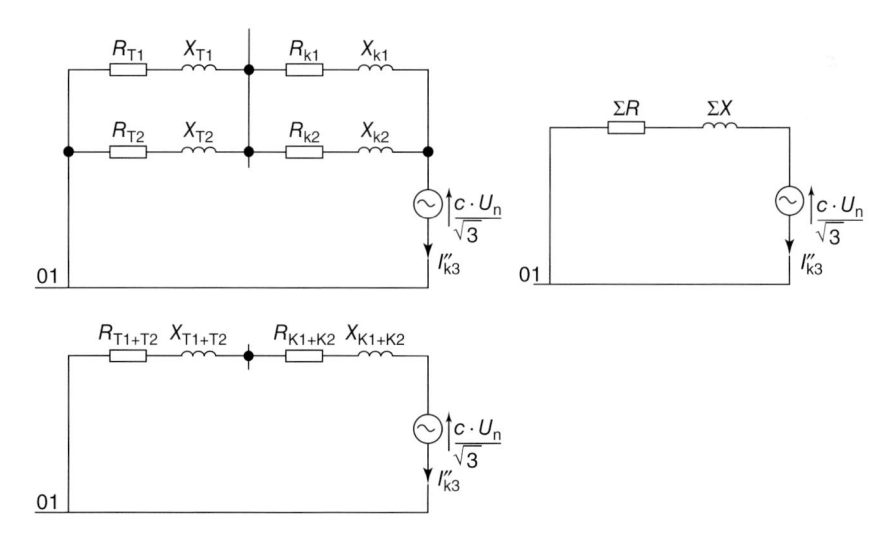

Figure 18.5 Equivalent circuit.

Cables:

$$R_L = \frac{l}{\kappa \cdot S \cdot n} = \frac{50\,\text{m}}{56\frac{\text{m}}{\Omega \cdot \text{mm}^2} \cdot 185\,\text{mm}^2 \cdot 2} = 2.4\,\text{m}\Omega$$

$$X_L = x'_L \cdot \frac{l}{n} = 0.08\,\text{m}\Omega \cdot \frac{50\,\text{m}}{2} = 2\,\text{m}\Omega$$

$$Z_k = \sqrt{R_k^2 + X_k^2} = \sqrt{4.15^2 + 9.57^2}\,\text{m}\Omega = 10.43\,\text{m}\Omega$$

$$I''_{k3} = \frac{c \cdot U_{rT}}{\sqrt{3} \cdot Z_k} = \frac{1.1 \cdot 400\,\text{V}}{\sqrt{3} \cdot 10.43\,\text{m}\Omega} = 24.35\,\text{kA}$$

18.5 Example 5: Connection of a Motor

A transformer as shown in Figure 18.6 supplies a motor through an overhead line, cables, and conductors.

1) Calculate the resistances and reactances.
2) Calculate the single-pole short-circuit current.

Transformer:

$$Z_T = 15.238\,\text{m}\Omega$$
$$R_T = 2.8\,\text{m}\Omega$$
$$X_T = 15\,\text{m}\Omega$$

Overhead line:

$$R_{L1} = 1.24 \cdot \frac{l}{\kappa \cdot S}$$

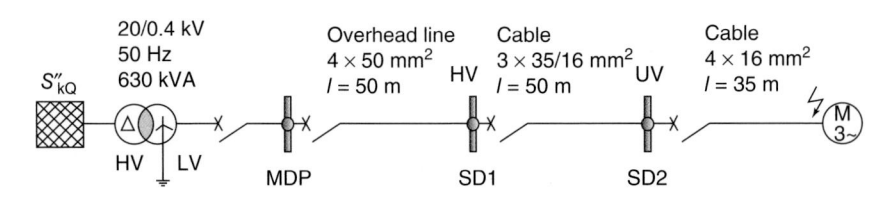

$$R_{L1} = 1.24 \cdot \frac{50\,\text{m}}{56\frac{\text{m}}{\Omega \cdot \text{mm}^2} \cdot 50\,\text{mm}^2} = 22.1\,\text{m}\Omega$$

$$R_{PEN} = 0.2\,\text{m}\Omega$$

$$X_{L1} = x' \cdot l = 0.33\frac{\Omega}{\text{km}} \cdot 50\,\text{m} = 16.5\,\text{m}\Omega$$

$$X_{PEN} = 16.5\,\text{m}\Omega$$

Figure 18.6 Example 5: connection of a motor.

Cables:

$$R_{L2} = 1.24 \cdot \frac{50\,\text{m}}{56\frac{\text{m}}{\Omega \cdot \text{mm}^2} \cdot 35\,\text{mm}^2} = 31.6\,\text{m}\Omega$$

$$R_{PEN} = 1.24 \cdot \frac{50\,\text{m}}{56\frac{\text{m}}{\Omega \cdot \text{mm}^2} \cdot 16\,\text{mm}^2} = 69.2\,\text{m}\Omega$$

$$X_{L2} = x' \cdot l = 0.08\frac{\Omega}{\text{km}} \cdot 50\,\text{m} = 4\,\text{m}\Omega$$

$$X_{PEN} = 4\,\text{m}\Omega$$

Conductors:

$$R_{L3} = 1.24 \cdot \frac{35\,\text{m}}{56\frac{\text{m}}{\Omega \cdot \text{mm}^2} \cdot 16\,\text{mm}^2} = 48.4\,\text{m}\Omega$$

$$R_{PEN} = 1.24 \cdot \frac{35\,\text{m}}{56\frac{\text{m}}{\Omega \cdot \text{mm}^2} \cdot 16\,\text{mm}^2} = 48.4\,\text{m}\Omega$$

$$X_{L3} = x' \cdot l = 0.08\frac{\Omega}{\text{km}} \cdot 35\,\text{m} = 2.8\,\text{m}\Omega$$

$$X_{PEN} = 2.8\,\text{m}\Omega$$

$$\sum R = 222.7\,\text{m}\Omega$$

$$\sum X = 61.6\,\text{m}\Omega$$

$$Z_k = \sqrt{R_k^2 + X_k^2} = \sqrt{227.7^2 + 61.6^2}\,\text{m}\Omega = 231.06\,\text{m}\Omega$$

$$I''_{k1} = \frac{c \cdot U_{nT}}{\sqrt{3} \cdot Z_k} = \frac{0.95 \cdot 400\,\text{V}}{\sqrt{3} \cdot 231.06\,\text{m}\Omega} = 949.5\,\text{A}$$

18.6 Example 6: Calculation for a Load Circuit

A grounding cable is connected to a transformer (Figure 18.7). The data for the cable are as follows: $r' = 0.482\,\Omega/\text{km}$, $x' = 0.083\,\Omega/\text{km}$, $\frac{R_{0L}}{R_L} = 4$, and $\frac{X_{0L}}{X_L} = 3.76$.

1) Calculate the resistances and reactances.
2) Calculate the three-pole and single-pole short-circuit currents.

Figure 18.7 Example 6: Calculation for a load.

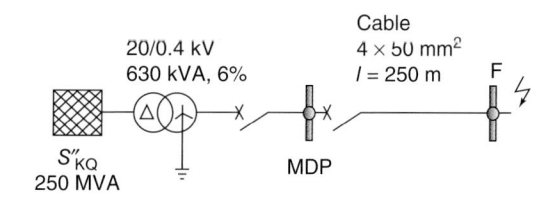

18 Examples: Calculation of Short-Circuit Currents

Calculation of I''_{k3}:

$$Z_{Qt} = \frac{c \cdot U_n^2}{S''_{kQ}} = \frac{1.1 \cdot (0.4\,\text{kV})^2}{250\,\text{MVA}} = 0.704\,\text{m}\Omega$$

$X_{Qt} = 0.995 \cdot Z_{Qt} = 0.7\,\text{m}\Omega$

$R_{Qt} = 0.1 \cdot X_{Qt} = 0.07\,\text{m}\Omega$

$$R_T = \frac{u_{Rr}}{100\%} \frac{U_{rT}^2}{S_{rT}} = \frac{1.1\%}{100\%} \cdot \frac{(400\,\text{V})^2}{630\,\text{kVA}} = 2.8\,\text{m}\Omega$$

$$Z_T = \frac{u_{kr} \cdot U_{rT}^2}{100\% \cdot S_{rT}} = \frac{6\%}{100\%} \cdot \frac{(400\,\text{V})^2}{630\,\text{kVA}} = 15.2\,\text{m}\Omega$$

$X_T = \sqrt{Z^2 - R_T^2} = \sqrt{15.2^2 - 2.8^2}\,\text{m}\Omega = 14.9\,\text{m}\Omega$

$R_{0T} = R_T = 2.8\,\text{m}\Omega$

$X_{0T} = 0.995 \cdot X_T = 0.995 \cdot 14.9\,\text{m}\Omega = 14.83\,\text{m}\Omega$

$R_1 = R' \cdot l = 0.482\,\Omega/\text{km} \cdot 0.25\,\text{km} = 120.5\,\text{m}\Omega$

$X_1 = X' \cdot l = 0.083\,\Omega/\text{km} \cdot 0.25\,\text{km} = 20.75\,\text{m}\Omega$

$R_{0l} = 4 \cdot R_1 = 482\,\text{m}\Omega$

$X_{0l} = 3.76 \cdot X_1 = 78.02\,\text{m}\Omega$

$R_k = (R_{Qt} + R_T + R_1) = 123.4\,\text{m}\Omega$

$X_k = (X_{Qt} + X_T + X_1) = 36.35\,\text{m}\Omega$

$Z_k = \sqrt{R_k^2 + X_k^2} = \sqrt{123.4^2 + 36.35^2}\,\text{m}\Omega = 128.6\,\text{m}\Omega$

$$I''_{k3} = \frac{c \cdot U_n}{\sqrt{3} \cdot Z_k} = 1.97\,\text{kA}$$

Calculation of I''_{k1} (Figure 18.8):

$2\underline{Z}_1 + \underline{Z}_0 = 2 \cdot (123.4 + j36.35)\,\text{m}\Omega + (484.8 + j92.85)\,\text{m}\Omega$

$2\underline{Z}_1 + \underline{Z}_0 = (731.6 + j165.6)\,\text{m}\Omega = 750.1\,\text{m}\Omega$

$$I''_{k1} = \frac{\sqrt{3} \cdot c \cdot U_n}{2 \cdot Z_1 + Z_0} = \frac{\sqrt{3} \cdot 0.95 \cdot 400\,\text{V}}{750.1\,\text{m}\Omega} = 877\,\text{A}$$

Simplified method for calculating the single-pole short-circuit current:
Transformer impedance:

$$Z_T = \frac{u_{kr}}{100\%} \frac{U_{rT}^2}{S_{rT}} = \frac{6\% \cdot (400\,\text{V})^2}{100\% \cdot 630\,\text{kVA}} = 0.015\,\Omega$$

Line impedance (outgoing and return lines):

$Z_1 = 2 \cdot z' \cdot l = 2 \cdot 0.486\,\Omega/\text{km} \cdot 0.250\,\text{km} = 0.243\,\Omega$

Total impedance:

$Z_{\text{total}} = Z_T + Z_L = 0.258\,\Omega$

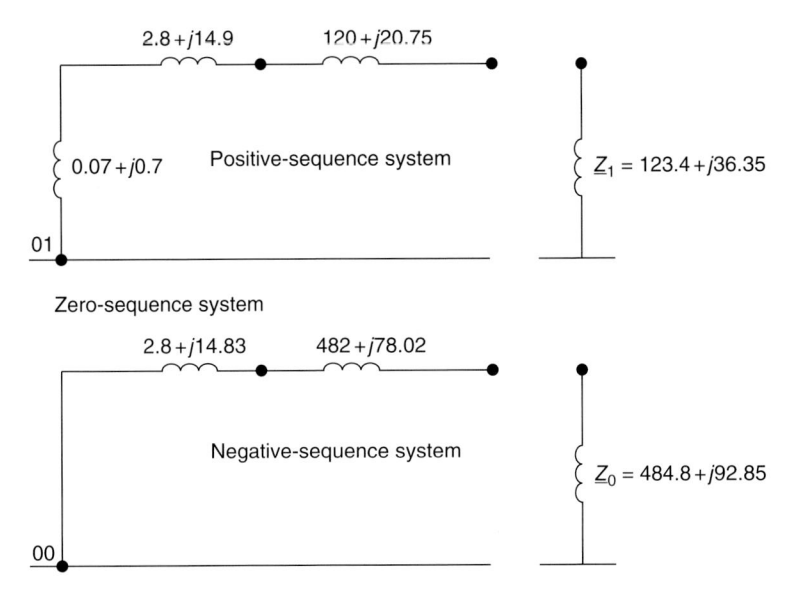

Figure 18.8 Equivalent circuit for I''_{k1}.

Single-pole short-circuit current:

$$I''_{k1} = \frac{c \cdot U_n}{\sqrt{3} \cdot Z_{Total}} = \frac{0.95 \cdot 400\,V}{\sqrt{3} \cdot 0.258\,\Omega} = 850.36\,A$$

The result is nearly the same by both methods.

18.7 Example 7: Calculation for an Industrial System

A transformer as shown in Figure 18.9 supplies the main distributor of an industrial system.

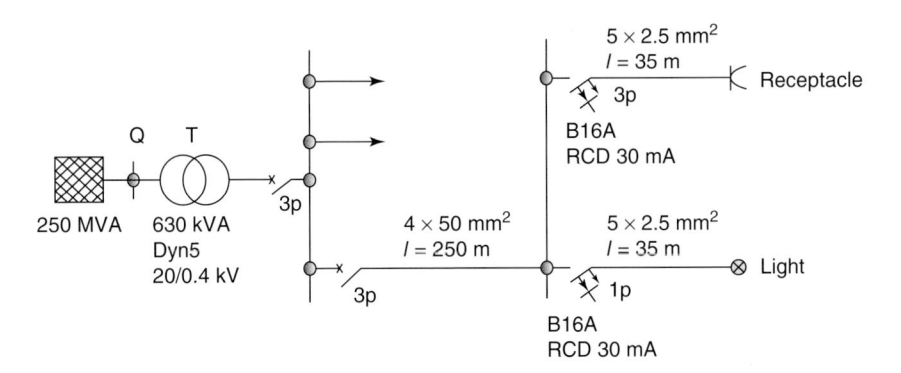

Figure 18.9 Example 7: network diagram.

1) Calculate the resistances and reactances.
2) Calculate the three-pole short-circuit current at the main distributor.
3) Calculate the single-pole short-circuit current at the outlet and at the light switch.

Determination of individual impedances:

$$Z_{Qt} = \frac{c \cdot U_n^2}{S_{kQ}''} = \frac{1.1 \cdot (400\,V)^2}{250\,MVA} = 0.704\,m\Omega$$

$$X_{Qt} = 0.995 \cdot Z_{Qt} = 0.7\,m\Omega$$

$$R_{Qt} = 0.1 \cdot X_{Qt} = 0.07\,m\Omega$$

$$Z_T = \frac{u_{kr} \cdot U_{rT}^2}{100\% \cdot S_{rT}} = \frac{6\% \cdot (400\,V)^2}{630\,kVA} = 15.24\,m\Omega$$

$$Z_{L1} = 2 \cdot z \cdot l_1 = 2 \cdot 0.486\,\Omega/km \cdot 0.25\,km = 243\,m\Omega$$

$$Z_{L2} = 2 \cdot z \cdot l_2 = 2 \cdot 9.02\,\Omega/km \cdot 0.035\,km = 631.4\,m\Omega$$

$$Z_{L3} = 2 \cdot z \cdot l_2 = 2 \cdot 15\,\Omega/km \cdot 0.015\,km = 450\,m\Omega$$

I_{k1}'' *at main distributor:*

$$Z_{L1} = 2 \cdot z \cdot l_1 = 2 \cdot 0.396\,\Omega/km \cdot 0.25\,km = 99\,m\Omega$$

$$Z_{HV} = Z_{Qt} + Z_T + Z_{l1} = 115\,m\Omega$$

$$I_{k1}'' = \frac{c \cdot U_n}{\sqrt{3} \cdot Z_{HV}} = \frac{0.95 \cdot 400\,V}{\sqrt{3} \cdot 115\,m\Omega} = 1.9\,kA$$

I_{k1}'' *at outlet:*

$$Z_{out} = Z_{Qt} + Z_T + Z_{l1} + Z_{l2} = 890.344\,m\Omega$$

$$I_{k1}'' = \frac{c \cdot U_n}{\sqrt{3} \cdot Z_{out}} = \frac{0.95 \cdot 400\,V}{\sqrt{3} \cdot 890.344\,m\Omega} = 246.4\,A$$

I_{k1}'' *at light switch:*

$$Z_{li} = Z_{Qt} + Z_T + Z_{l1} + Z_{l2} = 708.944\,m\Omega$$

$$I_{k1}'' = \frac{c \cdot U_n}{\sqrt{3} \cdot Z_{Li}} = \frac{0.95 \cdot 400\,V}{\sqrt{3} \cdot 708.944\,m\Omega} = 309.46\,A$$

18.8 Example 8: Calculation of Three-Pole Short-Circuit Current and Peak Short-Circuit Current

Given the network (Figure 18.10).

1) Calculate the resistances and reactances.
2) Calculate the three-pole short-circuit currents and the peak short-circuit currents at the fault locations.

18.8 Example 8: Calculation of Three-Pole Short-Circuit Current

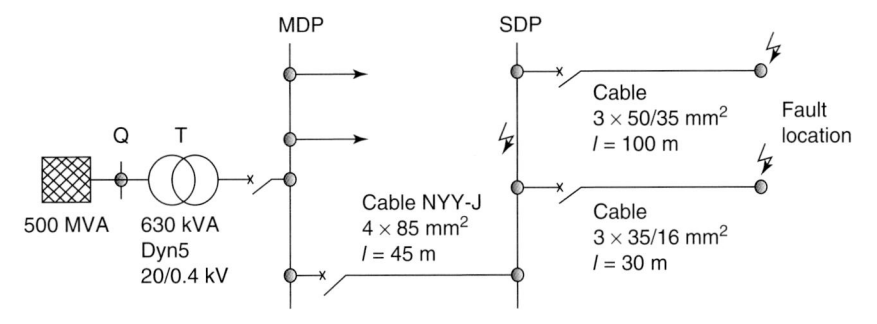

Figure 18.10 Example 8: network diagram.

Medium voltage:

$$Z_{Qt} = \frac{1.1 \cdot U_{rT}^2}{S_Q''} = \frac{1.1 \cdot (400\,\text{V})^2}{500\,\text{MVA}} = 0.352\,\text{m}\Omega$$

$$R_Q = 0.1 \cdot X_Q = 0.1 \cdot 0.35 = 0.0352\,\text{m}\Omega$$

Transformer:

$$u_x = \sqrt{u_{kr}^2 - u_{Rr}^2} = 5.9\%$$

$$R_T = \frac{u_{Rr} \cdot U_{rT}^2}{100\% \cdot S_{rT}} = 2.8\,\text{m}\Omega$$

$$X_T = \frac{u_{xr} \cdot U_{rT}^2}{100\% \cdot S_{rT}} = 15\,\text{m}\Omega$$

Supply cables:

$$R_1' = 0.101\,\Omega/\text{km}$$
$$X_1' = 0.08\,\Omega/\text{km}$$
$$R_{l1} = r' \cdot l = 4.545\,\text{m}\Omega$$
$$X_{l1} = x' \cdot l = 3.6\,\text{m}\Omega$$

Main low-voltage distributor (three-pole short circuit):

$$\underline{Z}_k = R_{Qt} + R_T + R_{l1} + j(X_{Qt} + X_T + X_{l1})$$
$$\underline{Z}_k = (0.0352 + 2.8 + 4.545)\,\text{m}\Omega + j(0.352 + 15 + 3.6)\,\text{m}\Omega$$
$$\underline{Z}_k = (7.38 + j18.952)\,\text{m}\Omega$$
$$Z_k = \sqrt{7.38^2 + 18.952^2}\,\text{m}\Omega = 20.34\,\text{m}\Omega$$
$$I_{k3}'' = \frac{c \cdot U_n}{\sqrt{3} \cdot Z_k} = \frac{1.1 \cdot 400\,\text{V}}{\sqrt{3} \cdot 20.34\,\text{m}\Omega} = 12.48\,\text{kA}$$
$$\frac{R_k}{X_k} = \frac{7.38\,\text{m}\Omega}{18.952\,\text{m}\Omega} = 0.389 \quad \text{from Figure 11.6 we get } \kappa = 1.32$$
$$i_p = \kappa \cdot \sqrt{2} \cdot I_{k3}'' = 1.32 \cdot \sqrt{2} \cdot 12.48\,\text{kA} = 23.29\,\text{kA}$$

Supply lines (cables and lines):

$$R_{12} = r \cdot l = 35.71 \, m\Omega$$
$$X_{12} = x \cdot l = 0.08 \, \Omega/km \cdot 0.100 \, km = 8 \, m\Omega$$
$$Z_k = \sqrt{35.71^2 + 8^2} \, m\Omega = 36.6 \, m\Omega$$

Subdistribution I (three-pole short circuit):

$$Z_k = (20.34 + 36.6) \, m\Omega = 57 \, m\Omega$$
$$I''_{k3} = \frac{c \cdot U_n}{\sqrt{3} \cdot Z_k}$$
$$I''_{k3} = \frac{1.1 \cdot 400 V}{\sqrt{3} \cdot 57 \, m\Omega} = 4.45 \, kA$$
$$i_p = \kappa \cdot \sqrt{2} \cdot I''_{k3} = 7.74 \, kA$$

Subdistribution II (three-pole short circuit):

$$R_{13} = \frac{l}{\kappa \cdot S}$$
$$R_{13} = \frac{30 \, m}{56 \frac{m}{\Omega mm^2} \cdot 35 \, mm^2} = 15.3 \, m\Omega$$
$$X_{13} = x \cdot l = 0.08 \, \Omega/km \cdot 30 \, m = 2.4 \, m\Omega$$
$$Z_k = R_{Qt} + R_T + R_{l1} + R_{l3} + j(X_{Qt} + X_T + X_{l1} + X_{l3})$$
$$= (22.68 + j21.352) \, m\Omega$$
$$Z_k = \sqrt{R^2 + X^2} = 31.15 \, m\Omega$$
$$I''_{k3} = \frac{c \cdot U_n}{\sqrt{3} \cdot Z_k}$$
$$I''_{k3} = \frac{1.1 \cdot 400 \, V}{\sqrt{3} \cdot 31.15 \, m\Omega} = 8.15 \, kA$$
$$i_p = \kappa \cdot \sqrt{2} \cdot I''_{k3} = 12.67 \, kA$$

18.9 Example 9: Meshed Network

Given a meshed network, as shown in Figure 18.11.

1) Calculate the impedances.
2) Carry out the network transformations.
3) Calculate the three-pole short-circuit currents and the peak short-circuit currents at the fault location F.

The following data are given:
With $S_{rT} = 160 \, MVA$, $u_{kr} = 12\%$, $U_n = 50 \, kV$, and $Z = 0.5 \, \Omega/km$ for all lines.

18.9 Example 9: Meshed Network

Figure 18.11 Meshed network.

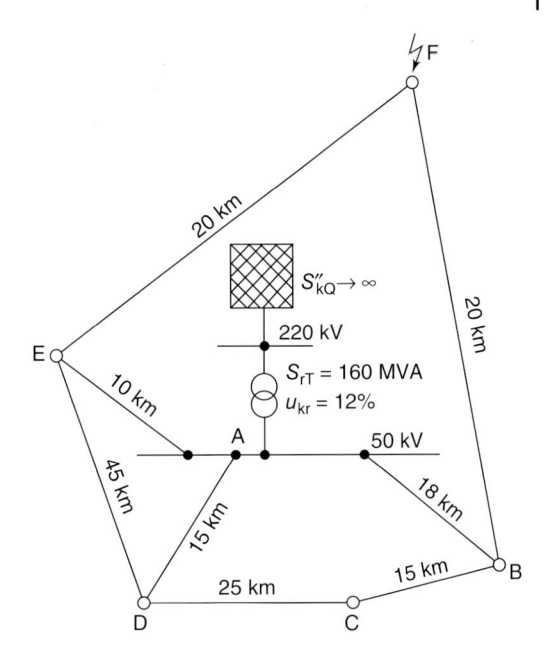

The impedance of the transformer:

$$Z_T = \frac{u_{kr} \cdot U_{rT}^2}{100 \cdot S_{rT}} = \frac{12\% \cdot (50\,\text{kV})^2}{100\% \cdot 160\,\text{MVA}} = 1.875\,\Omega$$

The impedances of the individual conductors:

$$Z_{AE} = 10\,\text{km} \cdot 0.5\frac{\Omega}{\text{km}} = 5\,\Omega$$

$$Z_{AD} = 15\,\text{km} \cdot 0.5\frac{\Omega}{\text{km}} = 7.5\,\Omega$$

$$Z_{AB} = 18\,\text{km} \cdot 0.5\frac{\Omega}{\text{km}} = 9\,\Omega$$

$$Z_{ED} = 45\,\text{km} \cdot 0.5\frac{\Omega}{\text{km}} = 22.5\,\Omega$$

$$Z_{EF} = 20\,\text{km} \cdot 0.5\frac{\Omega}{\text{km}} = 10\,\Omega$$

$$Z_{BD} = 40\,\text{km} \cdot 0.5\frac{\Omega}{\text{km}} = 20\,\Omega$$

$$Z_{BF} = 20\,\text{km} \cdot 0.5\frac{\Omega}{\text{km}} = 10\,\Omega$$

Transformation of the delta star impedances (Figure 18.12b):

$$Z_{AG} = \frac{Z_{AE} \cdot Z_{AD}}{Z_{AD} + Z_{AE} + Z_{ED}} = \frac{5 \cdot 7.5}{7.5 + 5 + 22.5} = 1.07\,\Omega$$

$$Z_{EG} = \frac{Z_{AE} \cdot Z_{ED}}{Z_{AD} + Z_{AE} + Z_{ED}} = \frac{5 \cdot 22.5}{7.5 + 5 + 22.5} = 3.21\,\Omega$$

$$Z_{DG} = \frac{Z \cdot Z}{Z_{AD} + Z_{AE} + Z_{ED}} = \frac{22.5 \cdot 7.5}{7.5 + 5 + 22.5} = 4.82\,\Omega$$

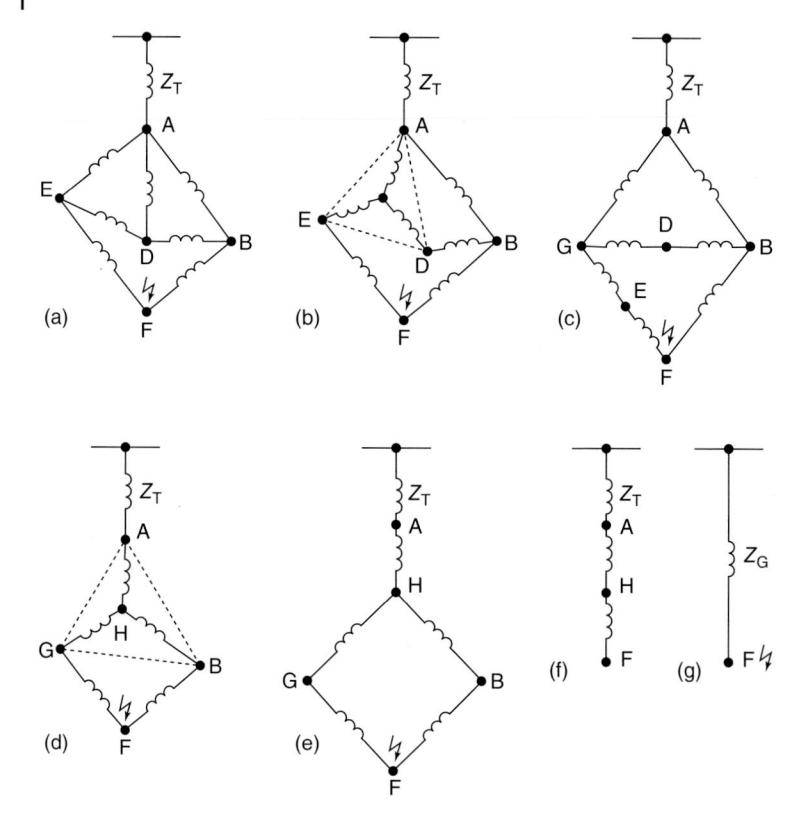

Figure 18.12 Delta star transformations.

Addition of series impedances (Figure 18.12c):

$$Z_{EG} + Z_{EF} = Z_{GEF} = 13.21\,\Omega$$
$$Z_{DG} + Z_{BD} = Z_{GDB} = 24.82\,\Omega$$

Transformation of the delta star impedances (Figure 18.12d):

$$Z_{AH} = \frac{Z_{AB} \cdot Z_{AG}}{Z_{AB} + Z_{AG} + Z_{GDB}} = \frac{9 \cdot 1.07}{9 + 1.07 + 24.82} = 0.276\,\Omega$$

$$Z_{BH} = \frac{Z_{AB} \cdot Z_{GDB}}{Z_{AB} + Z_{AG} + Z_{GDB}} = \frac{9 \cdot 24.82}{9 + 1.07 + 29.82} = 6.397\,\Omega$$

$$Z_{GH} = \frac{Z_{AG} \cdot Z_{GDB}}{Z_{AB} + Z_{AG} + Z_{GDB}} = \frac{1.07 \cdot 29.8}{9 + 1.07 + 24.82} = 0.76\,\Omega$$

Addition of series impedances (Figure 18.12e):

$$Z_{GH} + Z_{GEF} = Z_{HGF} = 0.76\,\Omega + 13.21\,\Omega = 13.97\,\Omega$$
$$Z_{BH} + Z_{BF} = Z_{HBF} = 6.7\,\Omega + 10\,\Omega = 16.7\,\Omega$$

Calculation of parallel impedances (Figure 18.12f):

$$Z_{HF} = \frac{Z_{HGF} \cdot Z_{HBF}}{Z_{HGF} + Z_{HBF}} = 7.54\,\Omega$$

Sum of all impedances (Figure 18.12g):

$$Z_k = Z_T + Z_{AH} + Z_{HF} = 9.69\,\Omega$$

$$I''_{k3} = \frac{c \cdot U_n}{\sqrt{3} \cdot Z_k} = \frac{1.1 \cdot 50\,\text{kV}}{\sqrt{3} \cdot 9.69\,\Omega} = 3.26\,\text{kA}$$

$$S''_{k3} = \frac{c \cdot U_n^2}{Z_k} = \frac{1.1 \cdot (50\,\text{kV})^2}{9.69\,\Omega} = 282.5\,\text{MVA}$$

$$i_p = 1.8 \cdot \sqrt{2} \cdot I''_{k3} = 1.8 \cdot \sqrt{2} \cdot 3.26\,\text{kA} = 8.29\,\text{kA}$$

18.10 Example 10: Supply to a Factory

A factory is supplied from an overhead line and cables with two transformers, as shown in Figure 18.13.

1) Calculate the impedances at the fault locations.
2) Calculate I''_{k3} at the fault locations.

Impedances at the fault locations:

$$Z_{Qt} = \frac{c \cdot U_{nQ}^2}{S''_{kQ}} = \frac{1.1 \cdot (0.4\,\text{kV})^2}{500\,\text{MVA}} = 0.352\,\text{m}\Omega$$

$$X_{Qt} = 0.995 \cdot Z_{Qt} = 0.995 \cdot 0.352\,\text{m}\Omega = 0.35\,\text{m}\Omega$$

$$R_{Qt} = 0.1 \cdot X_{Qt} = 0.1 \cdot 0.35\,\text{m}\Omega = 0.035\,\text{m}\Omega$$

$$Z_T = \frac{u_{kr}}{100\%} \cdot \frac{U_{rT}^2}{\sum S_{rT}} = \frac{6\%}{100\%} \cdot \frac{(400\,\text{V})^2}{4\,\text{MVA}} = 2.4\,\text{m}\Omega$$

$$R_T = \frac{u_{Rr}}{100\%} \cdot \frac{U_{rT}^2}{S_{rT}} = \frac{1.05\%}{100\%} \cdot \frac{(400\,\text{V})^2}{4\,\text{MVA}} = 0.42\,\text{m}\Omega$$

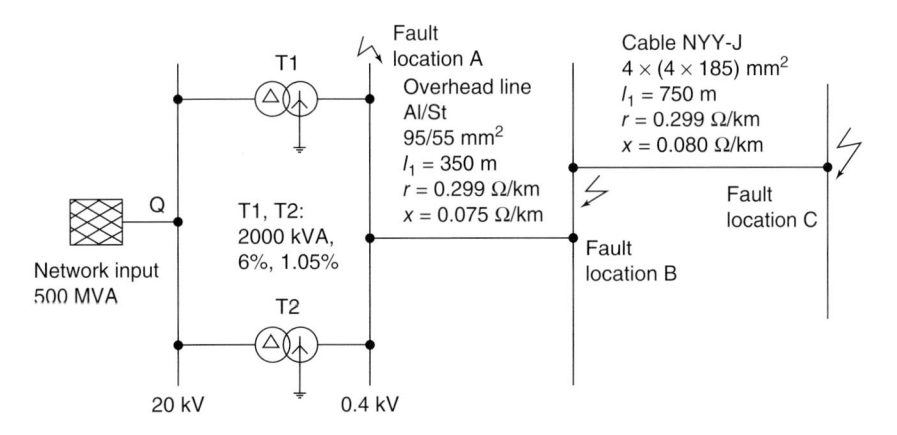

Figure 18.13 Supply to a factory.

Network input 500 MVA
Q
T1, T2: 2000 kVA, 6%, 1.05%
20 kV / 0.4 kV

Fault location A
Overhead line Al/St 95/55 mm²
$l_1 = 350$ m
$r = 0.299\,\Omega/\text{km}$
$x = 0.075\,\Omega/\text{km}$

Cable NYY-J
4 × (4 × 185) mm²
$l_1 = 750$ m
$r = 0.299\,\Omega/\text{km}$
$x = 0.080\,\Omega/\text{km}$

Fault location B

Fault location C

$$X_T = \sqrt{Z_T^2 - R_T^2}$$

$$X_T = \sqrt{2.4^2 - 0.42^2}\,m\Omega = 2.363\,m\Omega$$

Overhead line:

$$R_F = r' \cdot l = 0.299\,\Omega/km \cdot 0.350\,km = 104.7\,m\Omega$$
$$X_F = x' \cdot l = 0.075\,\Omega/km \cdot 0.350\,km = 26.3\,m\Omega$$
$$Z_F = \sqrt{R^2 + X^2} = \sqrt{104.7^2 + 26.3^2}\,m\Omega = 107.95\,m\Omega$$

Cables:

$$R_k = r' \cdot \frac{l}{n} = 0.101\,\Omega/km \cdot \frac{0.750\,km}{4} = 18.94\,m\Omega$$
$$X_k = x' \cdot \frac{l}{n} = 0.080\,\Omega/km \cdot \frac{0.750\,km}{4} = 15\,m\Omega$$
$$Z_k = \sqrt{R^2 + X^2} = \sqrt{18.94^2 + 15^2}\,m\Omega$$
$$Z_k = 24.16\,m\Omega$$
$$Z_{kA} = Z_{QT} + Z_T = 2.75\,m\Omega$$

Short circuit at position A:

$$I''_{k3} = \frac{c \cdot U_n}{\sqrt{3} \cdot Z_{kA}} = \frac{1.1 \cdot 400\,V}{\sqrt{3} \cdot 2.75\,m\Omega} = 92.37\,A$$

Three-pole short circuit at position B:

$$Z_{kB} = Z_{kA} + Z_F = 2.75\,m\Omega + 107.95\,m\Omega = 110.7\,m\Omega$$
$$I''_{k3} = \frac{1.1 \cdot 400\,V}{\sqrt{3} \cdot 110.7\,m\Omega} = 2.29\,kA$$

Short circuit at position C:

$$Z_{kC} = Z_{kB} + Z_K = 110.7\,m\Omega + 24.16\,m\Omega = 134.86\,m\Omega$$
$$I''_{k3} = \frac{1.1 \cdot 400\,V}{\sqrt{3} \cdot 134.86\,m\Omega} = 1.88\,kA$$

18.11 Example 11: Calculation with Impedance Corrections

Given the network diagram (Figure 18.14):

1) Calculate the impedances at the fault locations.
2) Calculate the impedance corrections.
3) Calculate the transferred short-circuit currents.
4) Calculate I''_{k3}, i_{p3}, and I_a.

18.11 Example 11: Calculation with Impedance Corrections

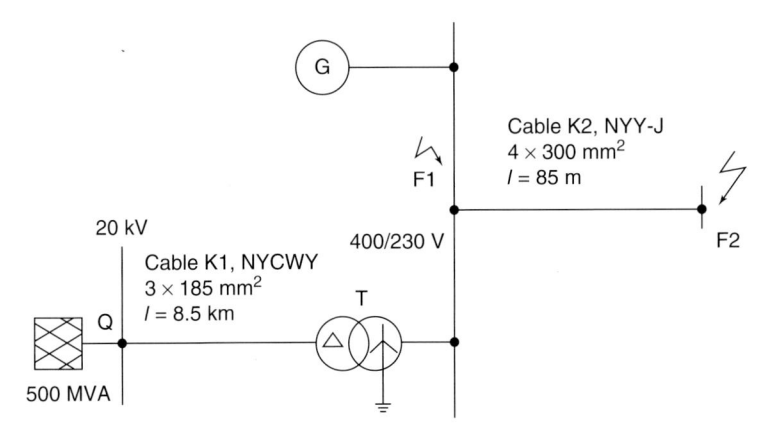

Figure 18.14 Calculation with impedance corrections.

The following data are known:
Transformer:
$S_{rT} = 1000\,\text{kVA}$, connection symbol: Dyn5, $u_{kr} = 6\%$, $u_{Rr} = 1.05\%$
Generator:
$S_{rG} = 600\,\text{kVA}$, $U_{rG} = 0.4\,\text{kV}$, $\cos\varphi_{rG} = 0.8$, $x''_d = 12\%$, $x''_{(0)G} = 8\%$

Cable K1:
$r' = 0.105\,\Omega/\text{km}$, $x' = 0.072\,\Omega/\text{km}$

Cable K2:
$r' = 0.066\,\Omega/\text{km}$, $x' = 0.079\,\Omega/\text{km}$, $\dfrac{R_{0L}}{R_L} = 4$, $\dfrac{X_{0L}}{X_L} = 3.66$

Impedances of the network feed-in:

$$Z_Q = \dfrac{1.1 \cdot U_n^2}{S''_{kQ}} = \dfrac{1.1 \cdot (20\,\text{kV})^2}{500\,\text{MVA}} = 0.88\,\Omega$$

$$Z_Q = \sqrt{R_Q^2 + X_Q^2}$$

$$X_Q = \dfrac{Z_Q}{1.005} = \dfrac{0.88\,\Omega}{1.005} = 0.8756\,\Omega$$
$R_Q = 0.1 \cdot 0.8756\,\Omega = 0.08756\,\Omega$
$\underline{Z}_Q = (0.08756 + j0.8756)\,\Omega$

Impedances of the supply cable:

$\underline{Z}_{K1} = l \cdot (r' + jx') = 8.5\,\text{km}(0.105 + j0.072)\,\Omega/\text{km} = (08925 + j0.612)\,\Omega$
$\underline{Z}_{G1} = \underline{Z}_Q + \underline{Z}_K = (0.98 + j1.4876)\,\Omega$

$$\underline{Z} = \underline{Z}_{G1} \left(\dfrac{U_{rTLV}}{U_{rTHV}}\right)^2 = (0.98 + j1.4876)\left(\dfrac{0.4\,\text{kV}}{20\,\text{kV}}\right)^2$$
$= (0.000392 + j0.000595)\,\Omega$

$$u_{xrT} = \sqrt{u_{kr}^2 - u_{Rr}^2} = \sqrt{(6^2 - 1.05^2)}\% = 5.9\%$$

$$R_T = \frac{u_{Rr}}{100\%} \cdot \frac{U_{LVT}^2}{S_{rT}} = \frac{1.05\%}{100\%} \cdot \frac{(0.4\,\text{kV})^2}{1000\,\text{kVA}} = 0.00168\,\Omega$$

$$X_T = \frac{u_{kr}}{100\%} \cdot \frac{U_{LVT}^2}{S_{rT}} = \frac{5.9\%}{100\%} \cdot \frac{(0.4\,\text{kV})^2}{1000\,\text{kVA}} = 0.00944\,\Omega$$

$$\underline{Z}_T = (0.00168 + j0.00944)\,\Omega$$

$$\underline{Z}_G = \underline{Z} + \underline{Z}_T = (0.002072 + j0.01)\,\Omega$$

$$Z_G = 0.01\,\Omega$$

$$X_{Gen} = X_d'' = \frac{x_d'' \cdot U_{rG}^2}{100\% \cdot S_{rG}} = \frac{12\%}{100\%} \cdot \frac{(0.4\,\text{kV})^2}{600\,\text{kVA}} = 0.032\,\Omega$$

$$R_{Gen} = 0.15 \cdot X_d'' = 0.15 \cdot 0.032\,\Omega = 0.0048\,\Omega$$

Correction for generator impedance:

$$K_G = \frac{U_n}{U_{rG}} \cdot \frac{c_{max}}{1 + X_d'' \cdot \sin\varphi_{rG}} = \frac{0.4\,\text{kV}}{0.4\,\text{kV}} \cdot \frac{1}{1 + 0.12 \cdot 0.6} = 0.93$$

$$\underline{Z}_{GK} = K_G \cdot \underline{Z}_G = 0.93 \cdot (0.048 + j0.0298)\,\Omega$$

$$Z_{GK} = \sqrt{0.0445^2 + 0.0298^2}\,\Omega = 0.0536\,\Omega$$

Calculation of initial symmetrical short-circuit current:
Contribution of network feed-in:

$$I_{k3}'' = \frac{c \cdot U_n}{\sqrt{3} \cdot Z_G} = \frac{1.1 \cdot 0.4\,\text{kV}}{\sqrt{3} \cdot 0.01\,\Omega} = 25.403\,\text{kA}$$

Contribution of generator:

$$I_3'' = \frac{c \cdot U_n}{\sqrt{3} \cdot Z_{Gen}} = \frac{1.1 \cdot 0.4\,\text{kV}}{\sqrt{3} \cdot 0.0536\,\Omega} = 4.739\,\text{kA}$$

Sum of transferred short-circuit currents:

$$\sum I_{k3}'' = I_{k3Net}'' + I_{k3Gen}'' = 30.142\,\text{kA}$$

Calculation of peak short-circuit currents:
Contribution of network feed-in:

$$i_{p3Net} = \kappa \cdot \sqrt{2} \cdot I_{k3}''$$

$$\frac{R}{X} = \frac{0.002072}{0.01} = 0.2 \quad \kappa = 1.58$$

$$i_{p3Net} = 1.58 \cdot \sqrt{2} \cdot 30.142\,\text{kA} = 67.35\,\text{kA}$$

Contribution of generator:

$$i_{p3Gen} = \kappa \cdot \sqrt{2} \cdot I_{k3Gen}''$$

$$\frac{R}{X} = 0.15 \quad \kappa = 1.03$$

$$i_{p3Net} = 1.03 \cdot \sqrt{2} \cdot 4.739\,\text{kA} = 6.9\,\text{kA}$$

Sum of transferred short-circuit currents:

$$\sum i_{p3} = i_{p3Net} + i_{p3Gen} = 74.25\,\text{kA}$$

Calculation of symmetrical cut-off current:
Contribution of network feed-in:

$$I_{aNet} = I''_{k3Net} = 23.09\,\text{kA} \quad (\text{far from generator})$$

Contribution of generator:

$$I_{aGen} = \mu \cdot I''_{k3Gen} = 0.755 \cdot 4.31\,\text{kA} = 3.25\,\text{kA}$$

$$I_{rG} = \frac{S_{rG}}{\sqrt{3} \cdot U_{rG}} = \frac{600\,\text{kVA}}{\sqrt{3} \cdot 0.4\,\text{kV}} = 866\,\text{A}$$

$$\frac{I''_{k3Gen}}{I_{rG}} = \frac{4.31\,\text{kA}}{0.866\,\text{kA}} = 4.98$$

Sum:

$$\sum I_a = I_{a3Net} + I_{a3Gen} = 26.34\,\text{kA}$$

Total impedance at fault location F1:

$$\underline{Z}_p = \frac{\underline{Z}_G \cdot \underline{Z}_{GK}}{\underline{Z}_G + \underline{Z}_{GK}} = \frac{(0.00207 + j0.01) \cdot (0.0445 + j0.0298)}{(0.00207 + j0.01) + (0.0445 + j0.0298)}\,\Omega$$

$$= \frac{-0.000206 + j0.000507}{0.0466 + j0.0398}\,\Omega$$

$$= 0.00282 + j0.00847\,\Omega$$

$$Z_G = \sqrt{0.00282^2 + 0.00847^2}\,\Omega = 0.00893\,\Omega$$

Impedances of cable K2:

$$\underline{Z}_{k2} = l(r' + jx') = 0.085\,\text{km} \cdot (0.066 + j0.079)\,\Omega/\text{km}$$

$$= (0.00561 + j0.006715)\,\Omega$$

$$\underline{Z}_k = \underline{Z}_p + \underline{Z}_{k2}$$

$$= (0.00282 + j0.00847)\,\Omega + (0.00561 + j0.006715)\,\Omega$$

$$= (0.00843 + j0.0152)\,\Omega$$

$$\underline{Z}_k = \sqrt{(0.00843 + 0.0152)^2}\,\Omega = 0.0174\,\Omega$$

The initial symmetrical short-circuit current is then

$$I''_{k3} = \frac{c \cdot U_n}{\sqrt{3} \cdot Z_k} = \frac{1.1 \cdot 0.4\,\text{kV}}{\sqrt{3} \cdot 0.0174\,\Omega} = 14.59\,\text{kA}$$

18.12 Example 12: Connection of a Transformer Through an External Network and a Generator

A transformer is connected through an external network and a generator as shown in Figure 18.15.

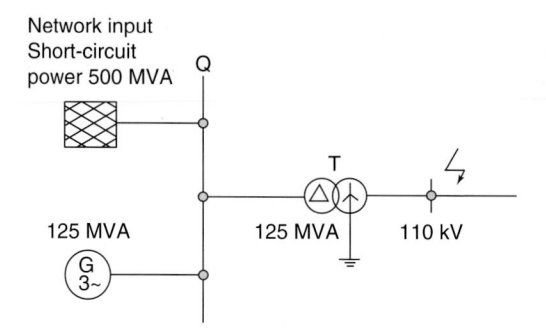

Figure 18.15 Example 12: connection of a transformer through an external network and a generator.

Network input Short-circuit power 500 MVA

125 MVA

125 MVA

110 kV

1) Calculate the impedances at the fault location.
2) Calculate I''_{k3}, I''_{k2E}, I''_{k2}, and I''_{k1} at the fault location.
3) Which is the largest current?

Generator:

$$X''_d = x''_d \cdot \frac{(1.05 \cdot U_n)^2}{S_{rG}} = 0.12 \cdot \frac{(1.05 \cdot 110\,\text{kV})^2}{500\,\text{MVA}} = 12.8\,\Omega$$

Network:

$$Z_{Qt} = \frac{1.1 \cdot U_n^2}{S''_{kQ}} = \frac{1.1 \cdot (110\,\text{kV})^2}{125\,\text{MVA}} = 26.62\,\Omega$$

Transformer:

$$Z_T = \frac{u_{Rr}}{100\%} \cdot \frac{U_{rT}^2}{S_{rT}} = 0.15 \cdot \frac{(110\,\text{kV})^2}{125\,\text{MVA}} = 14.52\,\Omega$$

Parallel impedance:

$$Z_p = \frac{Z_{Gt} \cdot Z_{Qt}}{Z_{Gt} + Z_{Qt}} = \frac{(12.8 \cdot 26.62)\,\Omega}{(12.8 + 26.62)\,\Omega} = 8.64\,\Omega$$

Three-pole current:

$$I''_{k3} = \frac{c \cdot U_n}{\sqrt{3} \cdot Z_G} = \frac{1.1 \cdot 110\,\text{kV}}{\sqrt{3} \cdot 23.16\,\Omega} = 3\,\text{kA}$$

Single-pole short-circuit current:

$$I''_{k1} = \frac{\sqrt{3} \cdot c \cdot U_n}{Z_1 + Z_2 + Z_0}$$

Positive-sequence system: $Z_1 = 23.16\,\Omega$
Negative-sequence system: $Z_2 = Z_1$
Zero-sequence system: $R_0 = 0$, $X_0 = 0.75 \cdot X_1 = 0.75 \cdot 14.52\,\Omega = 10.89\,\Omega$

$$I''_{k1} = \frac{\sqrt{3} \cdot 1.1 \cdot 110\,\text{kV}}{2 \cdot 23.16\,\Omega + 10.89\,\Omega} = 3.66\,\text{kA}$$

Two-pole short-circuit current with contact to ground:

$$I''_{k2E} = \frac{\sqrt{3} \cdot c \cdot U_n}{Z_0 + Z_1 + Z_0 \cdot \frac{Z_1}{Z_2}} = \frac{\sqrt{3} \cdot 1.1 \cdot 110\,\text{kV}}{\left(10.89 + 23.16 + 10.89 \cdot \frac{23.16}{23.16}\right)\Omega} = 4.66\,\text{kA}$$

Two-pole short-circuit current without contact to ground:

$$I''_{k2} = \frac{c \cdot U_n}{Z_1 + Z_2} = \frac{c \cdot U_n}{2 \cdot Z_1} = \frac{\sqrt{3}}{\sqrt{2}} \cdot I''_{k3} = 3.67\,\text{kA}$$

The two-pole short-circuit current with contact to ground is the largest.

18.13 Example 13: Motors in Parallel and their Contributions to the Short-Circuit Current

In a 20/6-kV network, as in Figure 18.16, there are four motors connected with the following data:

Transformer:

$$S_{rT} = 25\,\text{MVA}, \quad u_{kr} = 13\%,\ 20/6.3\,\text{kV}$$

Motors 1 and 2:

$$2 \times P_{rm} = 2.3\,\text{MW}, \quad U_{rG} = 6\,\text{kV}, \quad \cos\varphi_{rG} = 0.86,$$
$$p = 2, \quad I_a/I_{rm} = 5, \quad \eta = 0.97$$

Motors 3 and 4:

$$2 \times P_{rm} = 0.36\,\text{MW}, \quad U_{rG} = 6\,\text{kV}, \quad \cos\varphi_{rG} = 0.87,$$
$$p = 1, \quad I_a/I_{rm} = 5.5, \quad \eta = 0.98$$

1) Calculate the reactances.
2) Calculate the currents at the motors.
3) Calculate the cut-off currents at the motors.

Network input:

$$Z_{Qt} = \frac{c \cdot U_{nQ}^2}{S''_{kQ}} \cdot \frac{1}{t^2} = \frac{1.1 \cdot (20\,\text{kV})^2}{1000\,\text{MVA}} \cdot \frac{(6.3\,\text{kV})^2}{(20\,\text{kV})^2} = 0.044\,\Omega$$

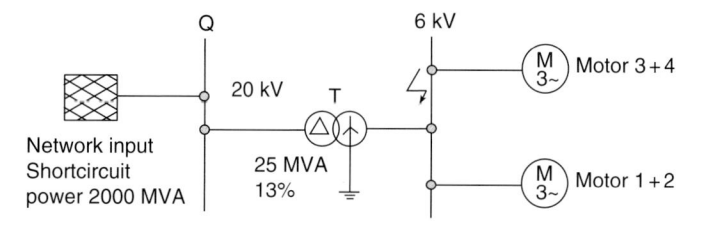

Figure 18.16 Example 13: influence of motors on the current.

18 Examples: Calculation of Short-Circuit Currents

Transformer:

$$Z_T = \frac{u_{kr}}{100\%} \cdot \frac{U_{LVT}^2}{S_{rT}} = \frac{13\%}{100\%} \cdot \frac{(6.3\,\text{kV})^2}{25\,\text{MVA}} = 0.206\,\Omega$$

Impedance:

$$Z_k = Z_{Qt} + Z_T = 0.25\,\Omega$$

Initial current without motors:

$$I_{k3}'' = \frac{c \cdot U_n}{\sqrt{3} \cdot Z_k} = \frac{1.1 \cdot 6\,\text{kV}}{\sqrt{3} \cdot 0.25\,\Omega}$$

Impedances for the asynchronous machines:

$$Z_{m1} = \frac{1}{2} \cdot \frac{\eta \cdot \cos\varphi}{I_{an}/I_{rm}} \cdot \frac{U_{rm}^2}{P_{rm}} = \frac{1}{2} \cdot \frac{0.86 \cdot 0.97}{5} \cdot \frac{(6\,\text{kV})^2}{2.3\,\text{MVA}} = 1.305\,\Omega$$

$$Z_{m2} = \frac{1}{2} \cdot \frac{\eta \cdot \cos\varphi}{I_{an}/I_{rm}} \cdot \frac{U_{rm}^2}{P_{rm}} = \frac{1}{2} \cdot \frac{0.87 \cdot 0.98}{5.5} \cdot \frac{(6\,\text{kV})^2}{0.36\,\text{MVA}} = 7.75\,\Omega$$

Transferred currents:

$$I_{km1}'' = \frac{c \cdot U_n}{\sqrt{3} \cdot Z_{m1}} = \frac{1.1 \cdot 6\,\text{kV}}{\sqrt{3} \cdot 1.305\,\Omega} = 2.92\,\text{kA}$$

$$I_{km2}'' = \frac{c \cdot U_n}{\sqrt{3} \cdot Z_{m2}} = \frac{1.1 \cdot 6\,\text{kV}}{\sqrt{3} \cdot 7.75\,\Omega} = 0.492\,\text{kA}$$

Initial symmetrical short-circuit current with influence of motors at the position:

$$\sum I_k'' = I_{k\,(\text{without motors})}'' + I_{km1}'' + I_{km2}''$$
$$= 15.24\,\text{kA} + 2.92\,\text{kA} + 0.492\,\text{kA} = 18.65\,\text{kA}$$

Short-circuit power:

$$S_k'' = \sqrt{3} \cdot U_n \cdot I_k'' = \sqrt{3} \cdot 6\,\text{kV} \cdot 18.65\,\text{kA} = 193.8\,\text{MVA}$$

Calculation of μ factors, with $t = 0.1$ s:

$$I_{rm1} = \frac{S_{rm1}}{\sqrt{3} \cdot U_{rm1}} = 0.221\,\text{kA}$$

$$I_{rm2} = \frac{S_{rm2}}{\sqrt{3} \cdot U_{rm2}} = 0.035\,\text{kA}$$

$$\frac{I_{km1}''}{I_{rm1}} = \frac{2.92\,\text{kA}}{0.221\,\text{kA}} = 11.2^{\text{TM}} \quad \mu = 0.64$$

$$\frac{I_{km2}''}{I_{rm2}} = \frac{0.492\,\text{kA}}{0.035\,\text{kA}} = 12.1^{\text{TM}} \quad \mu = 0.634$$

Calculation of q factors:

$$\frac{\text{Motor power}}{\text{Pole pair number}} = \frac{2.3\,\text{MVA}}{2} = 1.15^{\text{TM}} \quad q = 0.587$$

$$\frac{\text{Motor power}}{\text{Pole pair number}} = \frac{0.36\,\text{MVA}}{1} = 0.36^{\text{TM}} \quad q = 0.447$$

Cut-off current of motors:

$$I_{am1} = \mu \cdot q \cdot I''_{km1} = 0.64 \cdot 0.587 \cdot 2.92\,\text{kA} = 1.0997\,\text{kA}$$
$$I_{am2} = \mu \cdot q \cdot I''_{km2} = 0.634 \cdot 0.447 \cdot 0.492\,\text{kA} = 0.139\,\text{kA}$$
$$\sum I_a = 16.4\,\text{kA}$$

18.14 Example 14: Proof of the Stability of Low-Voltage Systems

For the selection and project management of electrical systems, it is necessary to check the short-current strength of the operational equipment against the mechanical and thermal stresses resulting from short circuits. This assumes the knowledge required for calculating short-circuit currents. In this section, the short-circuit currents are calculated and the operational equipment is dimensioned. After calculating the required short-circuit currents, it must be assessed whether the protection during indirect contact and stability against short circuits is ensured.

Tables 18.2–18.5 summarize the calculated short-circuit currents for Example 1 of this chapter.

Determination of the peak short-circuit current:

$$\kappa = 1.02 + 0.98 \cdot e^{-3(R/X)}$$
$$\kappa = 1.02 + 0.98 \cdot e^{-3(15.35/21.8)} = 1.138$$
$$i_p = \kappa \cdot \sqrt{2} \cdot I''_{k3} = 1.138 \cdot \sqrt{2} \cdot 9.14\,\text{kA} = 14.7\,\text{kA}$$

Determination of the thermal equivalent short-circuit current:

$$I_{thm} = I''_k \cdot \sqrt{m+n}$$

For far-from-generator short circuits we set $n = 1$.

$$m = \frac{1}{2 \cdot f \cdot t_k \cdot \ln(\kappa - 1)} [e^{4 \cdot f \cdot t_k \cdot \ln(\kappa - 1)} - 1]$$

$$m = \frac{1}{2 \cdot 50\,\text{Hz} \cdot 0.9\,\text{s} \cdot \ln(1.138 - 1)} [e^{4 \cdot 50\,\text{Hz} \cdot 1\,\text{s} \cdot \ln(1.138 - 1)} - 1] = 0.005$$

Table 18.2 Summary of results for I''_{k1min} at subdistributor.

Operational equipment	R (mΩ, 180 mm²)	R_{PE} (mΩ, 70 mm²)	X (mΩ)	X_{PE} (mΩ)
Primary network	—	—	—	—
Transformer	3.5		13	
Cable	15.68	26.88	6.8	6.8
	Total resistance = 46.06 mΩ		Total reactance = 26.6 mΩ	

Short-circuit impedance is 53.18 mΩ.
Short-circuit current $I''_{k1min} = 4.18\,\text{kA}$.

Table 18.3 Summary of results for I''_{k1min} at load.

Operational equipment	R (mΩ, 180 mm²)	R_{PE} (mΩ, 70 mm²)	X (mΩ)	X_{PE} (mΩ)
Primary network	—	—	—	—
Transformer	3.5		13	
Cable	15.68	26.88	6.8	6.8
Conductor	221.4	221.4	2	2
Sum of resistances	239.78	248.28	23.8	8.8
	Total resistance = 488.86 mΩ		Total reactance = 30.6 mΩ	

Short-circuit impedance is 489.8 mΩ.
Short-circuit current $I''_{k1min} = 448$ A.

Table 18.4 Summary of results for I''_{k3min} at distributor.

Operational equipment	R (mΩ)	X (mΩ)
Primary network	—	
Transformer	3.5	13
Cable	18.65	6.8
Total	16.15	19.8

Short-circuit impedance is 25.55 mΩ.
Short-circuit current $I''_{k3} = 9.14$ kA.

Table 18.5 Summary of results for I''_{k3min} at load.

Operational equipment	R (mΩ)	X (mΩ)
Primary network	—	
Transformer	3.5	13
Cable	18.65	6.8
Total	178.6	2

Short-circuit impedance is 204.25 mΩ.
Short-circuit current $I''_{k3} = 1.068$ kA.

Table 18.6 Checking the short-circuit strength.

400-V side	Short-circuit current calculation			Required short-circuit current strength			
	I''_{k3} (kA)	i_p (kA)	I_{thm} (kA)	I_{sc} (kA)	I_{ma} (kA)	I_{th} (kA)	I_{thz} (kA)
Main distribution	17.15	33.95		20	40		21.08
Subdistribution	8.58	13.8		18.5	20	8.6	17.17
Load	1			6	6		

so that the thermal equivalent short-circuit current is then

$$I_{thm} = 9.14\,\text{kA} \cdot \sqrt{0.005 + 1} = 9.16\,\text{kA}$$

Thermal short-circuit strength of subdistributor:

$$I_{thz} = 12.5\,\text{kA} \cdot \sqrt{\frac{s}{0.9\,s}} = 17.17\,\text{kA}$$

Thermal short-circuit strength of main distributor:

$$I_{thz} = 20\,\text{kA} \cdot \sqrt{\frac{s}{0.9\,s}} = 21.8\,\text{kA}$$

for $t_{th} = 1\,\text{s}$ and $t_k = 0.9\,\text{s}$.

The stability of the main and subdistributors against short circuits is ensured by means of the stress parameters (Table 18.6),

18.15 Example 15: Proof of the Stability of Medium-Voltage and High-Voltage Systems

The basic network design is shown in a single-phase representation with fault locations and with a 110-kV network input, supply lines, transformers, and busbars (Figure 18.17). Figure 18.18 illustrates the single-phase equivalent circuit, consisting of resistances and reactances, required for the calculation.

The short-circuit impedance of the network relative to the 110-kV side is

$$Z_Q = \frac{c(U_{nQ})^2}{S''_{kQ}} = \frac{1.1 \cdot (110\,\text{kV})^2}{5000\,\text{MVA}} = 2.662\,\Omega$$

$$X_Q = Z_Q = 2.662\,\Omega$$

The impedance is converted using the square of the transformation ratio t^2_{max} to give the value relative to the 20-kV side:

$$Z_{Qt} = 2.662\,\Omega \cdot \frac{1}{t^2_{max}} = 72.1\,\text{m}\Omega$$

$$X_{Qt} = 72.1\,\text{m}\Omega$$

Checking for far-from-generator short circuits:

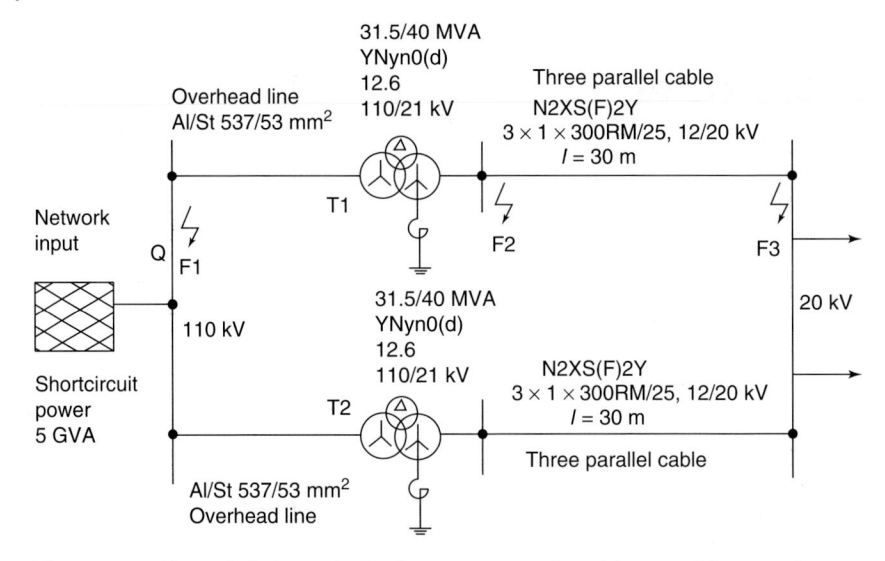

18 Examples: Calculation of Short-Circuit Currents

Figure 18.17 Network design – single-phase representation with network input and transformers.

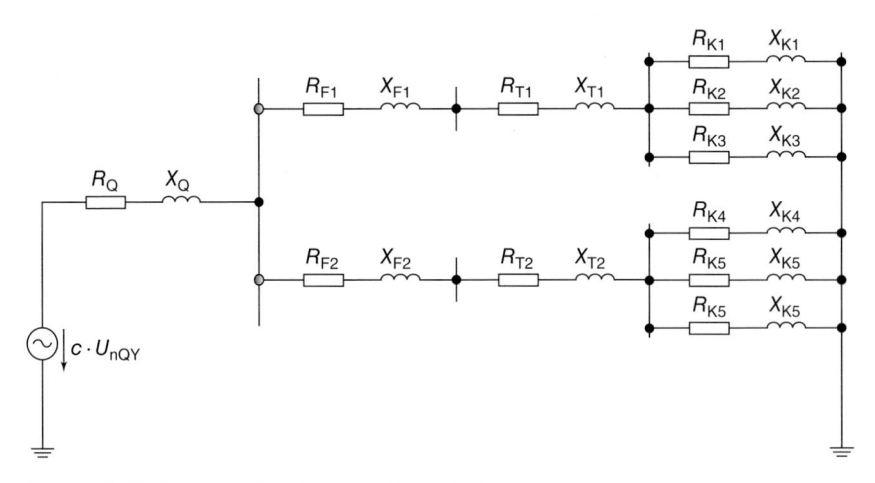

Figure 18.18 Operational equipment with equivalent circuit.

In accordance with IEC 60909:

$$X_{TSV} \geq 2X_{Qt} = 1.265\,\Omega \geq 2 \cdot 71.74\,m\Omega$$

The requirement is therefore satisfied.
Calculation of positive-sequence short-circuit impedances for the transformer:
The following data can be taken from the nameplate of the transformer:
11 000 V ± 16% in 13 steps (27 settings), 21, 31.5/110 kV.
Maximum setting (step 1):

$$Z_k = 68.8\,\Omega, \quad I_{TpV} = 124.5/181\,A, \quad U_{TpV} = 127.6\,kV$$

18.15 Example 15: Proof of the Stability of Medium-Voltage and High-Voltage Systems

Middle setting (step 14 m):

$$Z_k = 46.7\,\Omega, \quad I_{TpV} = 165.3/209.9\,\text{A}, \quad U_{TpV} = 110\,\text{kV}$$

Minimum setting (step 27):

$$Z_k = 30.7\,\Omega, \quad I_{TpV} = 196.8/249.9\,\text{A}, \quad U_{TpV} = 92.4\,\text{kV}$$

Here, it is necessary to clarify whether, in accordance with IEC 60909, the calculation can be performed only with the middle position. The following relationship holds true:

$$U_{TpV} = U_{nTpV}(1 \pm p_T)$$

p_T is obtained from the relationship:

$$p_T = 0.16 > 0.05$$

The step incrementation must be chosen so that the largest short-circuit current occurs. The transformation ratio is

$$t_{max} = \frac{U_{TpV}}{U_{TSV}} = \frac{127.6\,\text{kV}}{21\,\text{kV}} = 6.08$$

The transformer resistance is found from

$$R_T = \frac{P_{krT}}{3 I_{rTpV}} = \frac{136\,\text{kW}}{3 \cdot (209.9\,\text{A})^2} = 1.03\,\Omega$$

The impedances are converted to the values relative to the 20-kV side:

$$Z_T = X_T = 46.7\,\Omega \cdot \frac{1}{t_{max}^2} = 1.265\,\Omega$$

$$R_T = 1.03\,\Omega \cdot \frac{1}{t_{max}^2} = 27.84\,\text{m}\Omega$$

Calculation of cable impedances:
The following cable data [8] are given: N2XS(F)2Y 1 × 300 Rm/25, 18/20 kV:
From the resistance per unit length of the cable, we can calculate the resistance of the cable:

$R' = 0.0601\,\Omega/\text{km} \cdot 30\,\text{m} \cdot \frac{\text{km}}{1000\,\text{m}} = 1.8\,\text{m}\Omega$ and with the inductance per unit length of the cable $L' = 0.347\,\text{mH/km}$, we can calculate the inductance of the cable:

$$X' = 2\pi \cdot 50\,\text{Hz} \cdot 0.347\,\text{mH/km} = 109\,\text{m}\Omega/\text{km} \quad 30\,\text{m} = 3.27\,\text{m}\Omega$$

Overland lines:
The overland lines are of type Al/St 537/53 mm². Due to the large cross-section, the resistances expected can be neglected.

Calculation of short-circuit currents for different fault locations:
For the dimensioning of the operational equipment, the short-circuit currents are calculated according to Figure 18.19. First, a three-pole short circuit with simple input through parallel current paths is calculated.

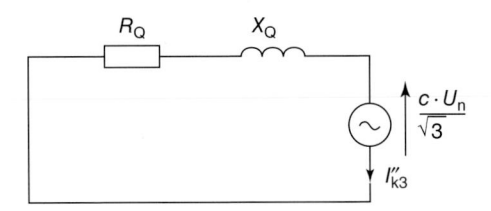

Figure 18.19 Equivalent circuit in the positive-sequence system at fault location F1.

The data for the resistance values in the short-circuit current path are

$$Z_Q = 2.662\,\Omega$$
$$X_Q = 2.662\,\Omega$$
$$R_Q = 0$$

The calculation of the initial symmetrical short-circuit current results from

$$I''_k = \frac{c(U_{nQ})}{\sqrt{3}Z_k} = \frac{1.1 \cdot 110\,\text{kV}}{\sqrt{3} \cdot 2.662\,\Omega} = 26.24\,\text{kA}$$

$$I''_k = I_k = I_a = 26.24\,\text{kA}$$

$$\frac{R_k}{X_k} = \frac{R_Q}{X_Q} = \frac{0}{2.662} = 0$$

κ is obtained from the relationship:

$$\kappa = 1.02 + 0.98 e^0 = 2$$

We can then determine the peak short-circuit current with

$$i_p = 2 \cdot \sqrt{2} \cdot 26.24\,\text{kA} = 74.23\,\text{kA}$$

Three-pole short circuit on transformer busbar:

The three-pole short circuit on the transformer busbar (Figure 18.20) is made up of the transferred short-circuit currents, which can be calculated as in the following.

The data for the operational equipment in the short-circuit current path are (the same values are used for transformers and cables)

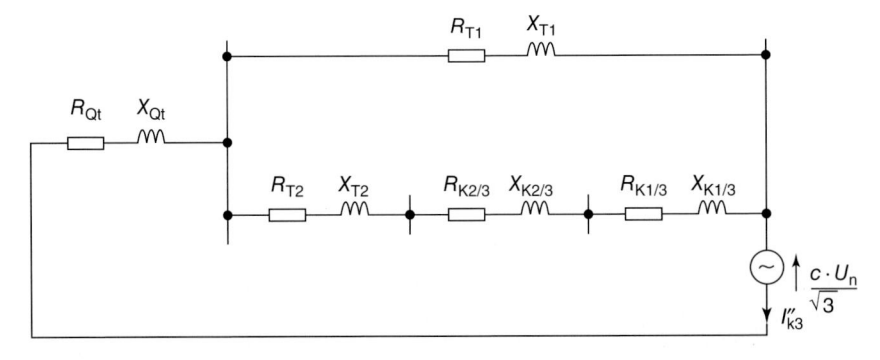

Figure 18.20 Equivalent circuit in the positive-sequence system at fault location F2.

18.15 Example 15: Proof of the Stability of Medium-Voltage and High-Voltage Systems

Network feed-in transformer cable

$Z_Q = 2.662\,\Omega \quad Z_T = 1265\,m\Omega \quad X_K = 1.09\,m\Omega$
$X_Q = 2.662\,\Omega \quad X_T = 1265\,m\Omega \quad R_K = 0.601\,m\Omega$
$R_Q = 0 \quad R_T = 27.87\,m\Omega$

Branch 1:

$$\frac{R_k}{X_k} = \frac{27.87\,m\Omega}{1265\,m\Omega} = 0.02203$$

$$R_k = 0.02203 \cdot X_k < 0.3 X_{k1}$$

Here, we can neglect R_k.

Branch 2:

$$\frac{R_k}{X_k} = \frac{27.87\,m\Omega + 0.601\,m\Omega + 0.601\,m\Omega}{1265\,m\Omega + 1.09\,m\Omega + 1.09\,m\Omega} = 0.02294$$

$$R_k = 0.02294 \cdot X_k < 0.3 X_k$$

Here again, we can neglect R_{kk}. For the calculation of the initial symmetrical short-circuit current, we use the reactances.

Short-circuit reactance:

$$X_k = X_{Qt} + \frac{X_{T1}\left(X_{T2} + \frac{X_{k1}}{2} + \frac{X_{k2}}{3}\right)}{X_{T1} + \left(X_{T2} + \frac{X_{k1}}{3} + \frac{X_{k2}}{3}\right)}$$

$$X_k = 72.1\,m\Omega + 0 \cdot \frac{1265\,m\Omega \cdot (1265\,m\Omega + 1.09\,m\Omega + 1.09\,m\Omega)}{1265\,m\Omega + (1265\,m\Omega + 1.09\,m\Omega + 1.09\,m\Omega)}$$

$$= 705.1\,m\Omega$$

The initial symmetrical short-circuit current is then:

$$I_k'' = \frac{c \cdot U_{nTSV}}{\sqrt{3}Z_k} = \frac{1.1 \cdot 20\,kV}{\sqrt{3} \cdot 705.1\,m\Omega} = 18.014\,kA$$

$$I_k'' = I_k = I_a = 18.014\,kA$$

The fictitious magnitude of the initial symmetrical short-circuit current is

$$S_k'' = \sqrt{3} \cdot 20\,kA \cdot 18.014\,kA = 624\,MVA$$

Calculation of the transferred short-circuit currents:

$$\frac{I_{k\text{-branch2}}''}{I_{k\text{-branch1}}''} = \frac{X_T}{X_T + X_k/3 + X_k/3} = \frac{X_T}{X_T + 2/3 \cdot X_k} = \frac{1}{1 + 2/3 \cdot \frac{X_K}{X_T}}$$

$$I_{k\text{-branch2}}'' = I_{k\text{-branch1}}'' \cdot \frac{1}{1 + 2/3 \cdot \frac{X_K}{X_T}}$$

18 Examples: Calculation of Short-Circuit Currents

The initial symmetrical short-circuit current results from the parallel current branches as follows:

$$I''_k = I''_{k\text{-branch1}} + I''_{k\text{-branch2}} + \cdots + I''_{k\text{-branch}n}$$

The same conditions, therefore, apply for the peak short-circuit, cut-off, and steady-state short-circuit currents. We can then determine the transferred short-circuit current for the first branch by rewriting the equations:

$$I''_{k\text{-branch1}} = I''_k \cdot \left[\frac{1}{1 + \dfrac{1}{1 + \dfrac{2}{3} \cdot \dfrac{X_K}{X_T}}} \right]$$

$$I''_{k\text{-branch1}} = 18.014\,\text{kA} \cdot \left[\frac{1}{1 + \dfrac{1}{1 + \dfrac{2}{3} \cdot \dfrac{1.09\,\text{m}\Omega}{1265\,\text{m}\Omega}}} \right] = 9.0095\,\text{kA}$$

For branch 2:

$$I''_{k\text{-branch2}} = I''_k - I''_{k\text{-branch1}} = 18.014\,\text{kA} - 9.015\,\text{kA} = 8.999\,\text{kA}$$

Determination of the peak short-circuit current for branch 1:

$$\kappa 1 = 1.02 + 0.98 \cdot e^{-3(R/X)} = 1.02 + 0.98 \cdot e^{-3 \cdot 0.02203} = 1.937$$
$$i_{p\text{-branch1}} = 1.937 \cdot \sqrt{2} \cdot I''_{k\text{-branch1}} = 1.937 \cdot \sqrt{2} \cdot 9.015\,\text{kA} = 24.7\,\text{kA}$$

Determination of the peak short-circuit current for branch 2:

$$\kappa 2 = 1.02 + 0.98 \cdot e^{-3(R/X)} = 1.02 + 0.98 \cdot e^{-3 \cdot 0.02294} = 1.935$$
$$i_{p\text{-branch1}} = 1.935 \cdot \sqrt{2} \cdot I''_{k\text{-branch1}} = 1.937 \cdot \sqrt{2} \cdot 8.999\,\text{kA} = 24.6\,\text{kA}$$

The total peak short-circuit current is the sum of the currents in branches 1 and 2:

$$i_p = i_{p\text{-branch1}} + i_{p\text{-branch2}} = 49.32\,\text{kA}$$

Here, it must be noted that the transferred short-circuit current per cable branch is only one-third of the calculated current.

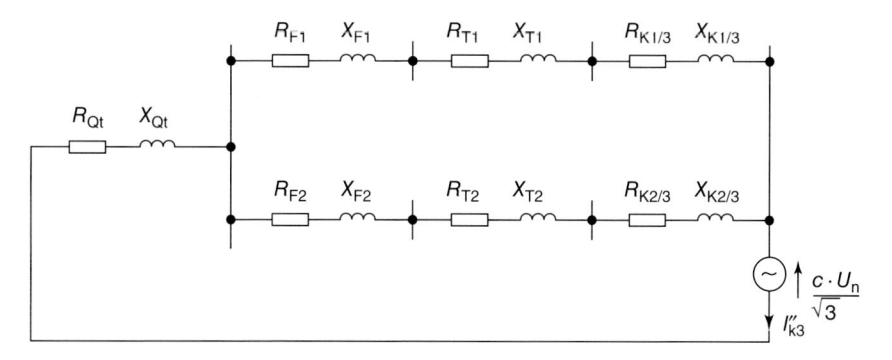

18.15 Example 15: Proof of the Stability of Medium-Voltage and High-Voltage Systems

Figure 18.21 Equivalent circuit in the positive-sequence system at fault location F3.

Three-pole short circuit on the 20-kV busbar (Figure 18.21):
Only the reactances are used here.

$$X_k = X_{Qt} + \frac{\left(X_{T1} + \frac{X_{k1}}{3}\right)\left(X_{T2} + \frac{X_{k2}}{3}\right)}{X_{T1} + \frac{X_{k1}}{3} + X_{T2} + \frac{X_{k2}}{3}}$$

$$X_k = 72.1\,\text{m}\Omega + \frac{(1265\,\text{m}\Omega + 1.09\,\text{m}\Omega)\cdot(1265\,\text{m}\Omega + 1.09\,\text{m}\Omega)}{(1265\,\text{m}\Omega + 1.09\,\text{m}\Omega)+(1265\,\text{m}\Omega + 1.09\,\text{m}\Omega)}$$

$$= 705.1\,\text{m}\Omega$$

$$I_k'' = \frac{c\cdot U_{LVT}}{\sqrt{3}Z_k} = \frac{1.1\cdot 20\,\text{kV}}{\sqrt{3}\cdot 705.1\,\text{m}\Omega} = 18.014\,\text{kA}$$

$$I_k'' = I_k = I_a = 18.014\,\text{kA}$$

The fictitious magnitude of the initial symmetrical short-circuit current is

$$S_k'' = \sqrt{3}\cdot 20\,\text{kA}\cdot 18.014\,\text{kA} = 624\,\text{MVA}$$

For the peak short-circuit current:

$$\frac{R_k}{X_k} = \frac{14.24\,\text{m}\Omega}{705.1\,\text{m}\Omega} = 0.02019$$

$$\kappa = 1.02 + 0.98e^{-3\cdot 0.02019} = 1.94$$

$$I_p = 1.94\cdot\sqrt{2}\cdot 18.04\,\text{kA} = 49.5\,\text{kA}$$

Stability of operational equipment against short circuits:
For the short-circuit strength of operational equipment, we must calculate the dynamic (i_p) and the thermal (I_{th}) stresses.
For far-from-generator short circuits:

$$\frac{I_k''}{I_k} = 1 \quad \text{and therefore } n = 1$$

In accordance with IEC 60909, we obtain for the thermal effect of the direct current (d.c.) aperiodic component m:

$$m = \frac{1}{2fT_k\ln(\kappa - 1)}[e^{4fT_k\ln(\kappa - 1)} - 1]$$

For fault location F3, we calculate m with $T_k = 1$ s:

$$m = \frac{1}{2 \cdot 50\,\text{Hz} \cdot 1\,\text{s} \cdot \ln(1.94-1)}[e^{4 \cdot 50\,\text{Hz} \cdot T_k \cdot \ln(1.94-1)} - 1] = 0.162$$

This yields the thermal short-time current:

$$I_{th} = I_k'' \sqrt{m+1} = 18\,\text{kA} \cdot \sqrt{0.162+1} = 19.4\,\text{kA}$$

For fault location F1, we calculate m with $T_k = 1$ s:

$$I_{th} = I_k'' \sqrt{m+1} = 27\,\text{kA} \cdot \sqrt{0.195+1} = 29.5\,\text{kA}$$

Dimensioning of the operational equipment:

The calculations (Table 18.7) are summarized in this section in order to make these available for dimensioning. For the dimensioning, the following operating conditions are assumed (Table 18.8). The standard value of the network frequency is 50 Hz, and the rated short-circuit duration T_k is assumed to be 1 s. The protection technology must be designed and set up in accordance with this. The dimensioning of the circuit breaker is taken from IEC 282 (Table 18.9). The rated short-circuit making current of the circuit breaker must be 2.5 times as large as the effective value of the rated short-circuit breaking current. If the peak short-circuit current is above 2.5 times this value, then the rated making current must have at least the value of the peak short-circuit current.

Table 18.10 gives the rated currents of the load interrupter switches. The rated steady-state current of the load interrupter switch is dimensioned according to the steady-state operating current.

The dimensioning of the disconnect switch and the grounding switch follows IEC 282 (Table 18.11).

Table 18.7 Current carrying capacities for the 110/20-kV level.

Rated voltage, U_r (kV)	Rated steady-state current, I_r (A)	Initial symmetrical short-circuit current, I_k'' (kA)	Steady-state short-circuit current, I_k (kA)	Peak short-circuit current, i_p (kA)	Thermal short-time current, I_{th} (kA)
123	420	27	27	75	30
24	2 200	18	18	50	20

Table 18.8 Rated voltages for the 110/20-kV level.

Highest voltage for operational equipment, U_m (kV)	Rated short-duration a.c. voltage, U_{rW} (kV)	Rated lightning impulse voltage, U_{rB} (kV)
24	50	95
123	230	550

a.c., alternating current.

18.15 Example 15: Proof of the Stability of Medium-Voltage and High-Voltage Systems

Table 18.9 Selection values and rated values for load interrupter switches in accordance with IEC 282.

Rated voltage (kV)	Rated short-circuit breaking current (kA)	Rated operating current (A)			
		800	1250	1600	2000
123	12.5	X	X		
	20		X	X	X
	25		X	X	X
	40			X	

Table 18.10 Selection values and rated values for load interrupter switches in accordance with IEC 282.

Rated steady-state current (A)	Rated short-time current (kA)	Rated peak current (kA)
630	31.5	78.75

Table 18.11 Selection values and rated values for disconnect switches and grounding switches in accordance with IEC 282.

Rated voltage (kV)	Rated initial symmetrical short-circuit current (kA)	Rated peak short-circuit current (kA)	Rated steady-state current (A)			
			800	1250	1600	2000
123	12.5	32	X	X		
	20	50		X	X	X
	25	63		X	X	X
	40	100			X	X

Dimensioning of the overvoltage surge arrester:
The dimensioning of the overvoltage surge arrester is accomplished with the help of IEC 60099-5 (1996). The use of a silicon carbide surge arrester is preferred in ground fault neutralizer grounded systems and is connected to the transformer between the conductors and ground. The required quenching voltage and its rated discharge current must be considered during dimensioning. With the load rejection factor $\delta_L = 1.1$, the quenching voltage is

$$U_L \geq \delta_L \delta \frac{U_m}{\sqrt{3}} = 1.1 \cdot \sqrt{3} \cdot \frac{123\,\text{kV}}{\sqrt{3}} = 135.3\,\text{kV}$$

In accordance with IEC 60099-5 (1996), the quenching voltage of 138 kV is chosen. A 10-kA arrester is recommended for the choice of arresters in accordance with the rated discharge current for networks with >60 kV. The stability against short circuits must be dimensioned, so that the short-circuit strength remains ensured at currents higher than the expected initial symmetrical short-circuit current.

Dimensioning of the current inverter:

The dimensioning of the current inverter follows from the short-circuit current to be expected, in order to ensure that the transformer does not become quickly saturated and the protection relays are no longer able to correctly sense the short-circuit current. The standard values in accordance with IEC 44-1 are as follows: 10, 15, 20, 30, 50, 75, and their decimal multiples or divisions. Here, the standard value 30 is chosen.

Dimensioning of the voltage transformer:

The standard values are provided in IEC 44-2. The information applies to inductive and capacitive transformers. The rated network voltage is the essential parameter for dimensioning. For conductor–conductor voltage transformers, 110 kV and a conductor–ground voltage transformer $\frac{110\,\text{kV}}{\sqrt{3}}$ must be selected. The rated voltage factors are obtained from IEC 44-2.

Dimensioning of the 20-kV switchgear:

1) Dimensioning of the supply line:
   - For the dimensioning of the conductors, the three-pole short circuit is used as the basis.
   - For the loading of the shielding, the neutral point connection and the asymmetrical short-circuit currents are of greatest importance. The maximum short-circuit current is calculated as the double ground fault.
2) Dimensioning of the circuit breakers:

The electrical data for the circuit breakers from the manufacturers' catalogs are compared with the data measured (Table 18.12). At this point an example will

Table 18.12 Dimensioning of the circuit breakers.

	Values from data sheet (kA)	Measured values (kA)
Rated short-duration power frequency withstand voltage	50	50
Rated lightning impulse withstand voltage	125	125
Rated short-circuit breaking voltage	25	18
Rated short-time current, 1 s	25	—
Thermal short-time current, 1 s	—	18
Rated short-circuit making current	63	—
Rated short-time current	25	—
Rated current of busbar	2500 A	2200 A
Rated current of branches	2000 A	—
Rated peak current	—	50

18.16 Example 16: Calculation for Short-Circuit Currents with Impedance Corrections

Table 18.13 Dimensioning the circuit breakers at different connection points on the busbar.

	Input field (kA)	Coupling field (kA)	Load field (kA)
Rated short-duration power frequency withstand voltage	50	50	50
Rated lightning impulse withstand voltage	125	125	125
Rated short-circuit breaking voltage	16	20	20
Rated duration of short circuit	3 s	3 s	3 s
Rated short-circuit making current	40	50	40
Rated current	1250 A	2000 A	1250 A

Rated current of transformer is 1000 A.
Rated current of vacuum circuit breaker is 1250 A.

Figure 18.22 Dimensioning of the circuit breakers; (1) input field, (2) load field, and (3) coupling field.

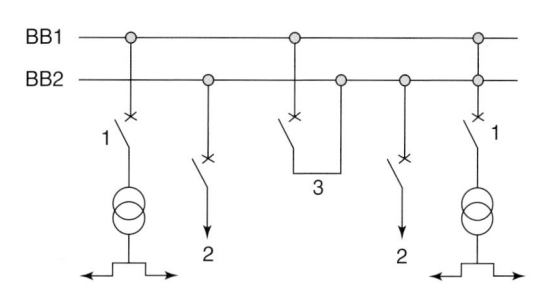

be given for the dimensioning of the circuit breakers (Table 18.13) at different connection points to the busbar (Figure 18.22).

The dimensioning parameters are the:

- rated short-circuit breaking capacity,
- rated operating current, and
- short-circuit current determined by measurement.

18.16 Example 16: Calculation for Short-Circuit Currents with Impedance Corrections

Given a 220-kV network with the data for the operational equipment as in Figure 18.23.

Calculate the short-circuit currents and the impedance corrections.

Network:

$$Z_Q = \frac{c \cdot U_{nQ}^2}{S_{kQ}''} = \frac{1.1 \cdot (220\,\text{kV})^2}{8000\,\text{MVA}} = 6.65\,\Omega$$

18 Examples: Calculation of Short-Circuit Currents

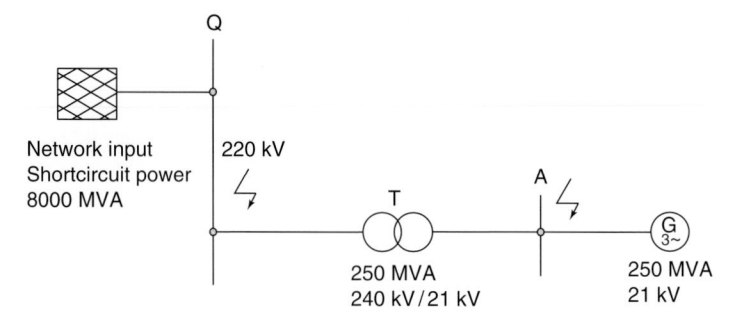

Figure 18.23 Example 16: Calculation of short-circuit currents with impedance corrections.

Generator:

$$Z_G = \frac{x_d'' \cdot U_{rG}^2}{100\% \cdot S_{rG}} = \frac{17 \cdot (21\,\text{kV})^2}{250\,\text{MVA}} = 0.30\,\Omega$$

Correction factor:

$$K_{G,KW} = \frac{c}{1 + x_d'' \cdot \sin\varphi_{rG}} = \frac{1.1}{1 + 0.17 \cdot 0.63} = 0.994$$

Corrected generator impedance:

$$Z_{G,KW} = K_{G,KW} \cdot Z_G = 0.994 \cdot 0.30\,\Omega = 0.298\,\Omega$$

Block transformer:

$$Z_{THV} = \frac{u_{kr}}{100\%} \frac{U_{rTHV}^2}{S_{rT}}$$

$$Z_{THV} = \frac{15\%}{100\%} \frac{(240\,\text{kV})^2}{250\,\text{MVA}} = 34.56\,\Omega$$

$$Z_{TLV} = \frac{u_{kr}}{100\%} \frac{U_{rTLV}^2}{S_{rT}}$$

$$Z_{TLV} = \frac{15\%}{100\%} \frac{(21\,\text{kV})^2}{250\,\text{MVA}} = 0.26\,\Omega$$

$$Z_{T,KW} = c \cdot Z_{TLV} = 1.1 \cdot 0.26\,\Omega = 0.286\,\Omega$$

Calculation of currents in Q:

$$I_k'' = I_{kQ}'' + I_{kKW}''$$

$$I_{kQ}'' = \frac{c \cdot U_{nQ}}{\sqrt{3} \cdot Z_Q} = \frac{1.1 \cdot 220\,\text{kV}}{\sqrt{3} \cdot 6.65\,\Omega} = 21\,\text{kA}$$

$$Z_{KW} = K_{KW} \cdot (t_r^2 \cdot Z_G + Z_{THV})$$

$$K_{KW} = \left(\frac{t_f}{t_r}\right)^2 \cdot \frac{c}{1 + (x_d'' - x_T) \cdot \sin\varphi_{rG}}$$

$$K_{KW} = \left(\frac{220\,\text{kV}}{21\,\text{kV}}\right)^2 \left(\frac{21\,\text{kV}}{240\,\text{kV}}\right)^2 \cdot \frac{1.1}{1 + (0.17 - 0.15) \cdot 0.63} = 0.913$$

18.16 Example 16: Calculation for Short-Circuit Currents with Impedance Corrections

$$Z_{KW} = 0.913 \cdot \left[\left(\frac{240\,kV}{21\,kV}\right)^2 \cdot 0.30 + 34.56\,\Omega\right] = 67.32\,\Omega$$

$$I''_{kKW} = \frac{1.1 \cdot 220\,kV}{\sqrt{3} \cdot 67.32\,\Omega} = 2.07\,kA$$

$$I''_k = 21\,kA + 2.07\,kA = 23.07\,kA$$

Calculation of currents in A:

$$I''_k = I''_{kG} + I''_{kT}$$

$$I''_{kG} = \frac{c \cdot U_{rG}}{\sqrt{3} \cdot Z_{G,KW}} = \frac{1.1 \cdot 21\,kV}{\sqrt{3} \cdot 0.298\,\Omega} = 44.75\,kA$$

$$I''_{kT} = \frac{c \cdot U_{rG}}{\sqrt{3} \cdot (Z_{T,KW} + \frac{1}{t_f^2} \cdot Z_Q)}$$

$$I''_{kT} = \frac{1.1 \cdot 21\,kV}{\sqrt{3} \cdot \left(0.286\,\Omega + \left(\frac{21\,kV}{220\,kV}\right)^2 \cdot 6.65\,\Omega\right)} = 38.48\,kA$$

$$I''_k = 44.75\,kA + 38.48\,kA = 83.23\,kA$$

Bibliography

1 DIN EN 60909-0:2016-10: *Kurzschlussströme in Drehstromnetzen*, Berechnung der Ströme.
2 EN 50522: Eathing of power installations exceeding 1 kV a.c. 2011-11.
3 Beiblatt 3 zu DIN VDE 0102:2003-07: Faktoren für die Berechnung von Kurzschlussströmen.
4 DIN VDE 0100: Elektrische Anlagen von Gebäuden mit Nennspannungen bis AC 1000V und DC 1500V.
5 IEC 909-3 (VDE 0102 Teil 3): 1997-06: Berechnung der Ströme in Drehstromanlagen, Doppelerdkurzschlussströme und Teilkurzschlussströme über Erde.
6 ABB: Switchgear manual 12th Edition, Cornelsen Verlag Düsseldorf, 2006.
7 I. Kasikci: *Analysis and Design of Low-Voltage Power Systems*. An Engineer's Field Guide, 1st edition. February 2004. XII, 387 Seiten, Hardcover 263 Abb., 79 Tab., Handbuch/Nachschlagewerk, ISBN 978-3-527-30483-7, Wiley-VCH, Berlin.
8 Heinhold: *Aufbaudaten und technische Werte für Kabel und Leitungen*, 3rd edn, Abschnitt 62, 1984.
9 Beiblatt 4 zu DIN VDE 0102:2003-02: Kurzschlussströme in Drehstromnetzen, Daten elektrischer Betriebsmittel für die Berechnungen von Kurzschlussströmen.
10 Nexans Deutschland GmbH – Energy Networks: Kabelkamp 20 30179 Hannover.
11 DIgSILENT: PowerFactory, Technical Reference Documentation, Version: 15.2, Edition: 1.
12 Beiblatt 2 zu DIN VDE 0102:1992-09: Berechnung von Kurzschlussströmen in Drehstromnetzen, Anwendungsleitfaden für die Berechnung von Kurzschlussströmen in Niederspannungsnetzen.
13 Entwurf Beiblatt 5 zu DIN VDE 0100: Errichten von Starkstromanlagen mit Nennspannungen bis 1000 V – Beiblatt 5: Maximal zulässige Längen von Kabeln und Leitungen unter Berücksichtigung des Schutzes, des Fehlerschutzes, des Schutzes bei Kurzschluss und des Spannungsfalls-Leitfaden für die Auswahl, die Dimensionierung und die Koordination der Schutzeinrichtungen und Kabel-, Leitungs- bzw. Schienenquerschnitte eines Stromkreises.

14 D. Oeding, B.R. Oswald: *Elektrische Kraftwerke und Netze*, 6th edn, Springer Verlag, Berlin-Heidelberg-New York, 2004, ISBN 3.540-00863-2.
15 A Hochreiner: *Symmetrische Komponenten in Drehstromsystemen*, Springer-Verlag Berlin, Göttingen-Heidelberg, 1957.
16 K. Heuck, K.D. Dettmann, D. Schulz: *Elektrische Energieversorgung, Erzeugung, Übertragung und Verteilung elekztrischer Energie für Studium und Praxis*, 7th edn, Vieweg, 2007, ISBN 978-3-8348-0217-0.
17 DIN VDE 0103:1994-11: Kurzschlussströme; Berechnung der Wirkung Begriffe und Berechnungsverfahren.
18 E. Handschin: *Elektrische Energieübertragungssysteme, Teil 1: Stationärer Betriebszustand*, Hüthig-Verlag, Heidelberg, 1983.
19 DIN EN 61660-1 (VDE 0102 Teil 10):1998-06: Kurzschlussströme in Gleichstrom–Eigenbedarfsanlagen, in Kraftwerken und Schaltanlagen.
20 E. Kreyszig: *Advanced Engineering Mathematics*, 7th edn, John Wiley, 1993.
21 L. Papula: Mathematik für Ingenieure und Naturwissenschaftler, Band 2, Viewegs Fachbücher der Technik, 2001, ISBN 3-528-94237-1.
22 A.J. Schwab: *Elektroenergiesysteme, Erzeugung, Transport, Übertragung und Verteilung elektrischer Energie*, Springer Verlag, Berlin, Heidelberg, 2006, ISBN-10 3-540-29664-6.
23 IEC 60 909:2002-07: Short-circuit current calculation in three-phase a-c systems.
24 Beiblatt 1 zu DIN VDE 0102:2002-11: Kurzschlussströme in Drehstromnetzen, Beispiele für die Berechnung von Kurzschlussströmen.
25 Beiblatt 1 zu DIN EN 61660-1:2003-02: Berechnung von Kurzschlussströmen in Drehstromnetzen, Beispiele.
26 DIN VDE 0102 Teil 3: Kurzschlussströme – Berechnung von Kurzschlussströmen in Drehstromanlagen Doppelerdkurzschlussströme und Teilkurzschlussströme über Erde.
27 E DIN IEC 73/89/CDV (VDE 0102 Teil 100):1997-08: Berechnung der Ströme in Drehstromanlagen; Teil 1: Begriffe und Berechnungsverfahren.
28 Beiblatt 1 zu DIN VDE 0103:1996-06: Kurzschlussströme; Berechnung der Wirkung, Begriffe und Berechnungsverfahren; Beispiele für die Berechnung.
29 Beiblatt 1 zu DIN VDE 0102 Teil 10:2002-11: Kurzschlussströme– Kurzschlussströme in Gleichstrom–Eigenbedarfsanlagen von Kraftwerken und Schaltanlagen Berechnung der Wirkungen.
30 IEC 61400-27-1 Ed.1: Wind Turbines – Part 27-1: Electrical simulation models – Wind turbines.
31 Dipl.Ing. Johannes Gester: *Starkstromleitungen und Netze*, VEB-Verlag, Berlin, 1977.
32 G. Seib: *Eelctrical Installations Handbook*, 3rd edn, John Wiley & Sons, 2000, ISBN 0-471-49435-6.
33 Switching, Protection and Distribution in Low-Voltage Networks: *Handbook with Selection Criteria and Planning Guidelines for Switchgear, Switchboards, and Distribution Systems*, Siemens, Germany, Publicis, 1994, ISBN-13: 978-3895780004, Publicis; 2nd revised edition.
34 I. Kasikci: *Projektierung von Niederspannungsanlagen*, Hüthig-Verlag, Heidelberg, 3rd edn, 2010, ISBN: 978-3-8101-0274-4.

35 R. Ayx, I. Kasikci: *Projektierungshilfe elektrischer Anlagen in Gebäuden, Praxiseinführung und Berechnungsmethoden. VDE-Schriftenreihe Bd. 148. 7. Aufl*, VDE Verlag, Berlin/Offenbach, 2012.
36 W. Knies, K. Schierack: *Elektrische Anlagentechnik, Kraftwerke, Netze, Schaltanlagen, Schutzeinrichtungen*, 5th edn, Hanser Verlag, 2000, ISBN-10 3-446-40574-7.
37 IEC 61400-27-1 Ed.1: Wind turbines - Part 27-1: Electrical simulation models – Wind Turbines.
38 B. Oswald: *Netzberechnung*, VDE-Verlag, Berlin Offenbach, 1992.
39 TIP: Planung der elektrischen Energieverteilung Technische Grundlagen, Siemens AG Energy Management, Medium Voltage & Systems, Artikel-Nr.: EMMS-T10007-00.

Standards

The following referenced documents are indispensable for the application of this document. For dated references, only the edition cited applies. For undated references, the latest edition of the referenced document (including any amendments) applies.

IEEE C 37.04	Standard Rating Structure for AC High-Voltage Circuit Breakers Rated on a Symmetrical Current including Supplements.
IEEE C 37.010	Standard Application Guide for AC High-Voltage Circuit Breakers Rated on a Symmetrical Current.
IEEE C 37.013	Standard for AC High-Voltage Generator Circuit Breakers Rated on a Symmetrical Current Basis.
IEEE 399	Power System Analysis – The Brown Book.
IEEE 141	Electric Power Distribution for Industrial Plants – The Red Book.
IEC 60027-7: 2010	Letter Symbols to be Used in Electrical Technology – Part 7: Power Generation, Transmission, and Distribution.
IEC 60038:2009	IEC Standard Voltages.
IEC 60050-131:2002	International Electrotechnical Vocabulary – Part 131: Circuit Theory.
IEC 60050-131:2002/A1:2002	International Electrotechnical Vocabulary – Part 131: Circuit Theory (Amendment 1).
IEC 60050-151:2001	International Electrotechnical Vocabulary – Part 151: Electric and Magnetic Devices.
IEC 60050-195:1998	International Electrotechnical Vocabulary – Part 195: Earthing and Protection Against Electric Shock.
IEC 60050-195:1998/A1.2001	International Electrotechnical Vocabulary – Part 195: Earthing and

Short Circuits in Power Systems: A Practical Guide to IEC 60909-0, Second Edition. Ismail Kasikci.
© 2018 Wiley-VCH Verlag GmbH & Co. KGaA. Published 2018 by Wiley-VCH Verlag GmbH & Co. KGaA.

	Protection Against Electric Shock (Amendment 1).
IEC 60071-1:2006	Insulation Co-ordination – Part 1: Definitions, Principles and Rules.
IEC 60071-1:2006/A.1:2010	Insulation Co-ordination – Part 1: Definitions, Principles and Rules.
IEC 60865-1:2011	Short-Circuit Currents – Calculation of Effects – Part 1: Definitions and Calculation Methods.
IEC 60909-0:2016-12	Short-circuit currents in three-phase a.c. systems – Part 0: Calculation of currents.
IEC TR 60909-1:2002-07	Short-circuit currents in three-phase a.c. systems – Part 1: Factors for the calculation of short-circuit currents according to IEC 60909-0.
IEC TR 60909-2:2008-11	Short-circuit currents in three-phase a.c. systems – Part 2: Data of electrical equipment for short-circuit current calculations.
IEC 60909-3:2009	Short-circuit currents in three-phase AC systems – Part 3: Currents during two separate simultaneous line-to-earth short circuits and partial short-circuit currents flowing through earth.
IEC TR 60909-4:2000	Short-circuit currents in three-phase a.c. systems – Part 4: Examples for the calculation of short-circuit currents.
IEC 62271-100:2008	High-Voltage Switchgear and Controlgear – Part 100: Alternating-Current Circuit-Breakers.
IEC 62428:2008	Electric Power Engineering – Modal Components in Three-Phase AC Systems – Quantities and Transformations.
IEC 60949:1988	Calculation of Thermally Permissible Short-Circuit Currents, Taking into Account Non-adiabatic Heating Effects.
IEC 60949:1988/A.1:2008	Calculation of Thermally Permissible Short-Circuit Currents, Taking into Account Non-adiabatic Heating Effects (Amendment 1).
IEC 60986:2000	Short-Circuit Temperature Limits of Electrical Cables with Rated Voltages from 6 kV ($U_m = 7.2$ kV) up to 30 kV ($U_m = 36$ kV).
IEC 60986:2000/A.1:2008	Short-Circuit Temperature Limits of Electrical Cables with Rated Voltages from 6 kV ($U_m = 7.2$ kV) up to 30 kV ($U_m = 36$ kV) (Amendment 1).

IEC 62271-100	High-Voltage Switchgear and Controlgear, Part 100: High-Voltage Alternating-Current Circuit Breakers.
IEC 62271-200	High-Voltage Switchgear and Controlgear, Part 200: AC Metal-Enclosed Switchgear and Controlgear for Rated Voltages above 1 kV and up to and including 52 kV.
IEC 62271-203	High-Voltage Switchgear and Controlgear, Part 203: Gas-Insulated Metal-Enclosed Switchgear for Rated Voltages above 52 kV.
IEC 60282-2	High-Voltage Fuses, Part 2: Expulsion Fuses.
IEC 60947-1	Low Voltage Switchgear and Controlgear, Part 1: General Rules.
IEC 60947-2	Low Voltage Switchgear and Controlgear, Part 2: Circuit Breakers.
IEC 61363-1	Electrical Installations of Ships and Mobile and Fixed Offshore Units, Part 1: Procedures for Calculating Short Circuit Currents in Three-Phase AC.

Explanations of Symbols

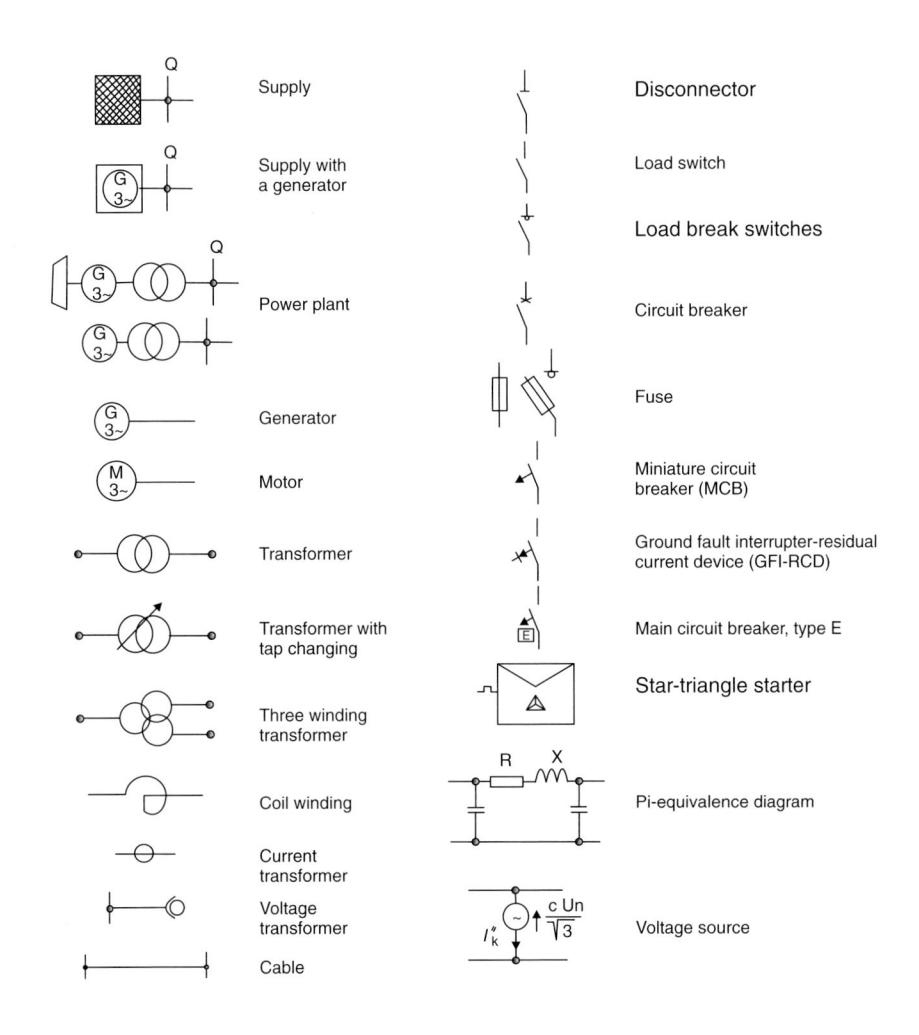

Symbol	Description
Supply	Supply
Supply with a generator	Supply with a generator
Power plant	Power plant
Generator	Generator
Motor	Motor
Transformer	Transformer
Transformer with tap changing	Transformer with tap changing
Three winding transformer	Three winding transformer
Coil winding	Coil winding
Current transformer	Current transformer
Voltage transformer	Voltage transformer
Cable	Cable
Disconnector	Disconnector
Load switch	Load switch
Load break switches	Load break switches
Circuit breaker	Circuit breaker
Fuse	Fuse
Miniature circuit breaker (MCB)	Miniature circuit breaker (MCB)
Ground fault interrupter-residual current device (GFI-RCD)	Ground fault interrupter-residual current device (GFI-RCD)
Main circuit breaker, type E	Main circuit breaker, type E
Star-triangle starter	Star-triangle starter
Pi-equivalence diagram	Pi-equivalence diagram
Voltage source	Voltage source

Explanations of Symbols

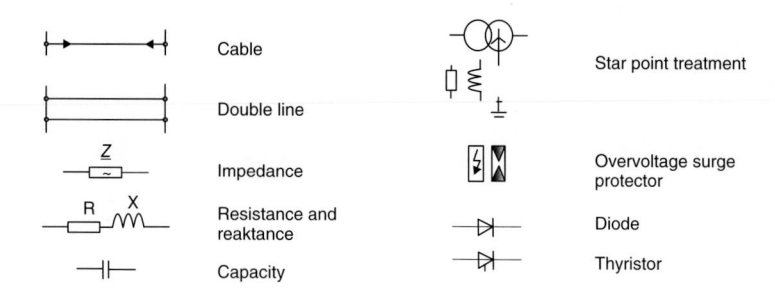

Symbol	Description
⊢──▶────◀⊣	Cable
═══════	Double line
—[Z]~—	Impedance
—[R]—/\/\/\—[X]	Resistance and reaktance
—‖—	Capacity
—(X)— with ⌷ ⌇ ⏚	Star point treatment
⚡ ⚡	Overvoltage surge protector
—▶⊢—	Diode
—▶⊬—	Thyristor

Symbols and Indices

A, Initial value of DC aperiodic component
A, Cross-section of conductor
a, Center-to-center distance between conductors
$\underline{a}, \underline{a}^2$, Rotational operators
b, Width of rectangular conductor
c, Voltage factor
C, Capacitance
C_E, Ground capacitance
E, Internal voltage of voltage source; source voltage
E_B, No-load voltage of battery
E'', Subtransient voltage of synchronous machine
f, Frequency
h, Height of conductor
L', Distributed inductance
I_a, Cut-off current
I_{an}, Starting current
I_B, Operating current
I_E, Grounding current
I_k, Steady-state short-circuit current
I''_k, Initial symmetrical short-circuit current
I''_{k1}, Single-pole short-circuit current
I''_{k2}, Two-pole short-circuit current
I''_{k3}, Three-pole short-circuit current
I''_{k2E}, Two-pole short circuit with contact to ground
I''_{kEE}, Double ground fault
I_{ma}, Rated short-circuit making current
I_n, Nominal current of protective equipment
i_p, Peak short-circuit current
I_r, Rated current
I_{rM}, Magnetic setting current
I_{LR}/I_{rM}, Ratio of locked rotor current to rated current of motor
I_{sc}, Rated short-circuit breaking current
I_{cm}, Rated short-circuit making current
I_{cu}, Rated ultimate short-circuit breaking

Short Circuits in Power Systems: A Practical Guide to IEC 60909-0, Second Edition. Ismail Kasikci.
© 2018 Wiley-VCH Verlag GmbH & Co. KGaA. Published 2018 by Wiley-VCH Verlag GmbH & Co. KGaA.

I_{rT}, Rated current of transformer on higher-voltage or lower-voltage side
I_{th}, Rated short-time current
K, Correction factor
L_B, Inductance of battery
L_{BBr}, Total inductance of battery
L_{BL}, Inductance of a battery conductor
L_C, Inductance of capacitor
L_{CBr}, Total inductance of capacitor
L_{CL}, Inductance of a capacitor conductor
L_{CY}, Inductance of coupling branch for capacitor
L_{DL}, Inductance of conductor in converter arm
L_M, Inductance of DC motor
L_{MBr}, Total inductance of DC motor
L_{ML}, Inductance of a DC motor conductor
I_{rG}, is the rated current of the asynchronous generator
I_{th}, Thermal equivalent short-circuit current
L_s, Inductance of saturated choke coil
L_Y, Inductance of coupling branch
m, Factor for the heat effect of the d.c. component
M_r, Rated load torque of motor
n, Factor for the heat effect of the a.c. component
p, Pole pair of asynchronous motor
p, Ratio I_k/I_p
P, Effective power
p_G, Range of generator voltage regulation
p_T, Range of transformer voltage adjustment
P_{krT}, Total winding losses of transformer at rated current
P_{rM}, Rated effective power of motor
r_T, Transformation ratio of transformer
Q, Idle power
q, Factor for the calculation of breaking current of asynchronous motors
r, Resistance, conductor radius, absolute, or relative value
R, Resistance
R_l, Resistance of conductor
R', Resistance per unit length
R_{BL}, Resistance of battery conductor
R_s, Resistance of saturated choke coil
R_Y, Resistance of coupling branch
R_{BY}, Resistance of battery coupling branch
R_C, Resistance of capacitor
R_{CBr}, Total resistance of capacitor
R_{CL}, Resistance of a capacitor conductor
R_{DL}, Resistance of conductor in converter arm
R_L, Resistance referred to the line
R_M, Resistance of DC motor
R_{ML}, Resistance of DC motor conductor

R_{MY}, Resistance of DC motor coupling branch
R_{Qt}, Resistance referred to the low-voltage side of the transformer
R_R, Resistance of choke coil
S, Apparent power, cross-section
S_{rG}, Rated apparent power of the asynchronous generator
S_{rT}, Rated apparent power of the transformer
S''_k, Initial symmetrical short-circuit power
t, Time
T_k, Duration of short circuit
t_p, Time until onset of peak short-circuit current
U_m, Highest voltage for equipment, line-to-line (root mean square, RMS)
U_{NB}, Nominal voltage of battery
U_{nQ}, Nominal voltage of the network
U_r, Rated voltage, line-to-line (RMS)
U_{rG}, Rated voltage of the asynchronous generator
U_{rM}, Rated voltage of motor
u_{Rr}, Rated value for resistive voltage drop in %
u_{kr}, Rated value for short-circuit voltage in %
U_{rT}, Rated voltage of transformer on higher-voltage or lower-voltage side
U_n, Nominal system voltage
U_{nHV}, Nominal voltage on higher-voltage side
$\underline{Z}_k$, Short-circuit impedance of network
X, Reactance
x''_d, Subtransient reactance of synchronous motor
X_{Qt}, Reactance referred to the low-voltage side of the transformer
X_R, Reactance of choke coil
Z, Impedance
Z_E, Grounding impedance
$Z_{(1)}$, Positive-sequence impedance
$Z_{(2)}$, Negative-sequence impedance
$Z_{(0)}$, Zero-sequence impedance
$Z_{(G)}$, Impedance of the asynchronous generator
Z_M, Impedance of motor
Z_{Qt}, Short-circuit impedance referred to the low-voltage side of the transformer
Z_T, Positive-sequence short-circuit impedances of two-winding transformers
X_T, Inductive resistance of transformer
X''_d, Subtransient reactance
X'_d, Transient reactance
X_d, Synchron reactance
X_0, Zero reactance
φ, Phase angle
ε, Coefficient of grounding
κ, Factor for the calculation of the peak short-circuit current
λ, Factor for the calculation of steady-state short-circuit current
μ_0, Absolute permeability in vacuum; $\mu_0 = 4\pi \cdot 10^{-4}$ H/km
η, Efficiency of AC motor

μ, Factor for the calculation of the symmetrical short-circuit breaking current
μ_{WA}, Factor for the calculation of the symmetrical short-circuit breaking current of a wind power station unit with an asynchronous generator
μ_{WD}, Factor for the calculation of the symmetrical short-circuit breaking current of a wind power station unit with doubly fed asynchronous generator
ρ, Specific resistance
δ, Decay coefficient, ground fault factor
ψ, Angular velocity
01, Positive-sequence neutral reference
02, Negative-sequence neutral reference
00, Zero-sequence neutral reference.

Indices

a, Cut off
A, B, C, Description of position, e.g., busbar
B, Battery
Br, Battery branch
a.c., AC current
AMZ, Maximum current-dependent time relay
ASM, Asynchronous machine
C, Capacitor
D, Converter
d.c., DC current
E, Ground
F, Short-circuit position
G, Generator
HV, High voltage
i, Internal
K, Cable
k, Short circuit
k1, Single-pole short-circuit current
k2, Two-pole short-circuit current
k2E, Two-pole short-circuit with contact to ground
k3, Three-pole short-circuit current
kEE, Double ground fault
l, Length
L, Conductor
L_1, L_2, L_3, Life (Line) conductor
LV, Low voltage
M, Motor
max, Maximum
min, Minimum
MV, Medium voltage
n, Nominal value

N, Neutral conductor, network
OPE, Overcurrent protective equipment
OV, Overvoltage
PE, Protective Earth (ground) conductor
pS, Limiting dynamic value
PV, Photovoltaic power station unit
Q, Network connection point
R, Short-circuit limiting reactor
s, Source current
S, Power station unit (generator and unit transformer with on-load tap changer)
SO, Power station unit (generator and unit transformer with constant transformation ratio or off-load taps)
r, Rated value
S, Smoothing choke
SP, Connection box to on-site power
T, Transformer
UMZ, Maximum current-independent time relay
WA, Wind power station unit with asynchronous generator
WD, Wind power station unit with doubly fed asynchronous generator
WF, Wind power station unit with full-size converter

Secondary Symbols, Upper Right, Left

″, Subtransient value
′, Transient value
′, Resistance or reactance per unit length
*, Relative magnitude.

American Cable Assembly (AWG)

American cable assembly "American Wire Gauge (AWG)" is given for bigger cable cross-sections.

AWG in mm^2 conversion table:

1CM = 1 Circ. mil = 0.0005067 mm^2
1MCM = 1000 Circ. mils = 0.5067 mm^2.

Index

a
admittance matrix 212–213
asymmetrical short circuits 152, 153
asynchronous generators (AG) 99–101
asynchronous machine 71, 101, 105
   impedance 106
asynchronous motors
   equivalent circuit 98
   overview of 97
   in plant engineering 97
   short circuit currents 161–163
automatic disconnection
   TN system 56
   TT systems 57

b
batteries 147
Bending moment 180
breaking current 127
busbar configuration 183
busbar systems 45

c
cables and overhead lines 58
   average geometrical distance between conductors 86
   calculation of 105
   copper cables and conductors
     resistances per unit length 93
     resistance values at 20°C 91
     resistance values at 80°C 90
   double line 85
   equivalent capacitive reactance 86
   equivalent circuit 86
   equivalent radius 86
   4x conductor bundle line 86
   inductive load reactance 85
   length-specific values 85
   mast diagram 86
   NAYY and NYY cables, resistances and inductive reactances 92
   outgoing and return lines, impedance for 93
   permeability 86
   positive-sequence system 85
     inductive reactances per unit length 92
     resistances per unit length 91
   PVC-insulated cables
     impedance 87
     resistance values 88–90
   2x conductor bundle line 86
   XLPE-insulated cables
     effective capacitances of 95
     ground fault currents of 96
     inductances of 95
     resistances of conductors 94
     resistances per unit length 94
   zero-sequence resistances 85
calculation tools 197
capacitors 98, 148
central earthing point (CEP) 47
choke coils 96–97
circuit breakers 112, 132, 186, 187
computer programs 151
controllable-power transformers 83
Cramer's rule, application of 229–230

current converters 146
current limiting 70
cut-off energy 131

d

DC aperiodic component 2, 3, 49
DC motors 149
DC systems 143
DC systems, short circuit currents
    batteries 203–4
    calculation procedure 200
    capacitors 204–205
    current converters 202
    DC motors 205
    equivalent circuit 201
    IEC 61660–1, 199
    largest short circuit current 199
    resistances of line sections 201
    smallest short circuit current 199
    standardized approximation functions 200
    three-phase synchronous generator 199
    typical paths 200
delta-star transformation 54
determinants 209–212
disconnectors 112
doubly fed asynchronous generator (DFAG) 101

e

earth fault compensation 64–66
earth-fault relays 24
earthing systems 48
electrical system, short circuits 23
electromagnetic compatibility (EMC) 47
EN 50522 60, 125, 126
equivalent circuit diagrams 36
equivalent circuits, for power flow calculations 227, 228
equivalent electrical circuit 2
equivalent voltage source 2, 7, 10–11

f

far-from-generator short circuits 5, 155, 157

fault current(s) 49–51, 56
    calculation 31
fault current analysis
    cable selection 26
    distributors 26
    equivalent voltage source 24
    final circuits 26
    high-fault current 24
    IEC 60909-0 23
    load flow condition 24
    medium-voltage networks 24
    multi-phase reclosure 24
    network planning and management processes 25
    network's generators 24
    power calculations and system planning 25
    reverse feed 24
    selectivity detection 26
    short circuit currents and short circuit impedances 27
    three-phase system 25
    transformer
        medium-voltage switchgear 26
        parallel network operation 26
fuses 112

g

Gauss–Seidel method 224
generators
    correction factor 106
    impedance correction factor K_G for 127–129, 131
    impedance of 105
    transient reactance of 50
ground fault 1
ground fault tripping 132
ground loop impedance 30

h

HH fuses 131
high and low voltage motors
    transformers, with different nominal voltages 163–165
    transformers, with two windings 163

high voltage power systems
  generation, transmission and distribution 46
  high-voltage substation 44
  380 kV/110 kV substation 44
  three-phase high-voltage systems 45
  transmission line 45
high-voltage transformers, characteristic values of 85
hybrid matrix 213, 214

i

IEC 60 909 51, 127, 152
IEC 60 909–4 85
IEC 60027 133
IEC 60364-1 47
IEC 60364-4-41 30
IEC 60364-7-710 47
IEC 60909 11, 12
IEC 60909-0 1, 23, 27, 109, 130
  "dead" short circuit 29
  effective voltage 30
  medium voltage networks 30
  neutral point design 30
  short circuit calculation, range of applicability 31
  short-circuit current selection 31
  single-phase equivalent voltage source method 30
  symmetrical and asymmetrical short circuits 30
  VDE 0102 29
  VDE 0670 switchgear regulations 29
IEC 60947 187
IEC 61363-1 102
impedance(s) 54, 235–237
  asynchronous machines 71
  capacitors 72
  network feed-ins 47
  non-rotating loads 72
  static converters 73
  symmetrical components 142, 144–145
  synchronous machines 49
  transformers 51
impedance corrections 75, 193
  generators 76, 128–129, 131
  power station 77, 127, 129–130
  transformers 79, 130–131
impedance matrix 213
induction motors 165
industrial load center network 39, 41
industrial system, short circuit current 243–244
in-phase voltage control 83
insulation, heat transfer 119
isolated network
  advantages and disadvantages 64
  equivalent circuit 63
IT system
  circuitry of 53
  exposed conductive parts, ground resistance of 54
  hospitals and production, applications in 47
  indirect contact, protection for 53
  in industrial sector 53
  overcurrent protective equipment 53
  power source, grounding conditions of 53
  RCDs, use of 48

j

Jacobian method 223–224

l

linear equations 229
linear equations systems 208–209
linear load flow equations 218–219
load-break switches 112
load circuit 241–243
load interrupter switches 112
load nodes 216
load types and complex power 216–218
loop impedance 49
low-resistance grounded network 66–67
low voltage network
  radial networks
    disadvantages 39
    individual load circuits 40
    load distribution 40

low voltage network (*contd.*)
    meshed network 39, 41
    with redundant inputs 39, 40
  TN system 47, 48
  transformers, equivalent resistances and reactances 84
  type of connection to earth 47
low voltage switchgear 186–187
low-voltage transformers 81, 82

m

magnet wheel 73
making current 127
mechanical short circuit strength
  bending stress 170
  busbar arrangement 171
  busbars and parallel conductors, force effects 169
  circuit breakers 168
  conductor elements 169
  correction factor k_{12} 170
  disconnectors 168
  effective spacings 169
  fuses 168
  laws of rigidity 170
  load-break switches 168
  load interrupter switches 168
  moments of resistance and moments of inertia 171
  natural mechanical oscillating frequency 171
  operational equipment 168
  parallel conductors 167
medium voltage network
  configuration 43
  industrial load center network 39, 41
  with remote station 42
  ring network 39, 42
  short circuit current 42, 43
  supporting structure 42
  transformers, equivalent resistances and reactances 84
medium voltage switchgear 185
mesh diagram 4
meshed network 19, 37–39, 41, 246
moments of inertia 171, 181
moments of resistance 171
motors
  asynchronous motor
    equivalent circuit 98
    overview of 97
    in plant engineering 97
    short circuit currents 161–163
  energy converters 97
  high and low voltage motors
    transformers, with different nominal voltages 163–165
    transformers, with two windings 163
  impedance of 106
  induction motors, short circuits 165
  LV motor, calculation of 106
%/MVA method 14
MVA system calculation 19–22

n

near-to-generator short circuits 2, 5, 6, 155–157
NEC 250 47
negative-sequence short circuit impedance 2
NEPLAN 22, 230, 231
network(s)
  grounding compensation 43
  isolated free neutral point 42
  low impedance neutral point 44
network feed-in 71–73
network matrices 212–231
  admittance matrix 212–213
  current iteration 223–224
  equivalent circuits for power flow calculations 227–228
  examples 228–231
  Gauss–Seidel method 224
  hybrid matrix 213–214
  impedance matrix 213
  linear load flow equations 218–219
  Newton–Raphson, load flow calculation by 219–223
  Newton–Raphson method 224–226
  node voltages and line currents calculation 214–215

node voltages calculation, at predetermined node power 215
power flow analysis, in low voltage power system 226–227
power flow calculation 215–218
network transformations 54–55
network types 21
  low voltage 21
  medium voltage 23
neutral conductor 30
neutral point, arrangement 45
neutral-point transformer (NPT)
  branch with 121, 122
  branch without 120, 121
  compensated network 124–125
  grounding systems 126–127
  insulated network 125
  maximal one-phase short circuit currents 121–124
  Y-Δ winding 120
  Z-Z winding 119, 120
neutral point treatment 39
Newton method, application of 228, 229
Newton–Raphson, load flow calculation by 219–223
Newton–Raphson method 224–226
node generator 216
nodes, types of 216
node voltages and line currents calculation 214–215
node voltages calculation, at predetermined node power 215

o

Ohm's law 135
operational equipment 189
overcurrent protection 131
overcurrent protective devices
  assessment of capacity 189
  circuit breakers
    characteristics of 190
    overloading and short circuit current protection 193
    uses 197
  control transformers 193
  cut-off current 189
  fuses applications, power systems 194, 197
  high voltage – high power fuses 189
  limit switch fuses, time-current characteristics of 189
  miniature circuit breakers 189
  motor protection device, tripping curves 196
  overview of 190
  principle of current limitation 190
  protective functions and setting possibilities 193
  thermal relays, tripping curves 195
  time-current characteristics
    circuit breakers 196
    HH fuses 193
    limit switch fuses 191, 192
    miniature circuit breakers 195
overcurrent protective equipment 34
overhead lines, see cables and overhead lines 86
overloading 131
overload tripping 132

p

parallel circuit 54, 55
peak short circuit current 104, 153–155, 244–246
peak value 78
PE-insulated cables 181
PEN conductor 29
per unit analysis 12–13
phase-angle control transformers 85
positive-sequence short circuit impedance 2
power flow analysis 207–231
  determinants 209–212
  linear equations systems 208–209
  in low voltage power system 226–227
  network matrices 212–231
power generator 143
power plant network, service panel 238
protective conductor (PE) 47, 48

protective functions   132
protective ground conductor   30
p.u. system   14–19

q
quadrature-control transformers   85

r
radial networks   18, 36, 153, 233–235
   disadvantages   39
   individual load circuits   40
   load distribution   40
   meshed network   39, 41
   with redundant inputs   39, 40
reactive power, calculation of   228
reference variables   10
   calculation with   12
residual current devices (RCDs)   34, 48, 53, 57
ring networks   18, 35, 39, 42

s
salient-phase generator   73
series circuit   54, 55
series-regulating transformers   83
short circuit   1, 19
   asynchronous motors   105
   calculation   7, 127
   far-from-generator   5
   impedance   2
   low voltage switchgear   128
   mechanical   111
   near-to-generator   2, 3, 6
   negative-sequence impedance   2
   positive-sequence impedance   2
   positive-sequence system   4
   single-pole   6, 7, 94
   symmetrical breaking current   99
   thermal   111, 112
   three-phase networks   6
   three-pole   4, 6, 7, 91
   two-pole   6, 7, 93
   types   5
   zero-sequence impedance   2
short-circuit calculation methods   7–22
   equivalent voltage source   10–11
   %/MVA method   14
   MVA system calculation   19–22
   per unit analysis   12–13
   p.u. system   14–19
   reference variables, calculation with   12
   short-circuit current characteristics   14
   superposition method   7–10
   switching process calculation   14–15
   transient calculation   11
short circuit current(s)   1
   asymmetrical short circuits   152, 153
   calculation   21, 151, 153
   capacitors   98
   choke coils   96–97
   initial symmetrical   1, 3
   limitation   120
   nonrotating loads   98
   peak   2, 3, 97
   peak short circuit current   153–154
   power grid   104
   self-quenching   42
   ship and offshore installations   102–104
   single-phase short circuit
      equivalent circuit   151
      positive, negative and zero-sequence systems   151–152
   static converters   98
   steady state   2, 102
   steady state short circuit current   157–159
   symmetrical breaking current   2, 155–157
   three-phase short circuit
      equivalent circuit   148
      fault conditions   147–148
      requirements   147
   time behavior of   2–3
   two-phase short circuit
      with earth contact   148–149
      without earth contact   149–150
short circuit current calculation
   connection of a motor   240–241
   factory, supply to   249–250

Index

impedance corrections 250–253, 269–271
industrial system 243–244
load circuit calculation 241–243
low voltage systems, proof of stability 257–259
medium and high voltage systems
  current inverter, dimensioning 268
  different fault locations 261–262
  network design – single-phase representation 260
  operational equipment, dimensioning 266–267
  operational equipment, equivalent circuit 260
  overvoltage surge arrester, dimensioning 267
  peak short circuit current 264
  positive-sequence short circuit impedances for transformer 260
  three-pole short circuit, 20 kV bus bar 265
  three-pole short circuit on transformer bus bar 262–263
  transferred short circuit currents 263
  voltage transformer, dimensioning 268
  vs. operational equipment stability 265
meshed network 246–249
motors in parallel and contributions 255–257
power plant network, on-site connection box 237–238
protective measures proof 235–237
radial network 233–235
three-pole short circuit current and peak short circuit current 244
transformer connection, external network and generator 253
transformers in parallel 238

short-circuit impedance 2
short-circuit path, positive-sequence system 3–5
short circuit strength
  choice of switchgear 185
  low voltage switchgear 186–187
  medium voltage switchgear 185
short-circuit types, classification of 5–7
short-time current 127
short-time delay release 132
single-phase short circuit
  equivalent circuit 151
  positive, negative and zero-sequence systems 151–152
single-phase short circuit current 51, 77
  symmetrical components 140–142
  TN system 47
single-phase short circuits between phase and N 6
single-phase short circuits between phase and PE 6
single source 17
slack node 216, 217
squirrel-cage motors 161
star-delta transformation 55
static converters 98
steady-state condition 9
steady state short circuit current 157–159
step voltages 40
superposition method 2, 7–10
supply networks 17
  calculation 34
  calculation variables 34–35
  concept finding 33
  dimensioning 34
  high-voltage levels 33
  lines supplied from a single source 35
  low-voltage levels 33
  medium-voltage systems 33
  meshed network 37–38
  modern dimensioning tools 33
  power plants and electricity consumer 33

supply networks (*contd.*)
  radial network   36
  ring network   35
surge arrester   191
switching process calculation   14–15
symmetrical breaking current
      155–157
symmetrical components   81, 82
  impedances   85
synchronous generators (SG)   99–101
synchronous machine   49, 99
  generator   73, 74
  inner-and outer-phase machines   73
  nonstationary operation   74
  positive sequence, equivalent circuit
      and phasor diagram   74, 75
  reactances   74–79
  salient-phase generator   73
  stationary operation   74
  turbo generator   73
systems
  IT   35
  TN   29
  TT   34

t

Terra–Terra (TT) systems
  automatic disconnection   57
  circuitry of   52
  exposed conductive parts, ground
      resistance of   52
  overcurrent protective equipment
      52
  RCDs, use of   48
  in rural supply areas   47
thermal short circuit strength   181
  current limitation   176
  Cu screening   182
  electrical operational equipment
      173
  high and medium voltage networks
      176
  IEC 76–1   173
  initial symmetrical short circuit
      current   173
  line-protection circuit breakers,
      house installations   176
  low voltage systems   176
  m and *n* factors   173
  mechanical short-circuit strength
      178–183
  paper-insulated cables
      1–10 kV   177
      12/20 kV   178
      18/30 kV   179
  PVC-insulated cables at 1–10 kV
      180
  rated short time current density
      174, 175
  transformer, feeder of   176
three-phase networks   39
  short-circuit types in   6–7
three-phase networks, neutral point
      treatment
  earth fault compensation   64–66
  grounding systems   61
  isolated network   63–64
  line interruptions   59
  low-resistance grounded network
      66–68
  neutral grounding   69
  neutral point arrangement
      application of   66
      high-voltage networks   66
  surface potential profile   60
  touch voltage   59, 60
  transformers   60
  transverse faults   59
three-phase power systems
  standardized method   23
  superposition method   23
three-phase short circuit(s)   6, 74
  current   77
  equivalent circuit   148
  fault conditions   147–148
  requirements   147
three-phase synchronous generator
      143
three-phase system
  delta and star connection, neutral
      point   133, 134
  symmetrical components
    asymmetrical faults, calculation of
      136

Index | 297

  impedances   142–145
  line-line voltages   134
  line-neutral voltages   134
  one-phase short circuit   140–142
  phase and line currents   135
  phase voltages   133
  positive-, negative-and zero-sequence systems   137–140
  rotational operators   136
  superposition, principle of   142
  three-phase Delta, star source and loads   145
three-pole short circuit current,   244–246
TN system
  automatic disconnection   56
  circuitry of   49
  fault current, calculation of   49–51
  fault protection, requirements on   49
  in industrial sector   48
  loop impedance   49
  low voltage networks   47, 48
  overcurrent protective equipment   49
  PEN conductor   48
  protective ground conductor   48
  single-phase short-circuit current   47
  TN-C-S system, circuitry of   48
touch voltage   39
transformation ratio   57
transformers
  correction factor, calculation of   105
  correction factor K_T for   130, 131
  equivalent circuit   80, 82
  equivalent resistances and reactances   81, 82, 84
  external network and a generator   253–255
  high and low voltage motors   163–164
  impedance calculation   104
  neutral point treatment   60
  overview of   80
  in parallel   238–239

  positive-sequence impedance   81, 83
  short-circuit voltage   80–81
  with three windings   81, 82
  voltage regulation   83, 85
transient calculation   11
transient method   10
turbo generator   73
two-phase short circuit(s)   6–7
  with earth contact   148–149
  with ground   6
  without earth contact   149–150

u

undelayed release   132

v

voltage factor   2, 8
voltage-regulating transformers   57, 83, 85

w

watt-metric relays   24
wind farm
  data   107–111
  grounding arrangement   108
  negative-sequence impedance   108
  positive-sequence impedance   108
  power transformer   108
  three-legged core transformers   109
  transformers, correction factor for   109
wind energy plant
  backup protection   116–117
  data   110–111
  generator   110
  maximal three-phase short circuit   111, 115
  minimal one-phase short circuit   111, 112, 115–116
  NPT, *see* neutral-point transformer (NPT)   119
  partial network, one-phase short circuit   113
  thermal stress of cables   118–119
wind power with full converter   106

Y-Y transformer, equivalent circuit 109
wind turbines
  asynchronous generator 99, 100
  DFAG 101
  full converter 101
  high-voltage power network 99
  synchronous generators 99, 100
  wind farm, *see* wind farm 99
wound rotor motors 161

z

zero-sequence short circuit impedance 2